Physics and Chemistry of III–V Compound Semiconductor Interfaces

Physics and Chemistry of III–V Compound Semiconductor Interfaces

Edited by
Carl W. Wilmsen
Colorado State University
Fort Collins, Colorado

Plenum Press • New York and London

Library of Congress Cataloging in Publication Data

Main entry under title:

Physics and chemistry of III-V compound semiconductor interfaces.

Includes bibliographies and index.
1. Semiconductors—Surfaces. 2. Surface chemistry. 3. Semiconductor-metal boundaries. 4. Dielectrics—Surfaces. I. Wilmsen, Carl W.
QC611.6.S9P48   1985   621.3815′2′0153   85-6598
ISBN 0-306-41769-3

©1985 Plenum Press, New York
A Division of Plenum Publishing Corporation
233 Spring Street, New York, N.Y. 10013

All rights reserved

No part of this book may be reproduced, stored in a retrieval system, or transmitted, in any form or by any means, electronic, mechanical, photocopying, microfilming, recording, or otherwise, without written permission from the Publisher

Printed in the United States of America

Contributors

D. K. Ferry ● Department of Electrical Engineering, Colorado State University, Fort Collins, Colorado 80523

S. M. Goodnick ● Department of Electrical Engineering, Colorado State University, Fort Collins, Colorado 80523

D. L. Lile ● Department of Electrical Engineering, Colorado State University, Fort Collins, Colorado 80523

L. G. Meiners ● Electrical Engineering and Computer Sciences Department, University of California, San Diego, La Jolla, California 92093

G. Y. Robinson ● Department of Electrical Engineering, Colorado State University, Fort Collins, Colorado 80523

J. F. Wager ● Department of Electrical and Computer Engineering, Oregon State University, Corvallis, Oregon 97331

R. H. Williams ● Physics Department, University College, Cardiff CF1 1XL, U.K.

C. W. Wilmsen ● Department of Electrical Engineering, Colorado State University, Fort Collins, Colorado 80523

Preface

The application of the III–V compound semiconductors to device fabrication has grown considerably in the last few years. This process has been stimulated, in part, by the advancement in the understanding of the interface physics and chemistry of the III–V's. The literature on this subject is spread over the last 15 years and appears in many journals and conference proceedings. Understanding this literature requires considerable effort by the seasoned researcher, and even more for those starting out in the field or by engineers and scientists who wish to apply this knowledge to the fabrication of devices. The purpose of this book is to bring together much of the fundamental and practical knowledge on the physics and chemistry of the III–V compounds with metals and dielectrics. The authors of this book have endeavored to provide concise overviews of these areas with many tables and graphs which compare and summarize the literature. In this way, the book serves as both an insightful treatise on III–V interfaces and a handy reference to the literature.

The selection of authors was mandated by the desire to include both fundamental and practical approaches, covering device and material aspects of the interfaces. All of the authors are recognized experts on III–V interfaces and each has worked for many years in his subject area. This experience is projected in the breadth of understanding in each chapter. It is hoped that this book will help accelerate the understanding of III–V interfaces and the establishment of a III–V device industry.

C. W. Wilmsen

Fort Collins, Colorado

Contents

1. III–V Semiconductor Surface Interactions
   R. H. Williams

   1. Introduction 1
   2. Interface States and Schottky Barriers 2
   3. Clean Surfaces of III–V Semiconductors 7
      3.1. Crystallographic Structures of Surface and Bulk . . . 7
      3.2. Bulk and Surface Electronic States 13
      3.3. Surface Imperfections and Defects 18
   4. Adsorption of Gases on Clean III–V Semiconductors . . . 21
      4.1. General Introduction 21
      4.2. Oxygen Adsorption 22
      4.3. Chlorine on III–V Semiconductors 28
      4.4. H_2, H_2S, and H_2O Adsorption 29
   5. Metal Films on Clean III–V Surfaces 31
      5.1. General Introduction 31
      5.2. Interactions at Very Small Coverages 35
         5.2.1. Cesium on GaAs (110) 35
         5.2.2. Al, Ga, and In on GaAs and InP 39
         5.2.3. Au and Ag on GaAs and InP 44
      5.3. Interactions with Thick Metal Films 46
   6. The Electrical Nature of Intimate Interfaces 51
      6.1. Introduction 51
      6.2. Abrupt Boundary Models 54
      6.3. Nonabrupt Boundary Theories 56
   7. Conclusions 61
      References 61

2. Schottky Diodes and Ohmic Contacts for the III–V Semiconductors
   Gary Y. Robinson

   1. Introduction 73

2. Electrical Properties of Metal–Semiconductor Contacts . . 74
  2.1. Classical Models of the Interface 74
    2.1.1. Schottky Model 74
    2.1.2. Bardeen Model 77
    2.1.3. General Case 77
  2.2. Mechanisms of Barrier Formation 80
  2.3. Current Transport 85
    2.3.1. Thermionic Emission: Rectification 86
    2.3.2. Field Emission and Thermionic-Field Emission: Ohmic Behavior 91
  2.4. Capacitance of a Schottky Diode 97
3. Schottky-Diode Technology 98
  3.1. Measurement of ϕ_B 99
    3.1.1. Photoresponse Measurements 99
    3.1.2. Current–Voltage Measurements 101
    3.1.3. Capacitance–Voltage Measurements 105
  3.2. Barrier Energies 108
    3.2.1. GaAs 108
    3.2.2. InP 112
    3.2.3. Other Binary Compounds 117
    3.2.4. III–V Alloys 123
4. Ohmic-Contact Technology 129
  4.1. Methods of Forming Ohmic Contacts 130
    4.1.1. Diffusion and Ion Implantation 131
    4.1.2. Epitaxy 132
    4.1.3. Alloying 133
    4.1.4. Heterojunctions 134
  4.2. Measurement of r_c 137
    4.2.1. Cox–Strack Method 137
    4.2.2. Four-Point Method 139
    4.2.3. Shockley Technique 140
    4.2.4. Transmission-Line Model 142
  4.3. Alloyed Ohmic Contacts 144
    4.3.1. GaAs 144
    4.3.2. InP 150
    4.3.3. Other Binary Compounds 152
    4.3.4. III–V Alloys 152
  References 154

3. The Deposited Insulator/III–V Semiconductor Interface
J. F. Wager and C. W. Wilmsen

1. Introduction 165

Contents

2. General Overview of the Deposited Insulator/III–V Interface . . 166
3. Choice of Insulator and Deposition Technique 170
4. Interfacial Properties 175
   4.1. Interfacial Reactions 175
   4.2. Interfacial Oxide 179
   4.3. Interdiffusion and Impurity Incorporation 182
   4.4. Surface Evaporation 184
   4.5. Energy of the Depositing Molecules 184
   4.6. Interfacial Trapping and Instabilities 186
5. Experimental Results 189
   5.1. InSb 189
   5.2. GaAs 194
   5.3. InP 198
6. Concluding Remarks 205
   References 206

4. Electrical Properties of Insulator–Semiconductor Interfaces on III–V Compounds
L. G. Meiners

1. Introduction 213
2. Theoretical Background 214
   2.1. Differential Surface Capacitance 215
   2.2. Surface States 224
   2.3. Surface Conductance 226
3. Gallium Arsenide 228
   3.1. Chemically Clean Surface 237
   3.2. Native Oxides 238
   3.3. Deposited Insulators 239
4. Indium Antimonide 242
5. Indium Phosphide 247
   5.1. Native Oxides 248
      5.1.1. Thermally Grown Oxides 248
      5.1.2. Anodically Formed Oxides 248
   5.2. Deposited Dielectrics 249
6. Indium Arsenide 260
7. Gallium Phosphide 261
8. Gallium Arsenide Phosphide 262
9. Whither Surface States 262
10. Low-Temperature Deposition of Dielectric Layers . . . 269
11. Conclusion 274
    References 275

5. **III–V Inversion-Layer Transport**
   S. M. Goodnick and D. K. Ferry

   1. Introduction 283
   2. Quantization 287
      2.1. Surface Subbands 287
      2.2. Approximate Solutions 288
      2.3. Effects of Nonparabolicity 294
   3. Surface Scattering Mechanisms 295
      3.1. Coulomb Scattering 296
      3.2. Surface Roughness Scattering 299
   4. Phonon Scattering 304
      4.1. The Acoustic Interaction 305
      4.2. Scattering by Polar Modes 306
      4.3. Remote Optical Phonons 308
      4.4. High Fields 310
   5. Experimental Results 311
      5.1. Subband Structure 311
      5.2. Transport Measurements 314
      Summary 318
      References 319

6. **Interfacial Constraints on III–V Compounds MIS Devices**
   Derek L. Lile

   1. Introduction 327
   2. Dielectric–Semiconductor Interfacial Phenomena . . . 331
      2.1. Trapping 332
      2.2. Scattering 338
      2.3. Recombination 339
   3. MIS-Device Characteristics 342
      3.1. Field-Effect Transistors 343
      3.2. Charge-Coupled Devices 347
      3.3. Integrated Circuits 348
      3.4. Optical Devices 351
      3.5. Memory Cells 355
   4. Device Results 356
      4.1. Gallium Arsenide 356
      4.2. Indium Phosphide 362
      4.3. Other Binary Compounds 378
      4.4. Ternary and Quaternary Alloys 383
   5. Epilogue 388
      References 390

7. Oxide/III–V Compound Semiconductor Interfaces
C. W. Wilmsen

1. Introduction 403
   1.1. Initial Oxidation 404
   1.2. Thermodynamics 405
   1.3. Vapor Pressure 408
2. The Chemically Cleaned Surface 408
   2.1. Polishing and Exposure to Air 408
   2.2. Chemical Etching and Growth of a Chemical Oxide . 414
3. Thermal Oxides 420
   3.1. General Overview 420
   3.2. InP 423
   3.3. GaP 426
   3.4. GaAs, GaSb, and InSb 426
   3.5. InAs 428
   3.6. $In_{0.53}Ga_{0.47}As$ 428
4. Anodic Oxides 430
   4.1. Anodic Oxidation Process 431
   4.2. Anodization Parameters 434
      4.2.1. The Electrolyte 434
      4.2.2. Viscosity 436
      4.2.3. Current Density 436
      4.2.4. pH 437
   4.3. Initial Growth 438
   4.4. Chemical Composition of Anodic Oxides and Interfaces 445
      4.4.1. GaAs 445
      4.4.2. InP 446
      4.4.3. GaP 447
      4.4.4. InAs 449
      4.4.5. GaSb 449
      4.4.6. InSb 449
   4.5. Thermal Annealing of the Anodic Oxides 450
5. Plasma-Grown Oxide 453
   References 457

INDEX 463

III–V Semiconductor Surface Interactions

R. H. Williams

1. Introduction

Central to the successful operation of almost all solid-state semiconductor devices is the need for reliable and well-controlled electrical contacts. However, in spite of several decades of research it is a surprising fact that our knowledge of the fundamental behavior of metal–semiconductor contacts on a microscopic level is still very far from complete. Over the past decade, though, considerable progress has been made and in this chapter we consider the current position relating to our understanding of the interaction of metals with atomically clean III–V semiconductor surfaces. Before dealing with metal–semiconductor interactions we consider in some detail the nature of selected atomically clean surfaces, concentrating in particular on the most thoroughly studied solid in the group, namely GaAs, and we also report recent work relating to the adsorption of selected gases onto these surfaces. Small amounts of an adsorbed gas on an otherwise clean surface can, in certain instances, severely influence the detailed interaction between a metal and semiconductor.

One of the thrusts of current research in the area of semiconductors surfaces and interfaces is aimed at obtaining a detailed understanding of the interplay between the crystallographic structure, the chemical composition, and the electronic properties of those regions, and at understanding the detailed electrical behavior of simple devices, such as

---

R. H. Williams • Physics Department, University College, P.O. Box 78, Cardiff CF1 1XL, U.K.

Schottky diodes, where the nature of these interfaces play a key role. During the last 15 years spectroscopic techniques have been developed which can probe the chemical and structural nature of the outermost atomic layers on solids with unprecedented accuracy, and these have been used very successfully to probe III–V semiconductors. They include low-energy electron diffraction (LEED), reflection high-energy electron diffraction (RHEED), Auger electron spectroscopy (AES), X-ray photoelectron spectroscopy (XPS or ESCA), ultraviolet photoelectron spectroscopy (UPS) and its angle-resolved version (ARUPS), and secondary ion mass spectrometry (SIMS). In this chapter we shall be dealing extensively with the application of these techniques to probe clean III–V semiconductor surfaces, and gases and metals on them. However, the basis of the techniques will not be extensively considered; the interested reader is referred to books by Ibach,[1] Brundle and Baker,[2] and Cardona and Ley,[3] and to articles by Haneman,[4] Williams,[5] and Brundle.[6]

In addition to advances on the experimental front, the past 10 years have seen considerable advances in the development and application of theoretical methods for calculating the electronic structure of semiconductor surfaces and metals on them, as well as for predicting the likely configuration of atoms on clean surfaces.[7] For a familiarization with these the reader is referred to articles by Cohen and co-workers,[8-10] Appelbaum and Hamann,[11] Chadi,[7,12] and others.[13,14] In spite of these major advances, our understanding of the detailed aspects of semiconductor surfaces and their interfaces is still at an elementary stage and indeed even for what should be one of the simplest of semiconductor surfaces, namely the Si(111), there is still no clear understanding of its crystallographic form, its electronic structure, or the driving forces associated with the various phase transitions which this surface displays.[15,16] Indeed in many senses certain surfaces of III–V semiconductors are better understood than those of silicon even though they are more complex particularly as a result of the fact that they are composed of two or more elements.

2. Interface States and Schottky Barriers

The conventional view of Schottky-barrier formation at metal–semiconductor interfaces is well known and is illustrated in Fig. 1 for the case of an n-type semiconductor. Suppose that the semiconductor has an electron affinity χ_s and has no surface states in the fundamental band gap, and that it is brought into contact with a metal of work function ϕ_m. In the simplest possible situation, shown in Fig. 1(a), ϕ_m is larger than the work function ϕ_s of the semiconductor so that electrons are transferred from the semiconductor into the metal leaving behind positively

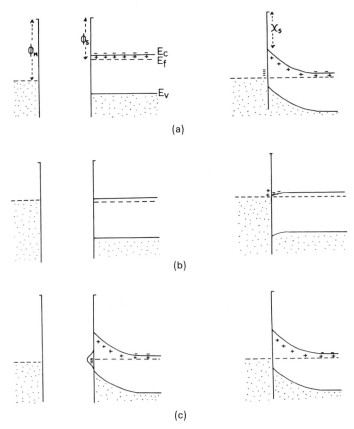

Figure 1. (a) Contact between a metal and an n-type semiconductor for the case where $\phi_m > \phi_s$ and there are no interface states. (b) Same as (a) but for the case where $\phi_m < \phi_s$. (c) A similar contact to (a) but with a high density of interface states.

charged donor impurity ions in the space-charge layer of the semiconductor. Provided that the surface dipoles that contribute to ϕ_m and χ_s remain unchanged, then the Schottky barrier ϕ_b is simply equal to $\phi_m - \chi_s$. The barrier should therefore vary for metals of differing work functions, and indeed if $\phi_m < \phi_s$ one would expect a zero height or an injecting contact as illustrated in Fig. 1(b). Similar diagrams may be drawn for the case of p-type semiconductors; in this case barriers analogous to that in Fig. 1(a) are formed when $\phi_m < \phi_s$.

Situations where electrical barriers are formed in accordance with the above model are referred to as the Schottky limit.[17] However, the assumption of complete absence of surface states in the fundamental band gap of the semiconductor is clearly not universally applicable, as pointed out by Bardeen.[18] Surface states may lead to the trapping of charge on the semiconductor surface and to the existence of band bending and a space-charge layer even in the absence of a metal contact. Consider again a very simple situation shown in Fig. (1c) where an n-type semiconductor has a band of acceptor-like surface states. Then, upon contact with a metal electrode the necessary charge transfer will only occur to or from the surface states, if the density of surface states is sufficiently large. In this limit of a large density of surface states, the interior of the semiconductor may be screened from the metal so that ϕ_b will be independent of ϕ_m. This situation is referred to as the Bardeen[18] limit.

Cowley and Sze[19] considered the situation where an n-type semiconductor has acceptor-like surface states of density D_s states cm^{-2} eV^{-1} uniformly distributed in the semiconductor band gap and in which the metal is separated from the semiconductor by a thin insulating layer of atomic dimensions. The width of this layer is δ and its relative permittivity is ε_1. It was shown that in this situation the Schottky barrier can be written as

$$\phi_b = \gamma(\phi_m - \chi_s) + (1 - \gamma)(E_g - \phi_0) \tag{1}$$

where E_g is the band gap of the semiconductor and ϕ_0 is the energy associated with a "neutral level" such that if the surface states are occupied up to ϕ_0 and are empty above it then the surface is neutrally charged. The quantity γ in Eq. (1) is given by

$$\gamma = \frac{\varepsilon_1 \varepsilon_0}{\varepsilon_1 \varepsilon_0 + e\delta D_s} \tag{2}$$

where e is the electronic charge and ε_0 the permittivity of free space. The various approximations involved in the derivation of Eqs. (1) and (2) have been considered in several articles[20,21] and in the book by Rhoderick.[22] Here it is sufficient to note that the condition corresponding to the Schottky limit is obtained for $D_s = 0$. On the other hand when D_s is large, γ is small and

$$\phi_b \simeq E_g - \phi_0 \tag{3}$$

In this latter case the Schottky barrier is independent of ϕ_m and the Fermi level is said to be "pinned" by the surface states. Assuming the interfacial layer to be a few angstroms of free space, we obtain $D_s \simeq 10^{13}(1-\gamma)/\gamma$ states cm^{-2} eV^{-1}, and for appreciable pinning, $D_s \gtrsim 10^{13}$ states cm^{-2} eV^{-1}. It is of interest to note that the number of surface atoms on a typical semiconductor is of the order of 10^{15} cm^{-2}, so that a very small relative density of surface states can have a very significant influence on Schottky-barrier formation. Other contributions to the measured values of ϕ_b, for example, image-force lowering, have been discussed in detail elsewhere[22] and will not be considered further here.

Until this point the detailed nature of the origin of the pinning states has not been considered. It is well known that surface states do occur on clean semiconductor surfaces and that they are allied to the termination of the periodic bulk lattice potential and to the unsaturated or "dangling" bonds at the surface. The nature of these states will be considered in detail for III–V compound semiconductors in Section 3. However, as pointed out by Heine,[23] it certainly is not clear that the states giving rise to Fermi-level pinning in Fig. 1(c) need be related in any simple way to the surface states in the band gap of the atomically clean surface. The very existence of the metal on the surface of the semiconductor will change the form of the potential and of the charge associated with the dangling bonds. One should therefore refer to the states responsible for pinning as "interface states." The detailed origin of these "interface states" and the way they relate to the microscopic structure of the interface region, will be a central theme of this chapter.

Although the linear interface potential theory of Cowley and Sze[19] has been extensively applied, it clearly is too simple, as it assumes a distribution of surface states which is highly unlikely to occur universally on all surfaces. Nevertheless, it has been extremely useful, with the common procedure being to evaluate D_s from the slope of the linear relationship between ϕ_b and ϕ_m for a range of metals on a given semiconductor. Unfortunately, the values of ϕ_m quoted in the literature, even for a metal such as gold, often vary over a range of up to 0.5 eV.[24] In view of these uncertainties, the application of the linear model has often been based on a rather unsound footing.

Some workers[25] have claimed that it is not the work function of the metal but rather its Pauling electronegativity X_m that should be considered in relation to Schottky-barrier height. The linear theory then yields the result

$$\phi_b = S(X_m - \chi_{sc}) + C \tag{4}$$

where C is a constant and the slope S is referred to as the "index of interface behavior. Kurtin *et al.*[25] have plotted the variation of S for a wide range of metals on many semiconductor systems and have suggested that $S \sim 0.1$ for covalent semiconductors such as Si, Ge, and the III–V compound semiconductors but that it has a value of around unity for ionic materials such as ZnS. It was reported that a sharp transition region separated the covalent ($S \sim 0$) and the ionic ($S \sim 1$) materials, but this now appears less certain following a reanalysis of the data by Schluter.[26] In addition, as will be demonstrated later, it is unlikely that the linear interface potential model is always applicable for describing interfaces formed between metals and a range of atomically clean semiconductor surfaces, including III–V materials.

Following the work of Heine[23] there have been a number of interesting and more sophisticated theoretical attempts to understand the basis of Schottky-barrier formation. Some have considered the influence and importance of many body effects,[27,28] whereas others[29-31] have considered the formation of new states at the interface due to the presence of the metal. These so-called "metal-induced gap states" (MIGS) will be briefly considered later when we discuss Fermi-level pinning by Al on GaAs (110) cleaved surfaces. The elegant self-consistent pseudopotential methods employed in these calculations ideally require the exact location and bond lengths associated with the atoms at the interface and beyond as an input. Unfortunately, however, at the present time such information is almost nonexistent for all but a few cases.

The precise growth mode of a metal film during its deposition onto any nonmetallic surface is a complex function of many parameters and some of these will be considered in more detail later. The metal atom on impact with the surface may stay stationary, may diffuse along the surface, or may diffuse into the nonmetallic solid. The latter case of interdiffusion across the boundary thus leads to an interface which is not atomically abrupt and there is much evidence that this situation arises when metals are deposited onto clean semiconductors even at toom temperature. For instance, it appears that gold diffuses into silicon in appreciable amounts when it is deposited on the clean (111) surface at room temperature.[32,33] The departure of the interface from the ideal atomically abrupt and ordered situation may have considerable implication in the formation of Schottky-barrier contacts. Accommodation of the contact atoms on substitutional or interstitial sites in the semiconductor leads to doping of the surface region, so that the assumption of a constant doping profile implicit in Fig. 1(a) is no longer applicable. In some cases, for example for Sn contacts on n-type InP or Sb on Si, the resulting near-surface doping leads to a barrier profile thin enough for electrons

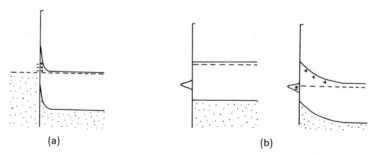

Figure 2. (a) An ohmic contact based on a thin barrier through which electrons can tunnel. (b) Defect states in the band gap and subsequent band bending for acceptor states on an n-type semiconductor.

to tunnel through, as illustrated in Fig. 2(a). The electrical contact may then show "ohmic" rather than rectifying behavior. In addition, constituent atoms from the semiconductor may diffuse into the metal contact leaving behind an excess of another constituent (e.g., As in GaAs) or generating point defects in the semiconductor. Such metal-induced defect levels may lead to electrically active sites at the interface and may lead to Fermi-level pinning as illustrated in Fig. 2(b). In a III-V solid, such as InP or GaAs, the departure of the surface from chemical stoichiometry by as little as a few percent may have a far-reaching influence on the Schottky-barrier formation. The role of interfacial defects forms the basis of the defect model[34-37] of barrier formation which we shall return to later in Section 6.

It may be seen, therefore, that the interface between a metal and a clean semiconductor may be a crystallographically and chemically complex region and one where a number of factors may contribute to the formation and profile of the electrical barrier. To gain an understanding of the relative importance of the various processes, we need to probe the microscopic atomic nature of the interface. The adsorption of gaseous adlayers on the clean semiconductor surface can generate new electronic states, either by direct interaction with the surface or by modifying the metal-semiconductor interaction. It is important, therefore, to gain an understanding of the nature of adsorbates on semiconductors and the detailed way in which they influence the electrical properties of the interface.

3. Clean Surfaces of III-V Semiconductors

In this section the crystallographic form and the associated electronic structures of selected III-V semiconductor surfaces will be considered. The most thorough studies have been carried out on the (110) cleaved face and of the various materials GaAs is the most extensively probed. For this reason the emphasis in this section will be biased towards GaAs surfaces, which will be used as a basis for discussing metal interaction with other III-V semiconductors.

3.1. Crystallographic Structures of Surface and Bulk

The majority of III-V compound semiconductors crystallize in the zinc-blende structure. This structure, illustrated in Fig. 3, can be considered as two interpenetrating face-centered-cubic lattices with group III and group V atoms on different sublattices. The lattice constants and structures associated with various compounds are given[38] in Table 1;

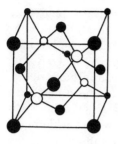

Figure 3. The zinc-blende structure of the III–V semiconductors.

note that GaN has a wurtzite structure. The bonding in the zinc blende compounds consists of sp^3 hybrid orbitals and, in contrast to the purely covalent group IV semiconductors, there is some charge transfer between the two elements and thus an ionic contribution to the bonding. The bonding and the ionicities have been considered in some detail by Phillips.[39] The ionic contribution to the bonding in III–V semiconductors manifests itself in its cleavage behavior. Crystals cleave along the nonpolar (110) plane although the least numbers of bonds would be broken by cleavage along (111) planes (which is why Si and Ge cleave along this plane). The ideal structures of the (001), (110), and (111) surfaces are illustrated in Fig. 4 [the (110) face is the relaxed configuration][55]. The polar (001) and (111) surfaces shown are composed of layers of all group III or all group V elements, whereas the (110) surface, in the ideal case, consists of equal numbers of each of the constituent atoms.

In general the clean free surfaces of solids do not retain the ideal structures illustrated in Fig. 4. The polar surfaces are often "reconstructed" and may adopt a range of chemical compositions. Some of the nonpolar cleaved (110) surfaces have been thoroughly studied by low-energy electron diffraction. In all cases studied the symmetry of the

Table 1. Lattice Constants and Structures of Some Compounds

Compound	Structure	Lattice constant at 300 K (Å)	Band gap at 300 K (eV)
GaN	Wurtzite	$a = 3.186, c = 5.176$	3.5
BN	Zinc blende	$a = 3.615$	~7.5
BP	Zinc blende	$a = 4.538$	6
AlSb	Zinc blende	$a = 6.1355$	1.63
GaSb	Zinc blende	$a = 6.0955$	0.67
GaAs	Zinc blende	$a = 5.6534$	1.43
GaP	Zinc blende	$a = 5.4505$	2.24
InSb	Zinc blende	$a = 6.4788$	0.16
InAs	Zinc blende	$a = 6.0585$	0.33
InP	Zinc blende	$a = 5.8688$	1.29

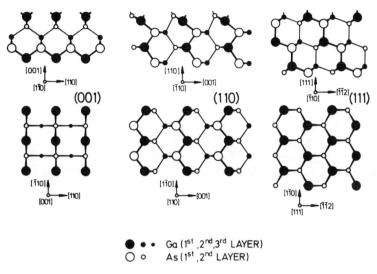

Figure 4. The structure of (001), (110), and (111) surfaces of a zinc-blende compound semiconductor (after Ref. 55).

surface unit mesh is that expected for a truncated bulk solid, that is, there is no "reconstruction" of the surface leading to unexpected diffraction beams. There are several reasons, however, for believing that the surface atoms are "relaxed" from their ideal bulk positions, adopting a configuration illustrated in Fig. 5. First, the LEED intensity profiles measured by

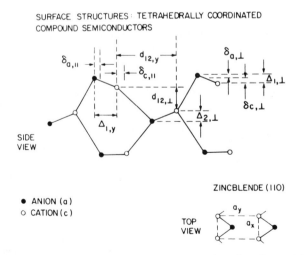

Figure 5. The relaxed (110) surface of a III–V compound semiconductor (after Ref. 44). Measured values of the parameters shown are given in Table 1.

Tong et al.[40] and Mayer et al.[41] on clean (110) GaAs surfaces suggest a tilting of the surface bonds so as to displace the As atom outwards and the Ga inwards by a small amount. Kahn et al.[41] also report a small movement associated with atoms in the second layer; the parameters determined by them are given in Table 2. Second, Chadi[7] has adopted a total energy minimization approach to the determination of the (110) surface atomic geometries. These calculations lead to the conclusion that the surfaces will be relaxed in a manner similar to that illustrated in Fig. 5, and that the magnitude of the relaxation may be appreciable and dependent upon ionicity. Third, it will be shown later that the absence of surface electron states in the band gap[42,43] of (110) GaAs cleaved surfaces demands a relaxation of the surface atoms as illustrated in Fig. 5.[8,9]

The structures derived by Mayer et al.,[41] Duke et al.,[44] and Kahn et al.[45] for (110) surfaces of GaAs, InP, and InSb are given in Table 2. These parameters are probably the best available at present but may require some modification in the light of improved and more sophisticated experiments. Slight differences may also be expected for (110) surfaces prepared by ion sputtering and annealing rather than by cleaving.[45] The precise structure illustrated in Fig. 5 is consistent with the tendency of the five valence electrons associated with the surface As atoms, in GaAs, to regroup into an s^2p^3 configuration. Likewise, the three electrons associated with the Ga atoms regroup to form a more planar sp^2 bond arrangement rather than the sp^3 configuration associated with bulk bonds.

The behavior of (001) and (111) surfaces differ substantially from that associated with (110) surfaces. As illustrated in Fig. 4, both (001) and (111) surfaces are polar due to the alternate arrangement of the anion and cation layer parallel to the surface, so that the ideal surface

Table 2. Parameters Determined for (110) Surfaced of GaAs, InSb, and InP

Material	Layer	$\delta_{a,\perp}$	$\delta_{c,\perp}$	$\delta_{a,\parallel}$	$\delta_{c,\parallel}$	$\Delta_{1,\perp}$	$\Delta_{2,\perp}$
GaAs (110)	1	0.144	0.506	−0.332	−0.486		
	2	0.06	0.06	0.0	0.0	0.650	−0.120
	3	0.0	0.0	0.0	0.0		
InSb (110)	1	0.177	0.604	−0.384	−0.584		
	2	0.090	0.090	0.0	0.0	0.781	−0.180
	3	0.0	0.0	0.0	0.0		
InP (110)	1	0.060	0.630	−0.344	−0.516		
	2	0.065	0.065	0.0	0.0	0.690	−0.130
	3	0.0	0.0	0.0	0.0		

may consist of all anions or all cations. In practice these surfaces are rarely ideal and display numerous reconstructed forms which are dependent on the method of preparation and the chemical composition of the surface. To date, most studies have been carried out on GaAs for which a variety of surface structures have been observed for (001) surfaces grown by molecular-beam epitaxy. The C(2 × 8), 2 × 4, and C(4 × 4) structures appear when the outermost atomic layer is rich in As (As stabilized), whereas the C(8 × 2), 4 × 2, 4 × 1, and 4 × 6 are generated when Ga is in excess (Ga stabilized). There have been several attempts to establish the precise chemical composition corresponding to the various surface reconstructions and the results presented by Bachrach[46] and Drathen et al.[47] are shown in Table 3. The column denoted "As coverage" refers to the fraction of the outermost surface layer composed of As atoms. There are some discrepancies in the two sets of measurements, in particular for the C(8 × 2) surface which is reported to have an arsenic coverage of 0.52 in one case and 0.22 in the other. This may result from the fact that different techniques were used to probe the structures; Drathen et al.[47] used Auger electron spectroscopy, whereas Bachrach[46] used photoelectron spectroscopy with synchrotron radiation as the exciting source. In view of these uncertainties, therefore, the values quoted in Table 3 should be taken as a guide rather than as definitive compositions.

The various structures can be generated by appropriate choice of substrate temperature and Ga to As flux ratio; the relevant phase diagram has been established by Massies et al.[48] At the present time, therefore, the precise crystallographic forms associated with the various (001) surface reconstructions are not well understood and the extension of LEED intensity studies to these faces is obviously desirable. In spite of the uncertainties, some models have been proposed. Neave and Joyce,[49] for example, favor a model of the surface which involves ordered vacancies, as suggested by Phillips,[50] rather than a model based on the distortion of the surface associated with Fermi-energy instabilities.[51] It may well

Table 3. Results of Bachrach[46] and Drathen et al.[47] for GaAs.

Surface structure	As coverage[46]	As coverage[47]
C(4 × 4)	1.0	0.86
C(2 × 8)	0.89	0.61
C(8 × 2)	0.52	0.22
1 × 6	0.42	0.52
4 × 6	0.31	0.27

be, of course, that the reconstructions involve many more atoms than just those which compose the surface layer so that establishing precise structures is likely to be a very difficult task. The possibility of facets must also be considered; Ludeke and Ley[52] found that the X-ray photoelectron spectra associated with (100) surfaces revealed many similarities to those associated with (110) surfaces and postulated that (110) microfacets may be playing an important role on (100) surfaces. Firm conclusions cannot be reached, however, at this stage.

Far less is known about the (001) surface structures of other III–V semiconductors. Bachrach[46] has studied the (001) surface of AlAs and reports that it is predominantly disordered, although a 3×2 phase was observed over a narrow composition range of 25% As surface coverage. For the case of InP, Farrow et al.[53] report that a $C(2 \times 8)$ phosphorus-stabilized surface could be reproducibly prepared and a 4×1 structure has also been reported.[54]

Finally, in this section, we consider briefly the (111) and ($\bar{1}\bar{1}\bar{1}$) surfaces of GaAs, which have been studied in some detail, particularly by Ranke and Jacobi.[55] The ideal (111) and ($\bar{1}\bar{1}\bar{1}$) surfaces should be composed of all Ga atoms and all As atoms, respectively, in the outermost atomic layer. However, the surfaces are not ideal and a number of reconstructed forms are observed. On the (111) surface a 2×2 form has been reported, whereas three forms have been described for the ($\bar{1}\bar{1}\bar{1}$) face.[55] These are (2×2), $(\sqrt{3} \times \sqrt{3})R30°$, and $(\sqrt{19} \times \sqrt{19})R23.2°$ and are reported to correspond to arsenic surface coverages of 0.87, 0.67, and 0.47 monolayer, respectively. It is possible that the 2×2 surface actually

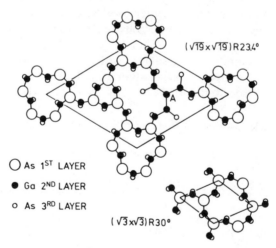

Figure 6. Models for the GaAs ($\bar{1}\bar{1}\bar{1}$) $(19^{1/2} \times 19^{1/2})R23.4°$ and $(3^{1/2} \times 3^{1/2})R30°$ surfaces proposed in Ref. 55.

corresponds to a coverage of one monolayer of As atoms and Ranke and Jacobi[55] suggest that it may result from a distortion of the surface involving alternatively raised and lowered rows of atoms on the surface, somewhat analogous to the early models proposed to account for the Si(111) 2×1 surface structure.[56] A short anneal at 770 K changes the 2×2 structure first into the $(\sqrt{3} \times \sqrt{3})R30°$ form and then into the $(\sqrt{19} \times \sqrt{19})R23.2°$ structure. Possible structures associated with these are shown in Fig. 6.[55] At the present time these can only be tentative but it is of interest to note that they do combine maximum bond saturation with minimum As content, as pointed out by Ranke and Jacobi.[55]

3.2 Bulk and Surface Electronic States

The electronic structure associated with the crystalline bulk form of III–V compound semiconductors has been widely studied both experimentally and theoretically and is now well understood. Figure 7(a) shows a calculated band structure for GaAs[57] for the various Brillouin-zone directions shown in Fig. 7(b). Details of the valence-band structure have been established by a range of optical techniques and in particular by X-ray photoelectron spectroscopy[58] and angle-resolved photoelectron spectroscopy.[59] The fundamental band gaps associated with the various III–V materials vary over a wide range and are presented in Table 1. Most of these, including GaAs and InP, are direct-gap semiconductors[38]; however, GaP displays an indirect band gap.

The termination of the crystalline lattice at the surface clearly leads to the generation of electronic states which differ from those in the bulk, that is, to surface states and resonances. These states are determined by the precise chemical composition and crystallographic structure associated with the surface atoms, and since the (110) surfaces are the most simple in this respect for III–V semiconductors, it is these surfaces that are the most widely studied and best understood. Early experimental studies of clean cleaved GaAs (110) surfaces[60–62] suggested that a band of unoccupied surface states did exist in the fundamental band gap and indeed theoretical estimates[8,10] for the unrelaxed surface also led to the conclusion that occupied and unoccupied states should be observed in the band gap. A series of more recent experiments, however, confirmed the earlier observation of Van Laar et al.[63] that no states exist in the gap provided the surface is of high quality. Van Laar et al.[63] observed the position of the Fermi level on n- and p-type GaAs crystals and did not find the Fermi-level pinning anticipated if surface states did exist in the band gap. However, they did find that surfaces with large defect densities resulting from poor cleaves did show strong pinning behavior. Huijser et al.[64] found that GaSb, InP, and InAs likewise show no Fermi-level

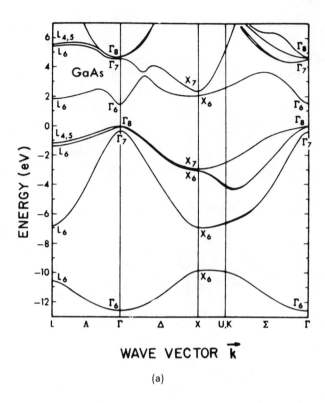

(a)

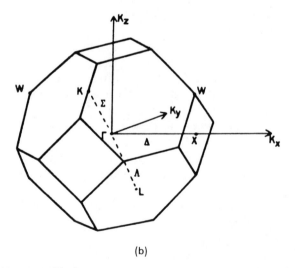

(b)

Figure 7. (a) The calculated electronic band structure of GaAs (after Ref. 57). (b) The reduced Brillouin zone of GaAs.

pinning on high-quality clean cleaved surfaces but, in contrast, some pinning behavior was observed for GaP, consistent with empty states in the gap about 1.7 eV above the valence-band maximum. The absence of gap states on GaAs (110) surfaces has been confirmed by optical-absorption techniques,[65,66] photoemission,[34] and contact potential difference,[63] as well as by electron-energy-loss spectroscopy.[64]

Surface states and resonances have been detected on GaAs (110) cleaved surfaces by several techniques and it has been shown that they lie outside the fundamental band gap. Theoretical calculations taking relaxation of the surface structure into account[9,12,67] are quite compatible with these observations. Chadi[7,12] used the tight-binding method to explore the influence on the gap surface states of distorting the surface as outlined in Section 3.1 and clearly showed the displacement of occupied and unoccupied surface states from the gap, as sketched in Fig. 8. Bertoni et al.[67] have also applied tight-binding methods to a range of III–V compounds and assumed that the relaxation of (110) surfaces is the same as that for GaAs (110). It was then shown that surface states on GaAs, GaSb, InP, InAs, and InSb (110) overlapped the valence and conduction bands and are absent from the gap. However, for GaP a band of empty states in the band gap was predicted and indeed there is experimental evidence that this does exist.

The occupied "dangling-bond" surface states on cleaved (110) surfaces of GaAs and InP have been detected by angle-resolved photoelectron spectroscopy,[68–70] and furthermore their dispersion has been measured. Figure 9(a) shows experimental photoemission spectra for cleaved InP, both for the clean surface and following adsorption of water. The peak labeled S_1 is particularly sensitive to adsorbates and has been attributed to emission from anion derived "dangling-bond" surface states.[70] A very similar behavior is observed for GaAs.[68,69] The surface state emission in Fig. 9(a) is particularly prominent for the larger angles, and the dispersion of the state may be plotted, as shown in Fig.

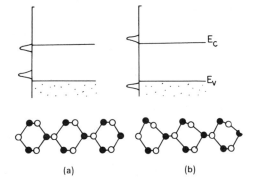

Figure 8. (a) Illustration of surface states in the band gap for an unrelaxed surface. (b) The surface states are driven from the gap by the relaxation of the surface atoms.

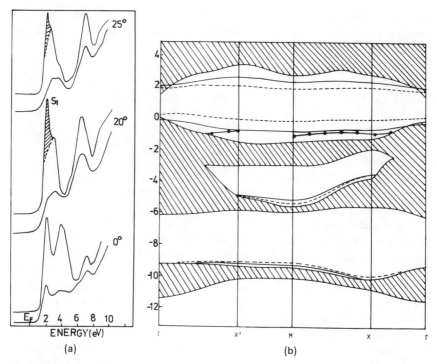

Figure 9. (a) Photoemission spectra from clean InP (110) surfaces at three polar angles of emission, and following exposure to 4×10^4 Langmuirs of water (lower). The enhanced attenuation of the leading peak is clear. The photon energy was 21.2 eV and the emission was along ΓX of the surface Brillouin zone (after Ref. 70). (b) The experimentally measured dispersion of the phosphorus dangling-bond surface state (closed circles). The solid lines are calculated surface states for a relaxed surface and the dotted line for an unrelaxed InP (110) surface. The hatched region is the bulk band structure projected onto the surface Brillouin zone (after Ref. 70). The vertical scale represents energy (in eV) relative to the valence-band edge.

9(b). Also shown in Fig. 9(b) are the theoretical estimates of Srivastava et al.[70] of the surface state energies for InP. These calculations employ both tight-binding and self-consistent pseudopotential methods and are similar to those carried out by Chelikowsky and Cohen[9] for GaAs. In the absence of surface relaxation occupied and unoccupied states are predicted in the gap. The results are shown by the dashed line in Fig. 9(b). For the relaxed surface, the parameters used are those deduced by Chadi[12]; the occuped anion "dangling-bond" derived surface state and the empty cation derived state are shown by the solid lines. The agreement between experiment and theory is satisfactory for these particular states. Similar calculations have also been carried out by Bertoni et al.[67]

Figure 10. Calculated charge-density contours associated with the upper occupied surface state on the InP (110) surface (after Ref. 70).

In Fig. 10 the charge-density contours associated with the upper occupied derived states of Fig. 9(b) are shown. It is very apparent that the occupied state is associated with the phosphorus surface atom; in the same way it was shown that the lowest unoccupied state is based on the surface indium atoms. The calculations of Chelikowsky and Cohen[9] for GaAs (110), making use of the surface parameters derived by Tong et al.,[40] lead to similar conclusions. Finally, we note that the empty surface states have been detected by techniques such as isochromat spectroscopy[72] and photoemission partial yield.[73] Care has to be taken in establishing the energies of states by these methods, however, because they can be severely influenced by excitonic effects.[74,75]

The nature of surface electronic states on the polar (001) and (111) faces of III–V semiconductors is, not surprisingly, not understood at present. Theoretical predictions of state energies are hampered by the lack of crystallographic information of the surface structures so that, to date, most authors have considered only the ideal 1 × 1 configuration. Appelbaum et al.[76] have carried out a self-consistent calculation for the Ga-terminated (001) GaAs surface and predict two bands of states in the fundamental gap for the ideal surface. One of these bands is derived from the "dangling bonds" whereas the other is associated with bridge bonds within the Ga plane. Ivanov et al.[77] and Pollmann and Pantelides[78] have applied the tight-binding method to study (100) GaAs surfaces and predict a number of surface states which overlap the occupied and unoccupied bands. Ludeke and Esaki[79] and Massies et al.[80] have used low-energy electron-loss spectroscopy (EELS) to probe these surfaces and interpret a loss peak at around 20 eV in terms of excitations from the Ga 3d core level to an empty state close to the conduction-band edge and related to the Ga "dangling bond." Ludeke and Esaki[79] also postulated the existence of an occupied As "dangling-bond" related state at around 1 eV below the valence-band edge, and an As–Ga back-bond related state at about 7–10 eV binding energy. The occupied states on the As-stable (001) 2×4 and $C(4 \times 4)$ surfaces prepared by molecular-beam epitaxy have been studied by Larsen et al.[81] using angle-resolved photoelectron spectroscopy. A "dangling-bond"-like surface state having largely sp_z character was identified on the 2×4 surface at a binding energy of around 1 eV. Furthermore, a model for the 2×4 structure was proposed based on the dimerization of adjacent surface atoms and with the bridge bond coupling the surface As atoms tilted so as to yield a p_z as well as a p_x component. Massies et al.[80] studied work-function differences on n- and p-type GaAs crystals and concluded that the Fermi level on these (001) surfaces was strongly pinned by a high density of gap states. It was suggested that these states were either due to defects such as Ga or As vacancies, or due to the intersection of facets. Ludeke and Ley[52] suggested that the (100) surface may be composed of (110) facets and atoms at the intersection of these facets may have the same dangling-bond configuration as for steps on the (110) surface. These steps are known to yield strong Fermi-level pinning.

Electronic states on the $(\overline{1}\overline{1}1)$ Ga-rich and As-rich surfaces have been studied by Ludeke and Esaki[79] and the (111) face by Ludeke and Koma[82] using EELS. These surfaces have also been probed by Jacobi et al.[83] using angle-resolved photoelectron spectroscopy. The latter observed emission from surface states located 1 to 2 eV below the valence-band edge on the (2×2) As-stabilized structure and this was considerably attenuated on conversion to the $\sqrt{19}$ As structure. This state was attributed to an As dangling bond which is symmetric about the surface normal for the former structure, that is, s or p_z derived, and which changed to purely s-like for the latter structure. The photoionization probability for s-states is low at the photon energy used (21.2 eV). At the present time, however, this interpretation is uncertain and must await further studies.

3.3. Surface Imperfections and Defects

As pointed out in the previous section clean cleaved high-quality (110) surfaces of III–V crystals such as GaAs do not lead to strong Fermi-level pinning in the fundamental band gap. The situation is different, however, if there exists a high density of steps on the surface. It has then been shown that gap states are generated which lead to strong pinning on GaAs (110) surfaces.[42,84] In Section 2 it was pointed out that strong pinning can be caused by as little as 10^{12}–10^{13} surface states cm^{-2}. Since the surface atom density is typically around 8×10^{14} cm^{-2} it may be seen that a very small deviation of the surface from the ideal may have a very pronounced effect, assuming of course that atoms in nonideal locations lead to gap states. Not a great deal is known about atoms at step edges on III–V cleaved surfaces but various configurations are possible. Figure 11(a) shows steps where the atoms on the edges of the step resemble those on a (111) or ($\overline{1}\overline{1}\overline{1}$) surface. Complex reconstruction effects may occur in these areas and without a detailed knowledge of the atomic positions it is difficult to predict whether resulting electronic states will lie in the band gap. Step-related defect levels have clearly been identified on vacuum cleaved InP crystals by Street et al.[85,86] using photoluminescence. Typical results are illustrated in Fig. 11(b). The peaks labeled 1 to 3 in Fig. 11(b) are associated with bulk recombination, but the peak labeled 4 was clearly associated with surface damage and was ascribed to luminescent recombination at step-induced defect levels just below the surface. The areas where these levels were prominent showed significant differences in band bending upon the adsorption of gases, such as oxygen, on the surface. Clearly, therefore, steps and other imperfections on cleaved surfaces may have a very significant influence on the interaction of the surface with metal and gaseous adsorbates.

The influence of heating a III–V crystal may also have significant effects on the nature of imperfections in the surface region. First, impurities from the surface may diffuse into the crystal and likewise dopants from the crystal may accumulate on the surface, giving rise to a complex distribution of interstitial and substitutional impurities in the surface region. Second, heating may lead to surfaces which are chemically nonstoichiometric. This is very often the case for III–V surfaces cleaned by ion sputtering and annealing and strong Fermi-level pinning is usually seen on surfaces prepared in this way.[34,87] Detailed studies of the chemical composition of these surfaces usually show them to be deficient in the anion component.[88] In addition, heating often leads to agglomerates of the cation species on the surface, for example, on InP the surface may be covered by small In droplets following sputtering and annealing.[89] Finally, surfaces prepared by sputtering and annealing often have a very large density of steps on the surfaces. Detailed studies of LEED spot profiles of GaAs (110) by Welkie and Lagally[90,91] show that sputtered and annealed surfaces are highly strained and that between 9 and 33% of the atoms on the surface may be at step edges. This density is certainly sufficiently high to significantly influence the properties of the surface.

Clearly, therefore, quite small deviations of a surface from the ideal, perfectly ordered, chemically stoichiometric situation may be of consider-

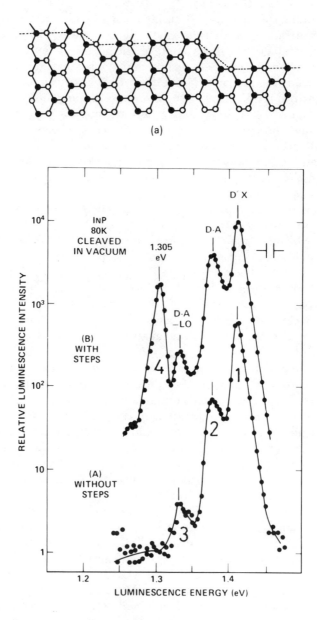

Figure 11. (a) Illustration of atoms on step edges, on a III–V semiconductor surface. (b) Photoluminescence spectra for a cleaved InP surface with and without steps (after Ref. 85).

able importance. Defects such as steps, vacancies, and antisite defects may influence the detailed interactions with gases and metals, the intermixing of the semiconductor and metal at an interface, as well as the Fermi-level pinning behavior. It is thus of considerable importance to understand the nature of these defects theoretically. Unfortunately, however, it is extremely difficult to calculate the energy levels and charge distributions associated with defects in a reliable and quantitative way, even when those defects are in the bulk. The current state of such calculations has been reviewed by Jaros[92] and Pantelides[93] and is outside the scope of this book. The most accurate calculations involve self-consistent pseudopotential approaches and indeed progress has been made by these means in our understanding of defects in silicon.[94] For the III–V's there have been attempts to estimate the energies associated with bulk and surface vacancies and antisite defects using the tight-binding technique.[95–102] Although the absolute accuracy of such calculations may be in doubt, they are useful as a qualitative guide and reference will be made to these calculations in Section 6.

4. Adsorption of Gases on Clean III–V Semiconductors

4.1. General Introduction

The adsorption of a gas on a clean semiconductor surface can lead to a drastic change in the nature of that surface and in its interaction with metal contacts. The fabrication of a wide range of semiconductor devices involves stages where metals are deposited onto surfaces covered by adsorbed contaminants so that obtaining a detailed understanding of adsorbate–semiconductor interactions is of technological as well as fundamental importance. Gases may interact with surfaces in various ways. Physisorption involves a weak interaction between the surface and the adsorbate, whereas chemisorption involves strong interactions. In addition, of course, the adsorbed gas may interact chemically with the semiconductor surface so as to yield new reaction products and dissociate the semiconductor, that is, break crystal bonds of the substrate. Both chemical reactions and chemisorption may lead to a modification of surface states on the clean semiconductor and may generate new donor or acceptor states in the energy gap. These states in turn may lead to band bending and a surface space-charge layer perhaps similar to that illustrated in Fig. 1(c). Again a density of surface states of around 10^{12} cm^{-2} may be sufficient to lead to appreciable effects so that an adsorbate surface coverage of 0.1–1% of a monolayer can have significant effects on the electronic properties of the whole surface region. The adsorbed species may also diffuse into the semiconductor and dope the surface region, and it can also alter the growth mode of a metal deposited on it.

In the following sections the adsorption of oxygen, chlorine, hydrogen, H_2S, and H_2O on clean III–V semiconductors is considered. The most thorough studies involve oxygen and chlorine adsorbed on cleaved GaAs and InP surfaces and for that reason the major part of this section will be devoted to these systems. The formation of thick oxide layers will not be considered in detail; this is considered in detail elsewhere in this volume.

4.2. Oxygen Adsorption

The most detailed studies of oxygen adsorption of III–V semiconductors has been on the (110) cleavage face of GaAs. Early studies of gas uptake on vacuum crushed crystals[103] revealed two phases of adsorption: an initial fast phase followed by a much slower phase which was more akin to physisorption. More recently, Dorn et al.[104] and Luth and Russell[105] studied adsorption on vacuum cleaved surfaces by LEED, Auger electron spectroscopy, and ellipsometry and estimated that the fast phase saturated following an adsorption of one oxygen atom for each pair of surface atoms (i.e., 0.5-monolayer adsorption). The initial sticking coefficients S_0 were estimated to be 3×10^{-5} and 1.4×10^{-6} for n- and p-type crystals, respectively. Mark and Creighton[106] reported similar values of S_0, but found no significant differences between n- and p-type crystals.

Extensive studies of oxygen adsorption on cleaved (110) GaAs surfaces have been carried out by Pianetta et al.[107] using photoelectron spectroscopy. Core-level spectra following excitation by radiation from a synchrotron source are shown in Fig. 12. Upon long exposures to oxygen, the As 3d emission suffers a large chemical shift and the resulting shifted satellite appears to saturate following an exposure of around 10^{12} Langmuirs. The sticking coefficients resulting from these studies are thus very much lower than previously measured ones. These differences in adsorption rate may be partly understood in terms of two experimental parameters. First, any hot filaments, such as ionization gauges in the vacuum system, can excite the oxygen gas before adsorption.[105,107] The gas molecules may be dissociated in this way or may be excited from the triplet ground state of O_2 to the singlet excited state.[108] In any case there is a large body of evidence showing that oxygen exposed to such filaments is more reactive and thus leads to higher adsorption rates. The differences in sticking coefficients reported by various workers may thus be partly due to experimental conditions pertaining at the time. Second, the state of perfection of the surface may also have a significant influence

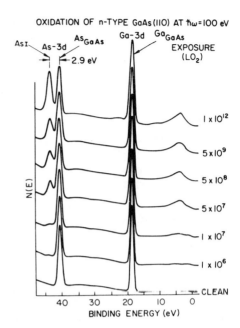

Figure 12. Photoelectron spectra for a vacuum cleaved GaAs surface and following exposure of the surface to oxygen (after Ref. 107).

on adsorption rates. A high density of steps or other imperfections such as those resulting from sputtering and annealing or poor cleaves leads to enhanced adsorption rates.[90,109,110] It has also been reported that the presence of small amounts of a metal such as Cs on the GaAs (110) cleaved surface may lead to oxygen adsorption rates many orders of magnitude higher than that appropriate for the clean surface.

The precise mechanisms involved in the adsorption process and subsequent oxidation have been the subject of much discussion and are not fully understood at the present time. The saturation of the initial adsorption stage at 0.5 monolayer observed by Dorn et al.[104] together with the chemical shift associated with the As 3d emission in Fig. 12, at first sight, suggest adsorption on As sites and charge transfer between oxygen and arsenic surface atoms. However, other experimental data lead to somewhat different conclusions. First, LEED studies[112] show that the absorbate-covered surface is not ordered, and the saturation coverage has been reevaluated as greater than 0.5 monolayer.[113] Second, the electron-energy-loss studies of Ludeke[114] show that oxygen adsorption has a significant influence on the transition from the Ga 3d core level to the empty Ga derived surface states, indicating the involvement of Ga sites in the adsorption process. In fact, the Ga 3d emission in Fig. 12 also does suffer a small chemical shift of around 0.8 eV, consistent with this view. Barton et al.[108] have used a cluster approach to calculate possible chemical shifts and show that chemisorption of oxygen to surface As can also lead to small chemical shifts of the Ga 3d emission, as observed. Theoretical calculations by Joannopoulos and Mele[115] also lead to the conclusion that the formation of O–As bonds may influence Ga derived surface states. And finally, the SEXAFS (Surface X-ray Extended Fine Structure Spectroscopy) measurements of Stohr et al.[116] are consistent with adsorption on As sites.

Detailed XPS studies of the adsorption of oxygen on (110) GaAs surfaces have been carried out by Brundle and Seybold.[117] Examination of emission from oxygen 1s and 2p orbitals leads to the conclusion that the oxygen is in a dissociated rather than molecular form. Brundle and Seybold also estimated the stoichiometry of the surface by detailed examination of the chemically shifted component of the Ga 3d and As 3d emission. It was suggested that the adsorbate phase is composed of the oxides Ga_2O_3 and As_2O_3. The data are also consistent with a very interesting model presented by Ludeke[114] in which the adsorption proceeds via an atomic oxygen bridge bonded to neighboring Ga and As atoms, as illustrated in Fig. 13. The surface resulting from this model has the desired disorder and has an effective stoichiometry equivalent to that of $GaAsO_3$.

The dependence of the adsorption process on surface perfection suggests that defects play a key role in the process. Barton et al.[108] have suggested that defect sites may serve to dissociate adsorbed molecular oxygen releasing atomic oxygen which then forms strong bonds with Ga and As sites. Mark and Creighton[106] suggested that the energy released in the exothermic adsorption process could generate further defect states. At the present time, however, no conclusive statements can be made regarding the kinetics of the adsorption process. It is of considerable interest to note that strong pinning of the Fermi level results from the adsorption of very small amounts of oxygen on cleaved surfaces of n- and p-type GaAs. Figure 14(a) shows the movement measured by Spicer and co-workers[34] as a function of oxygen exposure. The Fermi level of the n-type samples starts off by the conduction band, in the flat-band condition, and is then pinned around 0.7 eV below this when as little as 1% of the surface is adsorbed oxygen. There is also some photoemission evidence[118] that at these very low covereages the oxygen exists in a different chemical state from that observed at higher coverages. It may be that it is adsorbed at defect sites, or indeed it may be that the oxygen is in a largely molecular state, or it may be that surface defects are generated. The low coverages involved are close to the detection limit of the electron spectroscopies presently available.

The interaction of oxygen with (100) and (111) GaAs surfaces has not been as thoroughly studied. Recent studies have been reviewed by Ranke and Jacobi.[55] It seems that the oxygen sticking coefficient on

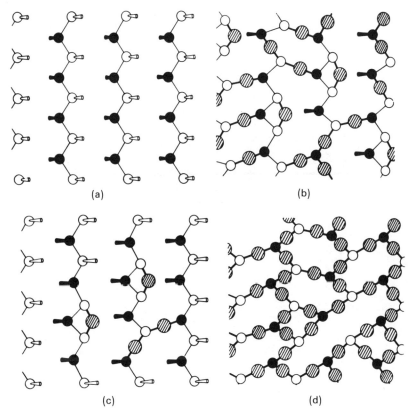

Figure 13. (a) The clean GaAs (110) surface and (b)–(d), with progressive oxidation (after Ref. 114). Open circles—Ga. Full circles—As. Shaded circles—O.

(100) surfaces is highly dependent on the chemical composition of the surface,[82] being appreciably larger on the Ga-rich 4 × 6 surface than on the As-stabilized C(2 × 8) face. It was assumed[82] that adsorption was mostly associated with gallium-rich sites but, as pointed out by Ranke and Jacobi,[55] this may not be so since the kinetics of the adsorption process may depend in a complex way on surface composition. The Ga-rich faces of the (111) surfaces also give rise to more rapid adsorption; for example, the sticking coefficient on the $(\sqrt{19} \times \sqrt{19})$ surface is around 10^3 smaller than on the 1 × 1 surface.[55] It is difficult to account for such large differences simply by assuming adsorption on Ga sites. Ranke and Jacobi[55] suggest that Ga atoms are actually displaced from the surface during the adsorption process, leaving highly active sites which may accelerate further adsorption. It was suggested that both molecular and atomic oxygen coexist in the adsorbed phase and interact with defect sites in a complex way.

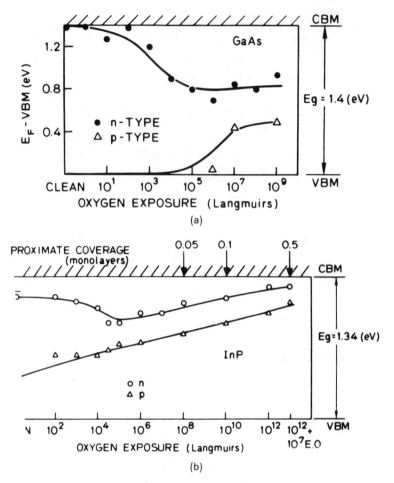

Figure 14. (a) The upper curve shows the shift of the Fermi level on an *n*-type cleaved (110) GaAs surface from its clean surface value, upon exposure to oxygen. The lower curve shows the shift for a *p*-type crystal. The vertical energy scale corresponds to the band gap (after Ref. 34). (b) Same as (a) but for cleaved InP (110) surfaces (after Ref. 34).

The adsorption of oxygen on other III–V semiconductors has not been widely studied. It appears that oxides are formed more readily on GaSb cleaved surfaces than on GaAs.[119] Exposure of InP cleaved surfaces to oxygen also leads to oxide formation providing there are hot filaments in the system leading to excited oxygen. The oxidation results in a chemical shift of over 4 eV in the phosphorus 2p photoemission,[119] whereas the shift associated with the In 4d emission is very much smaller (~0.3 eV). Adsorption of unexcited oxygen leads to a highly disordered surface layer[120] and the adsorption process may be significantly altered by electron beams such as those used in Auger electron spectroscopy.[120] Very small oxygen coverages again lead to strong pinning of the Fermi level on both n- and p-type crystals,[119,121] as illustrated in Fig. 14(b). On n-type crystals, the Fermi level at the surface first shifts by around 0.3 eV from near the conduction-band edge towards the center of the gap. The surface coverage at this point is less than 0.05 monolayers. For higher coverages the Fermi level shifts back towards the conduction-band edge as shown. This behavior is also reflected in the photoluminescence studies of Street et al.[85] At very low coverages it is likely that molecular oxygen is adsorbed forming acceptor states on the surface (yielding O_2^- species). At higher coverages defects are formed as the surface becomes disordered and these defect states are thought to pin the Fermi level close to the conduction band on n- and p-type crystals. These defect states will be considered further in Section 6.

Finally, in Table 4, a compilation of some recent important work relating to oxygen adsorption on clean III–V semiconductor surfaces is presented.

Table 4. Summary of Work Relating to Oxygen Adsorption

Material	Surface	Technique	References
GaAs	(110)	UPS	84, 90, 107, 123, 134
		XPS	117, 127, 132
		EELS	104, 105, 114
		SEXAFS	116
		LEED/AES	106
		Photovoltage	124, 126
		Photoyield	125
		Work function	122
		EPR	135
		Theory	108, 115
	(111) (100)	EELS	82
		UPS	55
		LEED/AES	106
InP	(110)	LEED/AES/CPD	120
		UPS/XPS	119, 123
		SIMS	128
GaP	(110)	Photoyield	129
		XPS	127, 132
		AES	133
	($\overline{111}$)	FEM	130
		Ellipsometry	136
InSb	(110)	XPS	127, 132
		LEED/Surface conductivity	131
GaSb	(110)	UPS/XPS	34, 119, 123

4.3. Chlorine on III–V Semiconductors

The adsorption of chlorine on silicon (111) surfaces has been widely studied and is now relatively well understood.[137,138] Studies have also been carried out on cleaved (110) surfaces of GaAs, GaSb, and InSb by Margaritando et al.[139] and on InP by Montgomery et al.[140] The latter used angle-resolved UPS and XPS as well as LEED and Auger electron spectroscopy. Low exposures of the clean cleaved InP surface to chlorine led to a UPS structure which was interpreted as molecular or atomic adsorption. Large exposures, however, leads to total surface disorder and the loss of some phosphorus from the surface. The electron affinity of the (110) surface increases appreciably even at low coverages, consistent with the transfer of electrons to the highly electronegative adsorbate.

Margaritando et al.[139] used angle-integrated photoemission and a synchrotron source to probe the saturation adsorption on cleaved surfaces. A typical spectrum for Cl on (110) InSb is shown in Fig. 15. Three strong chlorine-related features labeled A, B, and C are observed. Tight-binding calculations of the surface density of states were also carried out[139] for a number of possible adsorption geometries. The best agreement between experimental photoemission spectra and calculated local density of states was obtained for the geometry where chlorine atoms are attached to surface anions and where the III–V surface is still relaxed as illustrated in Fig. 5. A calculated density-of-states distribution for Cl on InSb is illustrated in Fig. 15 for the geometry shown. It was reported that the various peaks in the experimental spectra showed little dependence on the polarization of the exciting radiation in contrast to observations of Cl adsorption on silicon (111) surfaces.[138] This is almost certainly

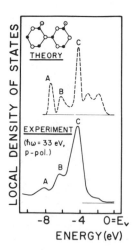

Figure 15. Comparison of the photoemission spectrum (lower curve) and calculated local density of states, for the geometry shown, of Cl on (110) InSb (after Ref. 139). The Cl is attached to the surface anions.

because the surface is disordered, as reported by Montgomery et al.[140] for InP. Unfortunately, the model used to calculate the local density of states in Fig. 15 assumes an ordered overlayer as indeed were all the models used in Ref. 138. The conclusion of adsorption on anion sites only must therefore be open to question, and further studies are required.

A brief report of the interaction of iodine with GaAs (111) surfaces has been presented by Jacobi et al.[141] It appears that As-stabilized (111) surfaces are etched continuously by iodine vapor, even at room temperature. The etching results in Ga_xI on the surface which can be desorbed by heating at 600 K.

4.4. H_2, H_2S, and H_2O Adsorption

Molecular hydrogen appears to adsorb only weakly on III–V compound semiconductor surfaces. Pretzer and Hagstrum[142] probed adsorption on the (110), (111), and ($\overline{1}\overline{1}\overline{1}$) surfaces of GaAs and Morgan and Van Velzen[143] used the gas volumetric method to study adsorption on crushed GaP crystals. More recently, the photoemission technique has been applied to study adsorption on cleaved (110) GaAs[134] and InP.[145] The adsorption rate is significantly influenced by the presence of hot filaments and the adsorption of atomic hydrogen leads to the attenuation of emission from occupied anion derived surface states.[145,68] Theoretical estimates of the surface electronic structure anticipated following hydrogen adsorption on GaAs and GaP (110) surfaces have been presented by Mangi et al.[146] A number of states are predicted with energy dependent on the adsorption site and bond lengths assumed. However, electron spectroscopic studies are not sufficiently complete for any conclusions to be drawn regarding adsorption sites.

The adsorption of water and hydrogen sulfide on cleaved (110) GaAs surfaces has been studied by Liehr and Luth[124] using surface photovoltage spectroscopy and by Buchel and Luth[147] using photoelectron spectroscopy. The adsorption of H_2S and H_2O on cleaved InP has also been studied by various methods[148,149] and the adsorption of H_2S on GaAs (100) surfaces grown by molecular-beam epitaxy (MBE) has been thoroughly probed by Massies et al.[150] Electron beams and hot filaments have an appreciable influence on the measured adsorption rates at room temperature, but this dependence is much less pronounced at a temperature of 700 K in the studies of H_2S adsorption on (100) GaAs.[150] This suggests that adsorption is molecular at room temperature but dissociative at the higher temperatures. The initial sticking coefficient associated with room-temperature H_2S adsorption showed a pronounced dependence on surface stoichiometry, decreasing markedly with increasing arsenic surface coverage. This dependence is shown in Fig. 16 for the 4 × 1, C(2 × 8), C(4 × 4), and (1 × 1) GaAs (100) surfaces which are reported to correspond to As coverages of about 0.2, 0.6, 0.9, and 1 monolayer, respectively. Since the saturation H_2S coverage does not differ for the various surfaces, Massies et al.[150] suggest that the different sticking coefficients are a reflection of differences in surface point defect concentration rather than gallium–sulfur preferential bonding. The defects are assumed to represent active sites for adsorption. The dissociative adsorption which takes place at higher temperatures is thought to lead to exchange reactions between As and S atoms in the outermost atomic layer of the (100) surface, and a 2 × 1 structure is observed. Surprisingly, this adsorbed layer does not alter the epitaxial relationship of Al contacts deposited on the clean or on the 2 × 1 adsorbate-covered surface. However, the adsorbed layer does have a very strong influence on the electrical character of the Al–GaAs contact. The Schottky barrier observed for the intimate interface is considerably reduced by the intervening adsorbed layer, consistent with the formation of new adsorbate-induced electronic energy levels near the conduction band of the semiconductor. Sulfur on arsenic sites leads to shallow donor levels in bulk GaAs.

Montgomery et al.[151] have also shown that the adsorption of H_2S on cleaved (110) InP surfaces can also influence Schottky-barrier formation at interfaces with metals such as Au and Ag. Photoemission and LEED studies[148,151] indicate that adsorption is largely molecular at room

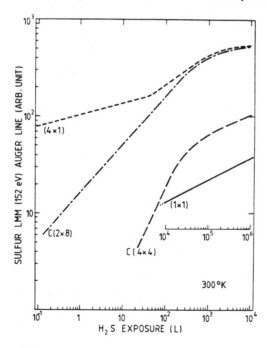

Figure 16. Sulfur Auger signal as a function of exposure to H_2S for adsorption on various reconstructed GaAs (100) faces (after Ref. 150).

temperature but some dissociation does occur, particularly for large exposures and surface coverages. The exposure of InP cleaved surfaces to water vapor leads to similar observations and conclusions. These observations are summarized in Fig. 31 where AES spectra, LEED patterns, and current-voltage characteristics following the deposition of Au contacts on the clean and adsorbate-covered surface are shown.[149] The sticking coefficients are again influenced by electron beams and it was suggested that H_2O adsorption is largely molecular at room temperature but that dissociation and exchange reactions occur at high exposures and lead to a disordered surface. The influence of H_2O adsorption on Schottky-barrier formation will be considered further in Section 6. The room-temperature adsorption of H_2O on clean cleaved GaAs surfaces has also been reported to be largely undissociative[147] and there is evidence that the presence of H_2O leads to enhanced oxidation of GaAs, InSb, and GaP cleaved surfaces.[132]

To summarize, therefore, the adsorption of gases such as H_2, H_2S, and H_2O on several faces of III–V solids appears to be largely undissociative for relatively low-exposure times at room temperature. Dissociative adsorption seems to be promoted by a high concentration of surface defects, by elevated temperatures, and by nearby electron beams and hot filaments. Dissociative adsorption is often accompanied by exchange reactions which in turn can have a dramatic effect on the surface electronic structure and on the electrical nature of contacts to these surfaces. Detailed aspects of the influence of adlayers on the electrical properties of III–V semiconductor–metal interfaces are considered further in Section 6.

5. Metal Films on Clean III–V Surfaces

5.1. General Introduction

The fabrication of electronic devices based on III–V semiconductors requires reliable and reproducible electrical contacts. In general the metal electrodes are deposited onto surfaces which have been processed in some way, perhaps subjected to chemical etching or to an oxidation cycle which leaves an oxide layer on the surface. This is often followed by annealing of the contact to ensure good adhesion and the appropriate electrical performance. There are many important physical and chemical processes which may take place at the interface and which may influence the detailed characteristics of the contact. Thus, the contact metal and semiconductor may chemically react to form new products. For example, the deposition of transition and refractory metals on silicon leads to a range of silicides,[152,153] many of which are of considerable technological interest.[154] Then, of course, interdiffusion may occur at the metal–semiconductor interface and the diffusion mechanism may be influenced by intermediate oxide or alloy layers at that interface. The electrical nature of the contact depends on many factors one of which is the possibility that the semiconductor surface itself can be doped by atoms from the contact. The whole area of contact formation, even on elemental semiconductors, is a complex one and is even more so for solids such as the III–V semiconductors which are composed of more than one element.

To illustrate some of these complexities we discuss Ag, Au, and Al contacts on chemically processed InP surfaces. In Fig. 17 the data of Kim et al.[155] are presented for Al–InP and Au–InP interfaces subjected to different annealing temperatures. The figure shows the elemental profile across the interface established by secondary ion mass spectrometry (SIMS). To within the depth resolution of the technique (some tens of angstroms) unannealed interfaces appear abrupt in both cases. However, following annealing at 400–500°C considerable interdiffusion has occurred at the Au–InP interface but not at the Al–InP one. In the former case not only has the gold diffused into the InP in appreciable quantities but also the indium has diffused preferentially into the gold contact. Studies of the resulting interface at a microscopic level reveal a complex morphology and new chemical products at the annealed interface.[156] Assuming a simple model with a constant diffusion coefficient D, the profile of the gold in InP may be expected to follow[159] a dependence of the form

$$C = C_0\left(1 - \text{erfc}\frac{x}{2(Dt)^{1/2}}\right) \quad (5)$$

where x is the distance from the interface, C_0 the gold concentration at $x = 0$, and t the annealing time. Detailed studies by McKinley et al.[157,158] of the interfacial layer formed by annealing Ag–InP contacts, using Auger electron spectroscopy and ion-depth profiling, could not be fitted to Eq. (5) with a constant value of D, although the general form of the results presented in Fig. 17 was reproduced for both Ag and Al contacts. Clearly, therefore, the interdiffusion processes are not simple and may be influenced by many variables.[160,161]

In a later section we shall see that complex interactions occur at metal–semiconductor interfaces even at room temperature and indeed interdiffusion can continue to change the detailed electrical nature of the contact days and months after the interface was formed. Experiments have been conducted on thin-film couples such as Au–Sn,[168] Ag–Ga,[169] and Au–In[170] which are influenced by the grain structure of the films, the existence of nucleation and defect sites, and electric fields. Even in

III–V Semiconductor Surface Interactions

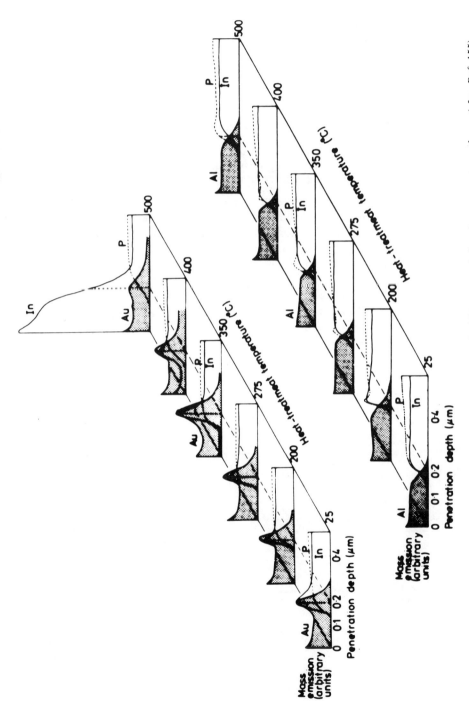

Figure 17. SIMS profiles across Au–InP and Al–InP interfaces which had been annealed at the temperatures shown (after Ref. 155).

simple metal couples the complex behavior is not well understood so that it is not surprising that metal III–V interactions are not well understood either.

In this section, therefore, we are interested in crystallographic structures, film growth, chemical reactions, and interdiffusion at interfaces between metals and III–V semiconductors. Since the presence of foreign layers on the semiconductor surface can significantly influence these processes, it is obviously desirable to start with surfaces which are atomically clean. We will deal, therefore, with clean intimate interfaces and concentrate in particular on the GaAs (110)–Cs system, the Al–GaAs (110), Al–GaAs (100) and Al-InP (110) systems, and the Au and Ag contacts to GaAs and InP. Aspects of the detailed interactions at some of these interfaces have already been reviewed by Matthews,[162] Phillips,[163] and Kern et al.[164] Kern et al., in particular, have concentrated on the mode of film growth. Provided that no compound formation takes place at the metal–semiconductor interface then surface and interfacial energies yield a good guide of the mode of film growth. For example, high-cohesive-energy metals such as Au, Ag, Al, and Ni usually grow as three-dimensional islands on low-cohesive-energy solids such as LiF and MgO. This mode of growth is referred to as the Volmer–Weber growth[165] and is illustrated in Fig. 18(a). In contrast some overlayers

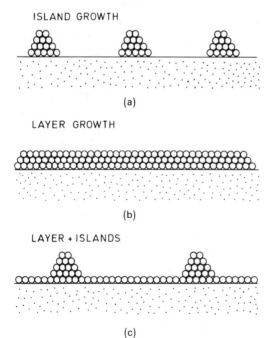

Figure 18. (a) Volmer–Weber growth mechanism of metals on a semiconductor surface. (b) Frank–Van der Merwe mechanism. (c) Stranski–Krastanov mechanism.

grow on solids by the layer-upon-layer, or Frank–Van der Merwe,[166] mechanism illustrated in Fig. 18(b). Rare-gas solids at low temperatures often grow on metals by this mode; the substrate–adlayer bonding is stronger than the adatom–adatom interactions. Finally, many metals on semiconductors grow by the Stranski–Krastanov mode[167] which is intermediate between the above two. It is illustrated in Fig. 18(c) and involves first a layer growth (strong substrate–adatom interactions) followed by island growth. Many metals grow on silicon surfaces in this way,[171] and although studies of metals on III–V semiconductors are incomplete and inadequate for firm conclusions to be drawn, there is evidence that the Stranski–Krastanov mechanism does often dominate in these systems also.

5.2. Interactions at Very Small Coverages

5.2.1. Cesium on GaAs (110)

The adsorption of Cs on GaAs surfaces has been relatively widely probed. This work is motivated in part by the technological application of electron emitters based on this system.[172,173] Extensive studies of the application of LEED, AES, and photoemission methods to probe (110), (111), ($\bar{1}\bar{1}\bar{1}$), and (100) surfaces of GaAs, GaP, and GaSb have been reported by Van Bommel and Crombeen[174–178] and others.[179–181] The most thorough study of Cs on GaAs (110) surfaces has been carried out by Derrien et al.[182,183] who applied a range of techniques to probe various aspects of the adsorption and desorption processes at different temperatures and for a range of surface coverages. Careful observation of the Cs Auger electron signal as a function of increasing layer thickness leads to the conclusion that the Cs film grows in a layer-upon-layer fashion and that every incident Cs atom sticks. For coverages below 4.6×10^{14} atoms/cm^2, defined as one monolayer ($\theta = 1$) in Ref. 182, a number of LEED patterns are observed, each one corresponding to a different value of θ. These patterns are illustrated in Fig. 19 as well as the corresponding surface structures proposed by Derrien and d'Avitaya.[183] Small coverages ($0.1 < \theta < 0.2$) lead to a p(3 × 2) pattern, whereas intermediate coverages ($0.3 < \theta < 0.5$) lead to a C(6 × 6) structure, followed at higher coverages by a C(4 × 4) structure. The adsorption leads to a large lowering of the electron affinity consistent with the fact that the Cs adsorbed atoms are partially ionized. The lateral interaction between the positively charged ions leads to the observed ordered surface structures at low coverages. The C(4 × 4) structure corresponds to a (110) plane of Cs atoms in which the spacing of the Cs atoms suffers a compression (~14%) with respect to that in bulk metallic Cs, and in this adsorption phase only eight-ninths of the Cs atoms are localized directly on Ga and As sites. It

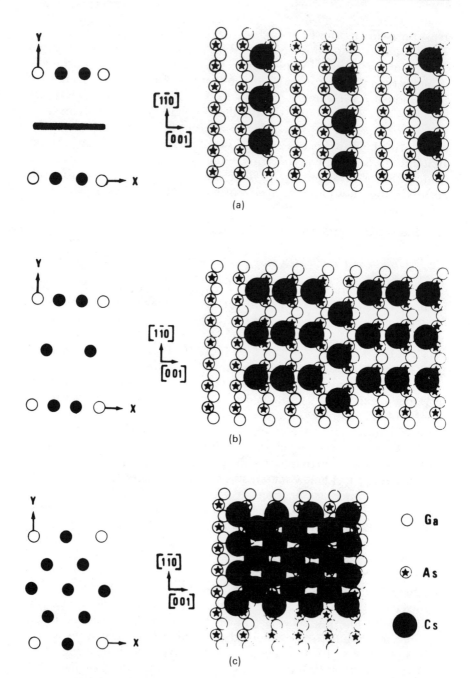

Figure 19. LEED patterns for Cs on GaAs (110) surfaces left for coverages of (a) ~0.15, (b) ~0.5, and (c) ~1 monolayer. The proposed structures are on the right (after Ref. 183).

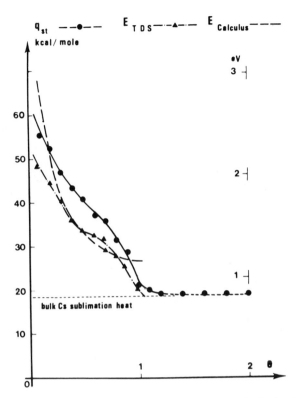

Figure 20. Isosteric heat of adsorption, the desorption energy, and calculated bond energy as a function of coverage (in monolayers) for Cs on GaAs (110) surfaces (after Ref. 183).

may be anticipated, therefore, that the bonding between the Cs atoms and the GaAs surface is weaker at monolayer coverages than at submonolayer coverages. The stronger bonding associated with coverages below half a monolayer is supported by thermal desorption spectra and measurements of the heats of adsorption.[183] Figure 20 shows the isosteric heat of adsorption, as well as the desorption energy and calculated bond energy as a function of adlayer coverage. Beyond a coverage of one monolayer the isosteric heat of adsorption levels out at a value approximately equal to the bulk sublimation heat. Thus the substrate has little influence on the Cs film growth for coverages in excess of one monolayer. Derrien and d'Avitaya[183] conclude that the layer-by-layer growth of layers other than the first is stabilized by entropic contributions.

The decrease in work function associated with the adsorption of Cs on the GaAs (110) surface is shown in Fig. 21.[183] It is assumed that the decrease is due to an induced electric dipole which decreases the effective electron affinity χ_{sc}.[183,191] Taking the electric dipole p to be $p = F(\theta)ed$,

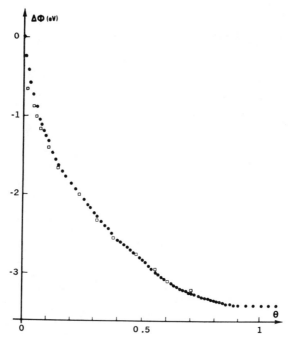

Figure 21. Work-function variations $\Delta\phi$ as a function of coverage (in monolayers) for Cs on GaAs (110) surfaces (from Ref. 183).

where e is the electronic charge and d the Cs adatom diameter, it may be estimated that the fractional charge transfer $F(\theta)$ varies from ~ 0.7 for $\theta \sim 0$ to ~ 0.2 for $\theta \sim 1$. Thus the Cs adatoms are practically ionized at low coverages and there is a large ionic contribution to the Cs–GaAs bond energy. This ionic component decreases with increasing coverage due to mutual depolarization, thus weakening the bond energies. This model, of course, assumes electron transfer from the adsorbed Cs atoms to the GaAs. One might therefore anticipate the formation of an accumulation layer near the semiconductor surface for Cs on an n-type crystal. Remarkably, however, the opposite seems to be the case; the adsorption of Cs leads to a depletion of electrons from the bulk of n-type GaAs and pinning of the Fermi level just below the middle of the band gap.[184] This is in accordance with the fact that large Schottky barriers have also been measured at GaAs alkali-metal interfaces.[185–190] This question will be considered further in Section 6. It is unfortunate that the measurements leading to Fig. 21 do not distinguish between contributions to $\Delta\phi$ due to Fermi-level shifts from those due to a reduction in the surface dipole. A plot similar to that shown in Fig. 21 has also been presented by Clemens et al.[191] but in this case the plot, at low coverages, was analyzed as a series of linear segments with breaks at coverages of approximately 0.08, 0.17, and 0.33 monolayer. The structures proposed corresponding to these breaks are similar to those in Fig. 19 at low coverages but differ somewhat at high coverages; however, they were not analyzed by LEED.

Ordered adlayer structures are also generated when Cs is deposited on (110) surfaces of GaP and GaSb at room temperature[177] consistent with a high degree of surface mobility. The adsorption on ($\overline{1}\overline{1}\overline{1}$), (111), and (001) surfaces, however, is disordered at room temperature but a number of different LEED structures are observed following heating. Fermi-level pinning energies on GaSb and InP (110) cleaved surfaces have also been established.[34]

5.2.2. Al, Ga, and In on GaAs and InP

There have been numerous studies, both experimental and theoretical, of the adsorption of group III atoms on cleaved III–V semiconductor surfaces.[192–204] The most thoroughly studied case is undoubtedly that of Al on GaAs (110) surfaces. Analysis of LEED intensity profiles by Duke et al.[192] yields the conclusion that annealing of Al–GaAs interfaces at 400°C leads to a replacement of Ga atoms by Al in the second layer of atoms from the surface. Indeed thermodynamical considerations lead one to expect that Al atoms would eventually replace Ga in the interface region, since the heat of reaction of AlAs is -28 kcal/mol compared to -17 kcal/mol for GaAs. However, in order to facilitate this exchange reaction several GaAs bonds must be broken and energy is necessary to activate the process. Elevated temperatures aid this process.

Interfaces generated by deposition of Al onto a GaAs (110) surface at room temperature are more complex than the abrupt ones generated by annealing. At room temperature the gradual weakening of the LEED pattern[193] indicates an appreciable amount of disorder which starts at coverages well below one monolayer. The evidence relating to Al:Ga exchange reactions at room temperature is somewhat mixed. Bachrach,[206] Skeath et al.[207] and Brillson et al.[194] report that Al does replace Ga in the outermost layer and that the reaction occurs for coverages well below one monolayer. Huijser et al.[208] however, maintain that this is not so and that the exchange processes observed were associated with surface defects. It seems certain that the initiation of exchange reactions on metals semiconductors in general is strongly influenced by surface defects and surface stoichiometry. This was shown quite clearly by Ludeke and Landgren[209] who studied the growth of Al on GaAs (100) surfaces grown by molecular-beam epitaxy. For the Ga-terminated (4 × 6) structure, Al did replace Ga for coverages in excess of 1.5 Å but for the As-stabilized

C(2 × 8) surface this did not happen. Likewise, the studies by Williams *et al.*[210] of Al on InP and other semiconductors show a clear variation in the metal thickness necessary to initiate the reactions in different systems. The mechanisms involved will be discussed in Section 5.3. At present, therefore, it is likely that aluminum does not replace Ga in the surface layer of GaAs (110) at room temperature provided the surface is of high quality and provided that the Al coverage is below around 0.5-1 monolayer.

Interesting theoretical estimates of the structural energies associated with this interface have been reported by Ihm and Joannopoulos.[203] A first-principles energy-minimization method was used to estimate the most stable configuration. They showed that the most stable one is that where Al replaces Ga atoms in the second layer beneath the surface. At temperatures where this cannot be activated, however, it was shown that two processes should exist. At very low coverages Al adsorption on twofold sites linking Ga and As is favored. For coverages in excess of 0.1 monolayers, it seems that the formation of Al–Al bonds is favored so that Al clusters are formed.

The adsorption of Al atoms in an ordered way on twofold sites, as outlined above, should lead to a sharp structure in the local density of states associated with the surface layer. In practice, however, this is not observed. Figure 22 shows valence-band photoemission spectra recorded by Skeath *et al.*[193] for a number of coverages of Ga on GaAs. The behavior of both Ga and Al is very similar; in both cases there is a gradual weakening of the substrate emission but without any sharp new features in the spectra. For a coverage as low as 0.3 monolayer, electron emission is observed from the Fermi edge of metallic Ga and Al, which led Skeath *et al.*[193] to suggest that the metal grows as two-dimensional rafts on the surface at these low coverages. This implies a weak-interactions model rather than the strong-interaction one. Indeed calculations by Zunger[200] suggest that Al should be weakly interacting with the GaAs (110) surface and that three-dimensional Al clusters should be formed at room temperature. On the other hand, angle-resolved photoemission and electron-loss spectroscopy studies by Huijser *et al.*[208,211] show beyond doubt that In and Al adsorption strongly influences the As derived dangling-bond surface state, thus suggesting strong interaction. At the present time, therefore, it appears that the evidence in support of either the strongly interacting or weakly interacting model is inconclusive and further detailed experimental studies are required.

The adsorption of aluminum on cleaved InP (110) interfaces has recently been the subject of a detailed study by McKinley *et al.*[157] Core-level photoemission, Auger electron spectroscopy, and LEED techniques show beyond doubt that exchange reactions leading to the release

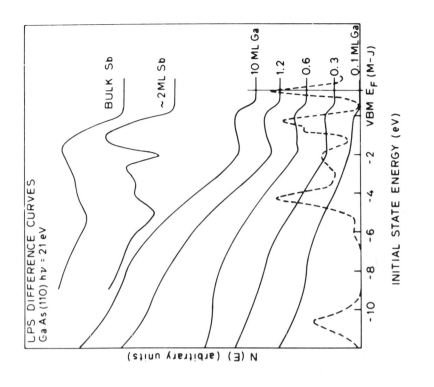
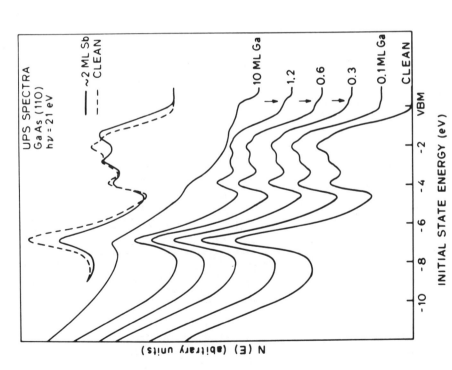

Figure 22. Valence-band photoemission spectra for clean GaAs(110) surfaces and following the deposition of Ga (lower left) and Sb (upper left). On the right difference spectra are shown (after Ref. 193). The dashed lines are calculated states for a strong bonding model.

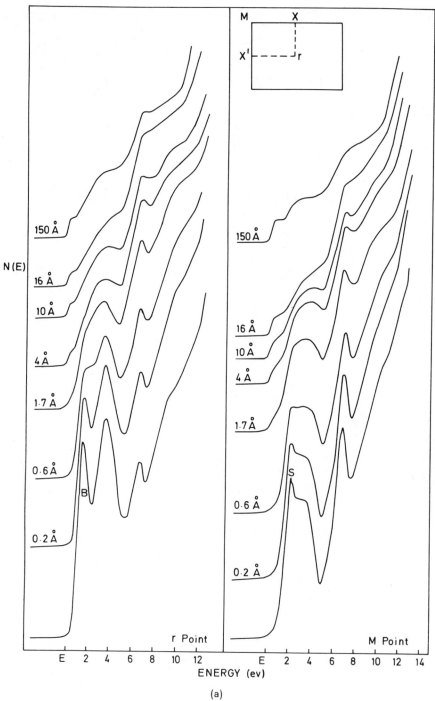

Figure 23. (a) Valence-band photoemission spectra for clean InP (110) surfaces and following deposition of aluminum (after Ref. 157). (b) The model proposed in Ref. 157 for the adsorption of Al on InP (110) surfaces for coverages below 0.5 monolayers.

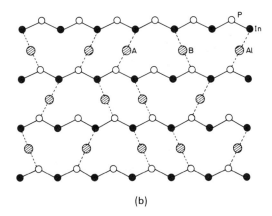

(b)

Figure 23. (*Continued*)

of In take place for coverages in excess of one monolayer. For coverages below 0.5 monolayer, adsorption is disordered and photoemission from the phosphorus derived dangling-bond state is severely attenuated but not eliminated. Figure 23(a) shows the spectral variation with increasing Al thickness. The lowest-binding-energy peak corresponding to emission from the M point in the surface Brillouin zone is associated with the phosphorus derived dangling-bond state and is totally attenuated when hydrogen is adsorbed. The fact that it is not totally attenuated for Al coverages of 0.5 monolayer led Williams *et al.*[157] to propose the structural model shown in Fig. 23(b) for small overlayer coverages. This involves adsorption on twofold sites and leads to appreciable disorder, as observed, and leaves a substantial number of unsaturated surface bonds. The disorder would have the effect of broadening any well-defined structure in the calculated density of states associated with ordered adsorption, so that sharp adsorption-induced peaks are not seen in the photoemission spectrum. For coverages in excess of 0.5 monolayer, Al–Al interactions are assumed to dominate and the Al film appears to grow in the form of islands so that a clear metallic Fermi-edge emission is observed in the photoemission spectra of Fig. 23(a). In many ways, therefore, the adsorption model for Al on InP (110) surfaces resembles that proposed by Ihm and Joannopoulos[203] for Al on GaAs (110) surfaces at low coverages, with the exception that the adlayer is highly disordered and that all dangling-bond surface states are not saturated. The exchange processes observed for coverages in excess of one monolayer is considered further in Section 5.3.

It is of interest to note that no room-temperature epitaxial growth of Al on GaAs or InP (110) surfaces has been observed although the epitaxial lattice-matching conditions are favorable. The behavior of the

GaAs (100) surface, however, is quite different. Here Al does grow epitaxially and Ludeke and Landgren[209] report that the film grows as Al (110) when deposited at elevated temperatures. Deposition at room temperature on the C(2 × 8) surface also leads to (110) growth but rotated 90° with respect to the high-temperature structure. On the predominantly Ga-terminated (4 × 6) surface a lattice-matched Al (100) 45° overgrowth was observed. For coverages in excess of 1.5 Å, however, an exchange of Ga by Al takes place as described earlier.

5.2.3. Au and Ag on GaAs and InP

There have been several studies of the growth of Al and Ag overlayers on III–V semiconductors and for thick overlayers there is a strong evidence for diffusion of Au and Ag atoms into the semiconductor and the diffusion of group III and group V atoms into the contact. This is considered in more detail in Section 5.3. Here we discuss the growth of films up to a few angstroms in thickness.

There have been detailed studies of Au adsorption on GaAs (110) carried out by Chye et al.[212] using photoelectron spectroscopy and by Bauer et al.[213] using the same method. There is no evidence that the adsorption leads to fully ordered structures at room temperature and much of the discussion has been involved with the question of whether or not the gold atoms diffuse into the GaAs. The emission from the Ga 3d and As 3d core levels are attenuated at approximately the same rate and more slowly than anticipated for layer-by-layer growth of the kind discussed for Cs growth on GaAs (110) surfaces. At the same time the growth of Au core-level emission is slower than for layer-by-layer growth, indicating either growth in the form of islands or diffusion of Au into the semiconducting crystal.

In an attempt to distinguish between these two, Chye et al.[212] looked closely at the photoemission spectra from the valence levels of the Au overlayer. A series of spectra for various coverages is presented in Fig. 24(a). The two well-defined peaks separated by 2.0 eV for $\theta = 5.8$ correspond to emission from the Au 5d levels. This splitting is 2.3 eV for a thick bulk Au layer and is below 2 eV for coverages well below one monolayer on GaAs, InP [see Fig. 24(b)], and GaSb (110) surfaces. It appears that this splitting is around 1.5 eV for gold in dilute alloys and where the Au atoms are well dispersed. The small value of the splitting for submonolayer Au coverages on III–V surfaces is then taken as an indication that the Au atoms are dispersed and do not form clusters on the surfaces. Unfortunately, however, it is not clear that this view is valid since similar differences in valence-band emission are observed when clusters of noble metals of different sizes are grown on inert substrates.[214,215] Indeed, McKinley et al.[158] have conducted careful studies

III–V Semiconductor Surface Interactions

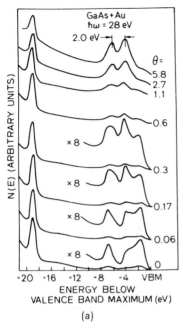

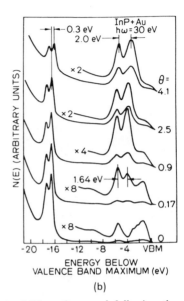

Figure 24. (a) Photoemission spectra for clean GaAs (110) surfaces and following the deposition of Au. The coverage θ is in monolayers (after Ref. 212). (b) Same as (a) but for Au on InP (110) (after Ref. 212).

of Ag growth on InP (110) cleaved surfaces and suggest that a two-dimensional island-growth mechanism is to be preferred, with nucleation at steps probable. The growth mechanism of the film, as well as the intermixing at the interface, is highly dependent on the perfection of the surface. For Ag on a high-quality cleaved surface no conclusive evidence for substantial intermixing was observed. However, if the InP surface was slightly roughened by argon ion bombardment the situation is drastically changed; in this case rapid diffusion of In into the Au overlayer is observed. Room-temperature interdiffusion mechanisms must involve defects since bulk diffusion constants of Au and Ag in III–V materials are too low to account for the observations. It is of interest to note that no evidence of interdiffusion across the interface was observed in recent detailed studies of Ag overlayers on GaAs (100) surfaces by Ludeke et al.[216] The same conclusion was reached by Massies and Linh[252] who also reported a layer-by-layer growth when Ag is rapidly deposited onto GaAs (100) surfaces at temperatures below 100°C.

The shift of the Fermi level at cleaved surfaces of III–V semiconductors has been widely studied as a function of increasing overlayer thickness. In many cases, provided the semiconductor surface is clean and of high quality, the substantial part of the band bending and Schottky-barrier

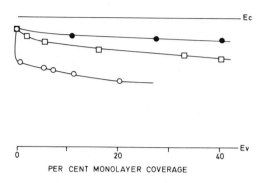

Figure 25. Shift of the Fermi level when Au is deposited on cleaved InP (110) surfaces. The lower curve represents deposition on a clean surface, whereas the middle and upper curves represent Au deposition on surfaces exposed to 10^3 and 10^9 Langmuirs of water vapor, respectively (after Ref. 149).

formation is complete for coverages of a few percent of a monolayer. This is illustrated in Fig. 25 for the case of Au on cleaved InP surfaces. The Fermi level at the clean surface lies close to the conduction band, in the flat-band situation. Small Au coverages lead to pinning of the Fermi level about 0.5 eV below the conduction band as shown. This pinning level is drastically influenced by the presence of impurities on the clean surface. Figure 25 shows the effect of exposing the clean surface to water for various times, before depositing gold. These Fermi-level shifts will be considered in detail in Section 6.

5.3. Interactions with Thick Metal Films

The deposition of metal coverages in excess of a few monolayers clearly leads to complex interdiffusion and reactions on many III–V semiconductor surfaces. Hiraki and co-workers[217–219] have studied metals on sputtered and annealed surfaces of several solids and observed copious diffusion of the cation species into thick Au overlayers, particularly for the lower-band-gap solids. Detailed XPS studies of the interaction of GaAs with several metals have also been reported by Waldrop and Grant.[220] To illustrate these effects we consider two examples, namely Au on GaSb and Ni on InP. Figure 26 shows a series of photoemission spectra following the deposition of various thicknesses of Au on GaSb.[221] The Ga 3d core level is rapidly attenuated but emission from the Sb 4d core level persists. This obviously means that the semiconductor is partly dissocated and that Sb is incorporated in the metal contact. The second example[224] is shown in Fig. 27. Here the In 4d and P 2p emission is shown for the clean surface and following the deposition of various amounts of Ni. With increasing Ni thickness the P 2p emission is rapidly attenuated but the In 4d emission persists. In addition the In 4d level emission is appreciably reduced. The InP surface is clearly dissociated so as to form nickel phosphide and release In atoms which

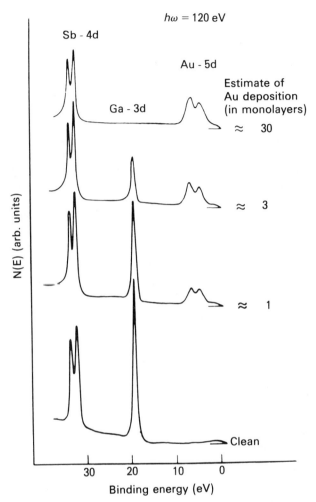

Figure 26. Photoemission spectra for a clean GaSb (110) surface and following the deposition of Au (after Ref. 221).

are incorporated in the growing Ni layer. For comparison, the corresponding behavior for a gold overlayer is also presented. In this case 5 Å of the metal leads to a shift of both lines, reflecting a Fermi-level shift at the interface. The attenuation of the emission lines is not in accordance with that expected for layer-upon-layer growth. The fact that emission from In 4d and P 2p core levels is still clearly seen following the deposition of some tens of angstroms indicates either considerable island growth, and substrate emission from the regions in between, or else a diffusion of In and P into the growing Au overlayer.

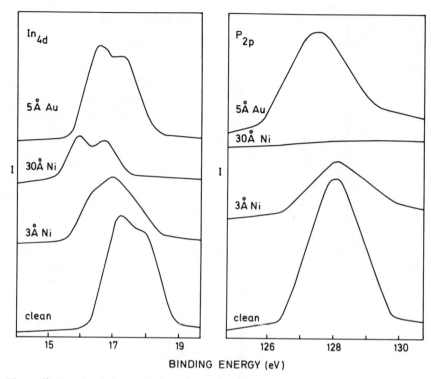

Figure 27. Core-level photoemission spectra when Ni and Au are deposited on cleaved InP (110) surfaces (after Ref. 224).

Many workers have assumed that data such as that in Fig. 27 reflect intermixing at the metal–semiconductor interface. If one simply considers emission from the phosphorus 2p core level in Fig. 27, it may be established that Ni attenuates the emission more rapidly than gold. Brillson et al.[222,223] have attempted to describe differences such as these in terms of an interface width T_0, defined as the overlayer thickness necessary to attenuate the anion core level emission by $1/e$. Brillson et al. studied a number of metals on a range of cleaved (110) III–V surfaces and values of T_0 for these are shown in Fig. 28. The horizontal scale represents the heat of the interface reaction, defined as the difference in semiconductor heat of formation and the heat of formation of the metal–anion complex which gives the most stable compound.[222] According to this picture the most abrupt interfaces are associated with those systems where the metal–anion interactions are strongest. It is not yet clear how useful this approach and the data presented in Fig. 28 will turn out to be. Unfortunately, the analysis assumes that the mode of metal film growth is identical for all metals on all III–V cleaved surfaces and that growth does not involve

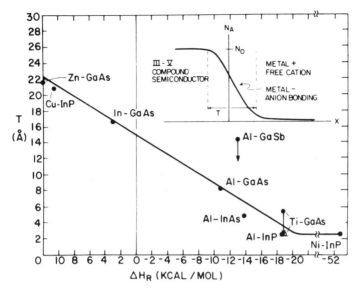

Figure 28. Plot of interface width as a function of heat of reaction (metal + anion) for a series of metal–semiconductor systems (after Ref. 222).

island formation. As emphasized previously these assumptions are highly questionable.

Bauer et al.[213] have recently used the photoemission technique to probe Au and Ge overlayers on MBE-grown AlAs and claim that bulk thermodynamical data are not appropriate as a guide to chemical processes occurring at the interface. It should also be noted that the nature of intermixing is highly dependent on the surface-defect structure as shown by McKinley et al.[158] and discussed in the previous section. Sputtered and annealed surfaces of InP led to greatly enhanced outdiffusion of In into Ag contacts and indeed on such surfaces indium droplets exist.[225] One must be careful therefore in comparing data on surfaces prepared in different ways in the absence of detailed information of surface-defect structure. However, it is certain that reactive metals such as Al and Ni do lead to a dissociation of the clean (110) III–V surfaces and in general it is the cation component which diffuses out into the contact. Unreactive metals such as Au and Ag may also diffuse into the semiconductor and anions and cations may diffuse into the metal. Usually, it is cation outdiffusion that dominates and is clearly observed for annealed contacts on most III–V semiconductors.[226]

It is of interest to examine a little further the mechanisms involved in the surface dissociation process. On a high-quality surface it appears that coverages of around one monolayer or more of the adlayer are necessary before the reactive metal initiates the surface exchange process. The heat of condensation of individual adsorbate atoms is not sufficient to activate the process. Careful experiments by Ludeke and Landgren lead to the view that the exchange process is initiated[209] by the formation of Al nuclei for the case of Al deposition on GaAs (100) surfaces at room temperature. As pointed out by Zunger[200] the energy released in the formation of clusters might be quite sufficient to trigger the exchange reactions. Thus, for Al on a GaAs surface the formation of a cluster with n bonds releases an energy of the order of nE_b, where E_b is the average binding energy of nearest neighbors. For Al, the value of $E_b \sim 0.5$ eV and for a four-atom cluster, $n = 6$ and the energy involved is ~ 3 eV, which is certainly sufficient to activate replacement reactions. It should be noticed, however, that this is only approximate and that the heat of adsorption of the Al atoms on the surface, which of course may be site dependent, must also be taken into account. The adsorption model proposed by Ihm and Joannopoulos[203] for Al on GaAs (110) cleaved surfaces indicates that the formation of nuclei or clusters will be favored for Al coverages in excess of 0.1 monolayer which is in accordance with the view that cluster formation may trigger Al:Ga exchange reactions. Clearly, though, this view must remain tentative until further more detailed experiments are conducted on a wide range of systems.

Further fascinating experiments have been carried out by Brillson et al.[222,223] whereby a thin layer of a reactive metal such as Al or Ni is deposited on the clean cleaved III–V surface and an unreactive metal such as gold is deposited on top of the reactive metal. Core-level photoemission spectroscopy was used to monitor the composition of the interface and the growing films. It appears that small quantities of the reactive metal can have a very profound influence on the interdiffusion between the semiconductor and the unreactive overlayer. Thus monolayer thickness of Al on GaAs (110) cleaved surfaces can increase the relative Ga to As diffusion into Au by over an order of magnitude. These processes have been considered in terms of "chemical trapping" of anions by the reactive metal and also in terms of electromigration due to electric fields at the interface.[223] However, the increased cation outdiffusion is analogous to that observed on semiconductor surfaces slightly disordered by light sputtering and on interfaces subjected to heating. Thus the disordering and exchange reaction generated by the Al overlayer is highly likely to dominate the mode of film growth of the unreactive film and the subsequent release of cations from the interface. The subsequent diffusion of Ga through the Au overlayer has a very low activation energy and is probably largely controlled by grain-boundary diffusion.[226]

6. The Electrical Nature of Intimate Interfaces

6.1. Introduction

In this section we concentrate in particular on the formation of ohmic and barrier contacts at interfaces between metals and clean semiconductor surfaces, as well as surfaces exposed to monitored and controlled contaminants. These interfaces, of course, are rather different to the "real contacts" used in electronic devices and considered elsewhere in this volume. Nevertheless, if an understanding of the true basis of Schottky-barrier formation is to be obtained it is of great importance to study interfaces which can be accurately controlled and monitored.

Schottky barriers at metal–semiconductor interfaces are normally measured by one of four methods:

1. The current–voltage method where the electrical transport properties of a diode are measured and the data fitted to appropriate theory.
2. The capacitance–voltage method where the capacitance of a diode is measured as a function of voltage applied.
3. The photoresponse method where the current through a diode is measured as a function of the wavelength of light illuminating the barrier contact.
4. The photoelectron spectroscopy method where the shift of electron emission peaks from well-defined levels (e.g., core level) in the semiconductor is measured during the deposition of the metal contact.

The application of these methods will not be considered here; they have been reviewed extensively elsewhere.[22,227] Nevertheless, it is important to point out that these methods do not necessarily measure exactly the same thing. The first three require quite thick metal contacts, whereas method 4 established the Schottky-barrier formation for metal coverages of the order of monolayers. For situations where the barrier is not uniform over the area of the interface methods, methods 1 and 2 tend to yield barrier values biased towards low-barrier patches, whereas methods 2 and 4 tend to yield average values. It is not surprising, therefore, that the various methods often yield Schottky-barrier values which differ by perhaps as much as 0.1 eV or more for a given contact.[228] In this section, therefore, we shall only be interested in relatively large differences in barrier values for different contacts to III–V semiconductors; the experimental data are not of sufficiently high quality at present to allow theories to seriously attempt to explain differences of less than about 0.1 eV for metals on these clean semiconductors. In the next section we summarize some of the major recent findings related to Schottky-barrier formation on cleaved III–V surfaces before considering various theoretical viewpoints in subsequent sections.

In Fig. 29 the various interfacial pinning levels for a range of metals on n- and p-type crystals of GaAs, InP, and GaSb, as measured by Spicer and co-workers[34,35] by photoelectron spectroscopy, are shown. These metals, as well as oxygen, were all deposited on the clean cleaved (110) surface. Let us consider the case of GaAs. Metals on n-type samples yield a pinning level at about midgap which is practically independent of the metal. Furthermore, a constant pinning level about 0.2 eV nearer the valence band, is observed for p-type crystals. Schottky barriers corresponding to similar values of pinning energies have been established for GaAs surfaces prepared in a host of ways,[87,228,229] that is, there is a remarkable independence of the barrier on the nature of the surface or any adlayer on it. The pinning of the Fermi level is achieved very rapidly; it seems that a metal coverage corresponding to a few percent of one atomic layer is sufficient to cause it.[34,35] Furthermore, the same pinning

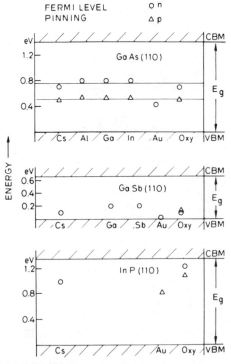

Figure 29. Fermi-level pinning positions for the adsorbates shown on cleaved surfaces of GaAs, GaSb, and InP (after Ref. 34).

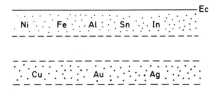

Figure 30. Fermi-level pinning energies for various metals on cleaved n-type InP crystals.

energy is observed on clean surfaces which have a large step density or which have been subjected to a sputtering and annealing cycle.[34]

The pinning energies on GaSb are seen to be located close to the valence-band edge for the range of overlayers shown, whereas in general on InP they appear localized in the upper part of the band gap. Detailed studies of contacts to clean InP have been carried out by Williams and co-workers,[36,231,232] using $I-V$, $C-V$, and photoemission techniques and by Hökelek and Robinson[233] using $I-V$ and $C-V$ methods. These measurements are summarized in Fig. 30, and in general the pinning levels fall into two regions, one around 0.1 to 0.3 eV from the conduction band and the other closer to mid gap. All the metals shown when deposited on etched surfaces yield barriers in the latter region.[37] The measured

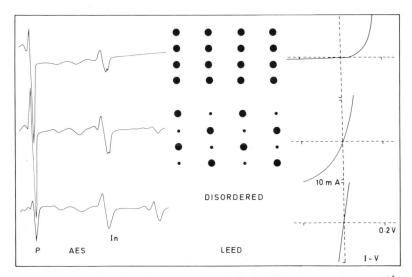

Figure 31. Auger spectra for clean InP (upper left) and following exposure to 10^4 L of water (center) and 10^9 L of water (lower). The middle panel illustrates the corresponding LEED, and the right panel shows $I-V$ characteristics following the deposition of Au electrodes (after Ref. 149).

Schottky barriers for metals such as Au and Ag on InP is drastically affected if the clean cleaved surface is subjected to a sputtering cycle, or if gases such as oxygen, chlorine, H_2S, or H_2O are adsorbed on the surface and the Au or Ag contact deposited on top. Figure 31 shows the diode behavior of Au on InP, both on clean cleaved surfaces and on surfaces subjected to different exposures of H_2O.[149] The corresponding LEED patterns and Auger electron spectra are also illustrated. The increased diode current reflects the decrease in the effective Schottky barrier, as illustrated in Fig. 25 previously.

An interesting observation of Schottky-barrier behavior for Au on a range of compound semiconductors has been found by McCaldin et al.[234] Analysis of a wide body of data suggested a direct correlation between the barrier height for holes, that is, on p-type material, and the electronegativity of the anion in the semiconductor. This observation is often referred to as the "common anion rule" and will be considered again later.

6.2. Abrupt Boundary Models

It is quite clear that the conventional Schottky model[17] in its most elementary form does not account well for the observations of metal-contact behavior on III–V semiconductors. There seems to be no clear simple dependence of ϕ_b on metal work function and indeed the data on InP are inconsistent with the linear interface potential model discussed in Section 2. The Bardeen model[18] involving intrinsic surface states in the semiconductor band gap is also inappropriate since there are no such states in the gap for clean cleaved GaAs and InP (110) surfaces. It could be, of course, that the adsorption of metal atoms leads to a reduction in the surface relaxation discussed in Section 3. It is the relaxation that drives the intrinsic states out of the gap on cleaved surfaces and the situation could be reversed if the relaxation were removed. However, detailed angle-resolved photoemission measurements[157] and partial-yield studies[34] on GaAs and InP show no evidence of this occurring, and indeed it would be somewhat surprising if such a wide range of metals led to the same pinning level on GaAs. Finally, the adsorption of hydrogen on InP (110) surface leads to a removal of the occupied dangling-bond state; yet the Schottky barrier measured on subsequent deposition of Au contacts is identical to that of Au on the clean surface.

It does not seem that intrinsic surface states dominate Schottky-barrier formation for these situations, therefore. However, metal-induced extrinsic states must be given consideration. Louie et al.[29] have considered theoretically the nature of the Al–GaAs contact. The Al was treated as a "jellium," that is, a structureless medium with the same average density as aluminum. The calculations show that metal-induced gap states (MIGS) may be generated at the interface and may pin the Fermi level. The metal wave functions do penetrate into the semiconductor in accordance with the views expressed by Heine.[23] These calculations have not been extended to other metals and obviously the assumption of a structureless "jellium" is rather unreaslistic. Chelikowsky *et al.*[235] have also calculated the interface states generated by Al adsorption on As sites, but, as outlined earlier, the validity of this structure is questionable. These models have not, so far, accounted for the variation of Schottky barriers observed on III-V semiconductors.

The importance of strong chemical interactions at metal–semiconductor interfaces has been emphasized in Section 5. Electrical barriers of transition metals on silicon were considered by Andrews and Phillips[236] who noticed a strong correlation of barrier height and the heat of formation of the appropriate silicide. Strong chemical bonding across

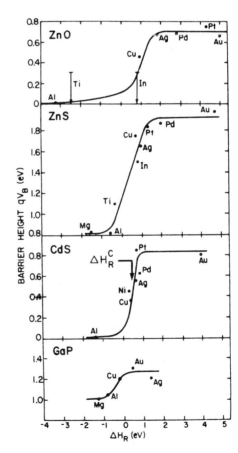

Figure 32. Barrier heights as a function of heat of reaction (metal + anion) for several metal–semiconductor systems (after Ref. 237).

the interface leads to a large dipole which can have a significant influence on the barrier, that is, the larger the heat of formation the lower the barrier. This model has been extended by Brillson to include metal–compound semiconductor interfaces. Figure 32 shows[237] an analysis of barriers for a range of metals on a number of semiconductors. The horizontal scale is the same as that used in Fig. 28. The model therefore considers that metals with large negative heats of reaction lead to strong bonds and a large interfacial dipole due to charge transfer between the adsorbate atom and semiconductor surface atoms; the Schottky barrier is correspondingly reduced. It is not clear that this model is in accordance with the fact that barriers are established for very small overlayer coverages or with the fact that the interfaces in many instances are totally disordered and not atomically abrupt.

6.3. Nonabrupt Boundary Theories

In view of the fact that considerable intermixing and disorder is generated at most metal III–V semiconductor interfaces fabricated at room temperature, it is clearly necessary to consider the influence of deviations from perfect order and abruptness on Schottky-barrier formation. As outlined in Section 2 the generation of ohmic contacts usually involves the doping of the semiconductor surface by at least one component from the contact; usually this is achieved by annealing the contact. The basis of the processes involved has been considered in several reviews[238–240] as indeed has the use of diffusion barriers to attenuate interdiffusion at metal–semiconductor interfaces. Oxide layers on III–V semiconductors may act as diffusion barriers and heat treatment is usually required before an ohmic contact can be formed in these cases. At intimate interfaces, however, ohmic contacts may be generated at room temperature. Thus, tin on clean cleaved n-type InP crystals gives good ohmic behavior[241] presumably due to doping of the surface region at room temperature and the replacement of In atoms by Sn. On an etched surface it is necessary to heat the contact so as to cause diffusion of Sn atoms through the oxide layer and into the semiconductor.

At the present time there are two models of Schottky barriers at imperfect interfaces which merit particular attention. The first is the "effective-work-function" model proposed by Freeouf and co-workers.[242–244] This model assumes that the Schottky picture outlined previously is appropriate but that the work functions that should be considered are not those of the pure contact metals. It is known that the annealing of Au contacts on clean or oxidized GaAs or InP leads to excess cations (Ga or In) in the gold contact so that the interface region is rich in the anion component. Thus, it is argued, the effective work function that should be used is not that of the metal but that of the anion. Then the barrier ϕ_{bp} on a p-type crystal is given by $\phi_{bp} = E_G + X_{sc} - \Phi_{anion}$. Since Φ_{anion} is directly related to the electronegativity of the anions, one can directly account for the "common anion rule" of McCaldin et al.[234] discussed in Section 6.1. The values of ϕ_{bp} measured for a range of semiconductors in contact with gold are accounted for well by this model. In addition the fact that barriers on GaAs are rather independent of contact-metal or surface treatment is taken as indicative of excess As at the interface.

This effective-work-function model is particularly interesting when mixed phases may be present at the interface. As mentioned previously, measurements of Schottky barriers on nominally identical systems by different workers show significant variations. It is also known that barriers measured by $I-V$ methods are invariably smaller than those determined by $C-V$, sometimes by as much as several tenths of an eV. Freeouf et al.[243] propose that these variations may be due to multiple phases at interfaces between semiconductors and thick metal films. Consider again the case of Au on GaAs. It has already been seen that upon slight heating Ga may diffuse through the gold, probably along grain boundaries. This will generate regions at the semiconductor interface which are depleted of cations and the interface will contain mixed phases as illustrated in Fig. 33. The $C-V$ technique tends to measure average values of barriers at such interfaces, whereas the $I-V$ method tends to give values associated with the lowest-barrier phases, since the current is exponentially dependent on ϕ_b. Freeouf et al.[243] suggest that the lack of reproducibility would follow from the kinetic aspects of the different metallurgical interactions.

The importance of multiple phases is certain for a whole range of metal–compound semiconductor systems. However, it is not clear that the model is appropriate for Schottky-barrier formation at very low metal coverages where consideration of multiple phases and anion-rich regions

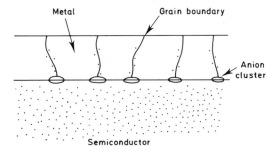

Figure 33. Diffusion of cations along grain boundaries leading to patches rich in anions at the interface.

is inappropriate. Provided the surfaces are of high quality, the Schottky barrier is largely formed on many cleaved surfaces when the metal coverage is only a few percent of an atomic layer.

The second important model which has been extensively considered is the so-called "defect model" of Schottky-barrier formation. This was first proposed by Spicer et al.[34,35] to account for the fact that a whole range of overlayers yield the same value of ϕ_b on cleaved GaAs. The model was also used by Williams et al.[37] in order to explain the influence of adsorbates on metal–InP contacts. Essentially, the model states that the Fermi level is pinned at the interface, not directly by the metal, but indirectly by defect states generated in the semiconductor.

The importance of defects in pinning the Fermi level at clean semiconductor surfaces is well known and has been referred to earlier. The most thoroughly studied case is again GaAs and it is known that surface steps,[63,64,84] as well as subsurface damage generated by sputtering and annealing, does lead to Fermi-level pinning at around midgap for both n- and p-type material.[87,245] It is also known that surfaces prepared by sputtering and annealing are nonstoichiometric chemically and it is natural, therefore, to attempt to relate pinning behavior to defects associated with subsurface nonstoichiometry. It is also of significance to observe that the pinning of the Fermi level observed at such a free surface does not appear to be removed when metals such as silver are deposited onto that surface.[87] Clearly, a high enough density of defects near the surface can have the same effect as surface states. If it is assumed that the deposition of small amounts of any metal on a clean GaAs (110) cleaved surface generated such defect states in the semiconductor, then it is easy to see why the Schottky barriers appear independent of the metal.

To illustrate the model we consider the case of contacts on indium phosphide surfaces. Suppose the region at and just below the interface contains phosphorus vacancies as illustrated in Fig. 34(a). These vacancies can act as donor or acceptor states and the pinning energy associated with them depends on their density and energies in the forbidden gap. The defect energy has been roughly calculated by Srivastava[101] using a semiempirical pseudopotential method and by Daw and Smith[95–98] using a tight-binding method. Both calculations lead to the conclusion that an anion vacancy will lead to an energy level in the upper part of the band gap as shown in Fig. 34(b). This level contains one electron and has a donor/acceptor character. It can clearly lead to the pinning behavior reported for InP in Fig. 30. Daw and Smith[95–98] have extended these calculations to other III–V compounds and their alloys and have considered anion and cation vacancies both in the bulk and at the surface. Remarkably, it appears that both vacancies would lead to Fermi-level pinning at around midgap in GaAs just as observed. It is clear that the type of calculations referred to above are only approximate and do not

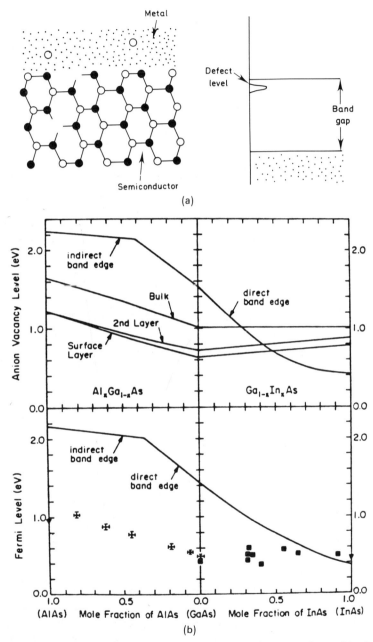

Figure 34. (a) The defect model of Schottky-barrier formation. Vacancies, antisite defects, or more complex defects in the semiconductor lead to interface gap states, illustrated on the right. (b) Measured pinning levels at metal interfaces with the alloys shown (after Refs. 245 and 246) relative to the valence band and conduction band—lower panels. The upper panels show predicted pinning energies (after Ref. 98) based on pinning by anion vacancies.

take fully into account important factors such as the relaxation of atoms around the vacancy. It would therefore be dangerous to place too much weight on calculated defect levels in one solid. However, it is perhaps more realistic to examine the trends of defect levels in a range of III–V semiconductors and their alloys and to explore whether this is in agreement with trends seen in the measurements of Schottky-barrier heights. An example of such a comparison is presented in Fig. 34(b) for alloys of AlAs–GaAs–InAs. The lower frame shows the band gap as a function of alloy composition and also the Fermi-level pinning energies in the gap.[245,246] The upper frame shows the calculated energies[95–98] associated with the upper occupied level of neutral anion vacancies in the bulk and close to the surface. The experimental trends are clearly well accounted for by the defect-level calculation and, similarly, good agreement has been obtained for other alloy systems. While this agreement is extremely encouraging it is clear that more complex defects will also have to be considered. The possibility of pinning by antisite defects (anions on cation sites) has been considered by Allen and Dow[99,100] and again trends of the kind illustrated in Fig. 34(b) have been well reproduced. Since it is cation outdiffusion into metal contacts that is usually observed, it is reasonable to seriously consider pinning by antisite defects of this kind.

The defect model, therefore, accounts reasonably well for many of the experimental observations of Schottky barriers at imperfect interfaces. Of course if there are defects at the surface, and in successive layers, as well as disorder then well-defined defect levels will be broadened out into a distribution of states. Such a distribution may well contribute to the nonideality of Schottky diodes.[247–249] There are other experimental data which also give strong support to the defect model. First, the strong pinning observed when the metal coverage is only perhaps 20% of one atomic layer is more easily understood. Second, studies involving the adsorption of gases on InP seem to indicate that surface disorder in some cases is accompanied by pinning of the Fermi level closer to the conduction band following subsequent metal deposition.[149,151] Finally, careful studies of Fermi-level pinning at Ge–GaAs interfaces appear to indicate that nonstoichiometry and defects are closely related to Fermi-level pinning at the interface.[250] Furthermore, the careful studies of Fermi-level pinning of Monch and Gant,[250] using the Kelvin probe method, and of Grant et al.,[251] using XPS, suggest that the two pinning levels observed in GaAs are in fact associated with one defect. The energy difference between the two levels then relates to the Coulomb repulsion energy associated with the two electrons in the state in one situation compared with no electrons in the other state. The "anion vacancy" model[95–98] accounts rather well for this energy difference; however, as noted previously the accuracy of the calculations is far from certain at the present time.

7. Conclusions

The precise nature of clean III–V compound semiconductor surfaces is complex and at the present time it is the cleaved (110) surfaces which are best understood. The atoms on these (110) surfaces are relaxed from their ideal bulk positions and in many cases this relaxation leads to the removal of empty and occupied surface states from the energy gap. Good experimental and theoretical data relating to the crystallography, chemistry, and electronic structures of these surfaces for a number of materials are gradually becoming available. Defects on these surfaces often lead to states in the band gap and these often have a dominating influence on many surface processes. The (100), (111), and ($\overline{1}\overline{1}\overline{1}$) of III–V semiconductors are relatively poorly understood but progress is being made at the present time, particularly as a result of the increasing use of molecular-beam epitaxy for material growth.

The interaction of gases and metals with III–V compound semiconductor surfaces is a complex function of many variables. Surface chemical stoichiometry and surface defects play a key role and the pinning of the Fermi level at free surfaces is highly sensitive to deviations of the surface from perfection. The adsorption mechanisms of metals also depend strongly on the chemical interactions between the adsorbed metal atoms and the surface. Highly reactive metals often disrupt the surface to a considerable extent, whereas weakly interacting metals often grow by island formation.

Interface states lead to strong pinning of the Fermi level at metal III–V semiconductor interfaces. Several mechanisms may contribute to the pinning which in turn depend strongly on the method of preparation of the interface and any subsequent heat treatment. Recent work suggests that the pinning is strongly influenced by metal-induced defects in the semiconductor and near the interface and models involving point vacancies and anion clusters have met with some success. The story is far from complete, however, and further theoretical and experimental studies on different surfaces of a range of materials are necessary. In particular the development of spectroscopic and microscopic methods of characterizing surface and interface defects and their influence on metal–semiconductor interactions needs to be strongly encouraged.

References

1. H. Ibach (ed.), *Electron Spectroscopy for Surface Analysis*, Springer-Verlag, Berlin (1977).
2. C. R. Brundle and A. D. Baker, *Electron Spectroscopy*, Vols. 1–3, Academic Press, New York (1978–1980).
3. M. Cardona and L. Ley, *Photoemission in Solids I and II*, Springer-Verlag, Berlin (1978 and 1979).

4. D. Haneman, in: *Surface Physics of Phosphors and Semiconductors* (C. G. Scott and C. E. Reed, eds.) Chapter I, Academic Press, New York (1975).
5. R. H. Williams, Electron spectroscopy of surfaces, *Contemp. Phys. 19*, 389–413 (1978).
6. C. R. Brundle, The application of electron spectroscopy to surface studies, *J. Vac. Sci. Technol. 11*, 212–224 (1974).
7. D. J. Chadi, Energy-minimization approach to the atomic geometry of semiconductor surfaces, *Phys. Rev. Lett. 41*, 1062–1065 (1978).
8. J. R. Chelikowsky and M. L. Cohen, (110) Surface states in III–V and II–VI zinc-blende semiconductors, *Phys. Rev. B 13*, 826–834 (1976).
9. J. R. Chelikowsky and M. L. Cohen, Self-consistent pseudopotential calculation for the relaxed (110) surface of GaAs, *Phys. Rev. B 20*, 4150–4159 (1979).
10. J. D. Joannopoulos and M. L. Cohen, Intrinsic surface states of (110) surfaces of group IV and III–V semiconductors, *Phys. Rev. B 10*, 5075–5081 (1974).
11. J. A. Appelbaum and D. R. Hamann, in: *Theory of Chemisorption* (J. R. Smith, ed.), Chapter 3, Springer-Verlag, Berlin (1980).
12. D. J. Chadi, (110) Surface atomic structures of covalent and ionic semiconductors, *Phys. Rev. B 19*, 2074–2082 (1979).
13. E. J. Mele and J. D. Joannopoulos, Electronic states of unrelaxed and relaxed GaAs (110) surfaces, *Phys. Rev. B 17*, 1816–1827 (1978).
14. E. J. Mele and J. D. Joannopoulos, Site of oxygen chemisorption on the GaAs(110) surface, *Phys. Rev. Lett. 40*, 341–346 (1978).
15. D. J. Chadi, Atomic structure of Si(111) surfaces, *Surf. Sci. 99*, 1–12 (1980).
16. D. J. Chadi, Origins of (111) surface reconstructions of Si and Ge, *J. Vac. Sci. Technol. 17*, 989–992 (1980).
17. W. Schottky, *Z. Phys. 113*, 367 (1939).
18. J. Bardeen, Surface states and rectification at a metal–semiconductor contact, *Phys. Rev. 71*, 717–727 (1947).
19. A. M. Cowley and S. M. Sze, Surface states and barrier heights of metal–semiconductor systems, *J. Appl. Phys. 36*, 3212–3220 (1965).
20. A. M. Cowley, Depletion capacitance and diffusion potential of gallium phosphide, *J. Appl. Phys. 37*, 3024–3032 (1966).
21. E. H. Rhoderick, The physics of Schottky barriers, *J. Phys. D 3*, 1153–1167 (1970).
22. E. H. Rhoderick, *Metal–Semiconductor Contacts*, Clarendon Press, Oxford (1978).
23. V. Heine, Theory of surface states, *Phys. Rev. A 138*, 1689–1696 (1965).
24. *Handbook of Physics*, 50th Edition (Ed. R. C. Weast), The Chemical Rubber Co., Cleveland, Ohio (1970).
25. S. G. Kurtin, T. C. McGill, and C. A. Mead, Fundamental transition in the electronic nature of solids, *Phys. Rev. Lett. 22*, 1433–1436 (1969).
26. M. Schluter, Chemical trends in metal–semiconductor barrier heights, *Phys. Rev. B 17*, 5044–5047 (1978).
27. J. C. Inkson, Many-body effects at metal–semiconductor junctions, *Phys. Rev. B 17*, 5044–5047 (1978).
28. J. C. Ikson, Schottky barriers and plasmons, *J. Vac. Sci. Technol. 11*, 943–946 (1974).
29. S. G. Louie, J. R. Chelikowsky, and M. L. Cohen, Ionicity and the theory of Schottky barriers, *Phys. Rev. B 15*, 2154–2162 (1977).
30. H. E. Zang and M. Schluter, Studies of the Si(111) surface with various Al overlays, *Phys. Rev. B 18*, 1923–1935 (1978).
31. J. R. Chelikowsky, Electronic structure of Al chemisorbed on the Si(111) surface, *Phys. Rev. B 16*, 3618–3627 (1977).
32. A. Hiraki, M. A. Nicolet, and J. W. Mayer, Low-temperature migration of silicon in thin layers of gold and platinum, *Appl. Phys. Lett. 18*, 178–181 (1971).

33. I. Abbati, L. Braichovich, A. Franciosi, I. Lindau, P. R. Skeath, C. Y. Su, and W. E. Spicer, Photoemission investigation of the temperature effect on Si–Au interfaces, *J. Vac. Sci. Technol.* 17, 930–935 (1980).
34. W. E. Spicer, I. Lindau, P. Skeath, and C. Y. Su, Unified defect model and beyond, *J. Vac. Sci. Technol.* 17, 1019–1027 (1980).
35. W. E. Spicer, I. Lindau, P. R. Skeath, C. Y. Su, and P. W. Chye, Unified mechanism for Schottky-barrier formation and III–V oxide interface states, *Phys. Rev. Lett.* 44, 420–426 (1980).
36. R. H. Williams, Surface defect effects on Schottky barriers, *J. Vac. Sci. Technol.* 18, 929–936 (1981).
37. R. H. Williams, R. R. Varma, and V. Montgomery, Metal contacts on silicon and indium phosphide cleaved surfaces and the influence of intermediate absorbed layers. *J. Vac. Sci. Technol.* 16, 1418 (1979).
38. S. M. Sze, *Physics of Semiconductor Devices*, Wiley–Interscience, New York (1969).
39. J. C. Phillips, *Bonds and Bands in Semiconductors*, Academic Press, New York (1973).
40. S. Y. Tong, A. R. Lubinsky, B. J. Mrstik, and M. A. Van Hove, Surface bond angle and bond lengths of rearranged As and Ga atoms on GaAs(110), *Phys. Rev. B 17*, 3303–3309 (1978).
41. R. J. Mayer, C. B. Duke, A. Paton, A. Kahn, E. So, J. L. Yeh, and P. Mark, Dynamical calculations of low-energy electron diffraction intensities from GaAs(110): Influence of boundary conditions, exchange potential, lattice vibrations and multilayer reconstructions, *Phys. Rev. B 19*, 5194–5202 (1979).
42. J. J. Van Laar and J. J. Scheer, Influence of volume dope on Fermi level position at gallium arsenide surface, *Surf. Sci.* 8, 342–356 (1967).
43. A. Huijser and J. J. Van Laar, Work function variations of gallium arsenide cleaved, single crystals, *Surf. Sci.* 52, 202–210 (1975).
44. C. B. Duke, R. J. Meyer, and P. Mark, Trends in surface atomic geometries of compound semiconductors, *J. Vac. Sci. Technol.* 17, 971–977 (1980).
45. R. J. Meyer, C. B. Duke, A. Paton, J. C. Tsand, J. L. Yeh, A. Kahn, and P. Mark, Dynamical analysis of low energy diffraction intensities from InP(110), *Phys. Rev. B 22*, 6171–6183 (1980).
46. R. Z. Bachrach, in: *Progress in Crystal Growth and Characterization*, (P. Pamplin, ed.) Vol. 2. p. 115, Pergamon Press, Oxford (1979).
47. P. Drathen, E. Ranke, and K. Jacobi, composition and structure of differently prepared GaAs(100) surfaces studies by LEED and AES, *Surf. Sci.* 77, L162–L166 (1978).
48. J. Massies, P. Devoldere, and N. T. Linh, Silver contacts on GaAs(001) and InP(001), *J. Vac. Sci. Technol.* 15, 1353 (1978).
49. J. H. Neave and B. A. Joyce, Structure and stoichiometry of (100) GaAs surfaces during molecular beam epitaxy, *J. Cryst. Growth* 44, 387–397 (1978).
50. J. C. Phillips, Excitonic instabilities, vacancies, and reconstruction of covalent surfaces, *Surf. Sci.* 40, 459–469 (1973).
51. J. A. Appelbaum, G. A. Baraff, and D. R. Hamann, GaAs(100): Its surface effective charge, and reconstruction patterns, *Phys. Rev. B 14*, 1623–1632 (1976).
52. R. Ludeke and L. Ley, *Physics of semiconductors*, *Inst. Phys. Conf. Ser.* 43, 1069 (1979).
53. R. F. C. Farrow, A. G. Cullis, A. J. Grant, and J. E. Pattison, Structural and electrical properties of epitaxial metal films grown on argon ion bombarded and annealed (001) InP, *J. Cryst. Growth* 45, 292–301 (1978).
54. C. R. Bayliss and D. L. Kirk, The compositional and structural changes that accompany the thermal annealing of (100) surfaces of GaAs, InP and GaP in vacuum, *J. Phys. D* 9, 233–244 (1976).
55. W. Ranke and K. Jacobi, *Progr. Surf. Sci.* 10, 1 (1981).

56. D. Haneman, Surface structures and properties of diamond-structure semiconductors, *Phys. Rev. 121*, 1093–1100 (1961).
57. J. R. Chelikowsky and M. L. Cohen, Nonlocal pseudopotential calculations for the electronic structure of eleven diamond and zinc-blende semiconductors, *Phys. Rev. B 14*, 556–582 (1976).
58. L. Ley, R. A. Pollak, F. R. McFeely, S. P. Kowalszyk, and D. A. Shirley, Total valence-band densities of states of III–V and II–VI compounds from X-ray photoemission spectroscopy, *Phys. Rev. B 9*, 600–621 (1974).
59. F. C. Chiang, J. A. Knapp, D. E. Eastman, and M. Aono, Angle-resolved photoemission and valence band dispersion $E(k)$ for GaAs: Direct vs. indirect models, *Solid State Commun. 31*, 917–920 (1979).
60. J. H. Dinan, L. K. Galbraith, and T. E. Fischer, Electronic properties of clean cleaved (110) GaAs surfaces, *Surf. Sci. 26*, 587–604 (1971).
61. W. E. Spicer, P. W. Chye, P. E. Gregory, T. Sukegaw, and I. A. Babalola, Photoemission studies of surface and interface states on III–V compounds, *J. Vac. Sci. Technol. 13*, 233–240 (1976).
62. D. E. Eastman and W. D. Grobman, Photoemission densities of intrinsic surface states of Si, Ge, and GaAs, *Phys. Rev. Lett. 28*, 1378–1381 (1972).
63. A. Huijser and J. Van Laar, Work function variations of gallium arsenide cleaved single crystals, *Surf. Sci. 52*, 202–210 (1975).
64. A. Huijser, J. Van Laar, and T. L. Van Rooy, Electronic surface properties of UHV-cleaved III–V compounds, *Surf. Sci. 62*, 472–486 (1977).
65. P. Chiaradia, G. Chiarotti, I. Davoli, S. Nannarone, and P. Sassaroli, Physics of semiconductors, *Inst. Phys. Conf. Ser. 43*, 195 (1978).
66. P. Chiaradia, G. Chiarotti, F. Ciccacci, R. Momeo, S. Nannarone, P. Sasaroli, and S. Selci, Optical detection of surface states in GaAs (110) and GaP (110), *Surf. Sci. 99*, 70–75 (1980).
67. C. M. Bertoni, O. Bisi, C. Calandra, and F. Mangi, Physics of semiconductors, *Inst. Phys. Conf. Ser. 43*, 191 (1978).
68. J. A. Knapp and G. J. Lapeyre, Angle resolved photoemission studies of surface states on (110) GaAs, *J. Vac. Sci. Technol. 13*, 757–760 (1976).
69. A. Huijser, J. Van Laar, and T. L. Van Rooy, Angular-resolved photoemission from GaAs (110) surfaces with adsorbed Al, *Surf. Sci. 102*, 264–270 (1981).
70. V. Montgomery, G. P. Srivastava, I. Sing, and R. H. Williams, The electronic structure of cleaved indium phosphide (110) surfaces: experiment and theory, *J. Phys. C 16*, 3627–3640 (1983).
71. V. Dose, H. J. Gossmann, and D. Straub, Investigation of intrinsic unoccupied surface states at GaAs (110) by isochromat spectroscopy, *Phys. Rev. Lett. 47*, 608–610 (1981).
72. R. S. Bauer, R. Z. Bachrach, S. A. Flodstrom, and J. C. McMenamin, Empty semiconductor surface states: Core-level photoyield studies, *J. Vac. Sci. Technol. 14*, 378–382 (1977).
73. R. S. Bauer, Ionicity effects on compound semiconductor (110) surfaces, *J. Vac. Sci. Technol. 14*, 899–903 (1977).
74. G. J. Lapeyre and J. Anderson, Evidence for a surface-state exciton on GaAs (110), *Phys. Rev. Lett. 35*, 117–120 (1975).
75. R. E. Allen and J. D. Dow, Theory of Frenkel core excitons at surfaces, *Phys. Rev. B 24*, 911–914 (1981).
76. J. A. Appelbaum, G. A. Baraff, and D. R. Hamann, Theoretical study of the GaAs (100) surface, *J. Vac. Sci. Technol. 13*, 751–756 (1976).
77. I. Ivanov, A. Mazur, and J. Pollmann, The ideal (111), (110) and (100) surfaces of Si, Ge and GaAs; A comparison of their electronic structure, *Surf. Sci. 92*, 365–384 (1980).

78. J. Pollmann and S. T. Pantelides, Scattering-theoretic approach to the electronic structure of semiconductor surfaces: The (100) surface of tetrahedral semiconductors and SiO_2, *Phys. Rev. B 18*, 5524–5544 (1978).
79. R. Ludeke and L. Esaki, Electron energy-loss spectroscopy of GaAs and Ge surfaces, *Phys. Rev. Lett. 33*, 653–656 (1974).
80. J. Massies, P. Etienne, F. Dezaly, and N. T. Linh, Stoichiometry effects on surface properties of GaAs (100) grown in situ by MBE, *Surf. Sci. 99*, 121–131 (1980).
81. P. K. Larsen, J. H. Neave, and B. A. Joyce, Angle resolved photoemission from As-stable GaAs (001) surfaces prepared by MBE, *J. Phys. C 14*, 167–192 (1981).
82. R. Ludeke and A. Koma, Selection-rule effects in electron-loss spectroscopy of Ge and GaAs surfaces, *Phys. Rev. Lett. 34*, 817–821 (1973).
83. K. Jacobi, C. Von Muschwitz, and W. Ranke, Angular resolved UPS of surface states on GaAs (111) prepared by molecular beam epitaxy, *Surf. Sci. 82*, 270 (1979).
84. W. Gudat and D. E. Eastman, Electronic surface properties of III–V semiconductors: Excitonic effects, band bending effects and interactions with Au and O adsorbate layers, *J. Vac. Sci. Technol. 13*, 831–837 (1976).
85. R. A. Street, R. H. Williams, and R. S. Bauer, Influence of the surface on photoluminescence from indium phosphide crystals, *J. Vac. Sci. Technol. 17*, 1001–1004 (1980).
86. R. A. Street and R. H. Williams, The luminescence of defects introduced by mechanical damage of InP, *J. Appl. Phys. 52*, 402–406 (1981).
87. J. M. Palau, E. Testemale, A. Ismail, and L. Lassabatere, Silver Schottky diodes on Kelvin, AES and LEED characterized (100) surfaces of GaAs cleaned by ion bombardment, *Solid State Electron. 25*, 285–294 (1982).
88. J. Tsand, A. Kahn, and P. Mark, Comparison of LEED and Auger data from cleaved and sputtered-annealed InP (110) surfaces, *Surf. Sci. 97*, 119–127 (1980).
89. R. F. C. Farrow, A. G. Cullis, A. J. Grant, and J. E. Pattison, Structural and electrical properties of epitaxial metal films grown on argon ion bombarded and annealed (001) InP, *Cryst. Growth 45*, 292–301 (1978).
90. D. G. Welkie and M. G. Lagally, Correlation of short-range order and sputter dose in GaAs (110) using a vidicon-based LEED system, *J. Vac. Sci. Technol. 16*, 784–788 (1979).
91. M. G. Lagally, T. M. Lu, and D. G. Welkie, Surface defects and thermodynamics of chemisorbed layers. *J. Vac. Sci. Technol. 17*, 223–230 (1980).
92. M. Jaros, Deep levels in semiconductors, *Adv. Phys. 29*, 409–525 (1980).
93. S. T. Pantelides, The electronic structure of impurities and other point defects in semiconductors, *Rev. Mod. Phys. 50*, 797–858 (1978).
94. G. A. Baraff, E. O. Kane, and M. Schluter, Silicon vacancy: A possible "Anderson negative-U" system, *Phys. Rev. Lett. 43*, 956–959 (1979).
95. M. S. Daw and D. L. Smith, Vacancies near semiconductor surfaces, *Phys. Rev. B 20*, 5150–5156 (1979).
96. M. S. Daw and D. L. Smith, Surface vacancies in InP and GaAlAs, *Appl. Phys. Lett. 36*, 690–692 (1982).
97. M. S. Daw and D. L. Smith, Energy levels of semiconductor surface vacancies, *J. Vac. Sci. Technol. 17*, 1028–1031 (1980).
98. M. S. Daw and D. L. Smith, Relation between the common anion rule and the defect model of Schottky barrier formation, *Solid State Commun. 37*, 205–208 (1981).
99. R. E. Allen and J. D. Dow, Unified theory of point-defect electronic states, core excitions and intrinsic states at semiconductor surfaces, *J. Vac. Sci. Technol. 19*, 383–387 (1981).
100. R. E. Allen and J. D. Dow, Role of surface antisite defects in the formation of Schottky barriers, *Phys. Rev. B 25*, 1423–1426 (1982).

101. G. P. Srivastava, *Procceedings of the Conference on Semi-Insulating III–V Materials*, Nottingham (1980).
102. M. Jaros, Deep levels in semiconductors, *Adv. Phys.* **29**, 409–525 (1980).
103. A. J. Rosenberg, J. N. Butler, and A. A. Meena, Oxidation of intermetallic compounds, adsorption of oxygen on III–V compounds and germanium at 78°K, *Surf. Sci.* **5**, 1–16 (1966).
104. R. Dorn, H. Luth, and G. J. Russell, Adsorption of oxygen on clean cleaved (110) gallium–arsenide surfaces, *Phys. Rev. B* **10**, 5049–5956 (1974).
105. H. Luth and G. J. Russell, Electron loss spectroscopy of clean and oxygen covered GaAs (110) surfaces, *Surf. Sci.* **45**, 329–341 (1974).
106. P. Mark and W. F. Creighton, The effects of surface index and atomic order on the GaAs–oxygen interaction, *Thin Solid Films* **56**, 19–38 (1979).
107. P. Pianetta, I. Lindau, C. M. Garner, and W. E. Spicer, Photoemission studies of the initial stages of oxidation of GaSb and InP, *Surf. Sci.* **88**, 439–460 (1979).
108. J. J. Barton, W. A. Goddard, and T. C. McGill, Reconstruction and oxidation of the GaAs (110) surface, *J. Vac. Sci. Technol.* **16**, 1178–1185 (1979).
109. P. Mark, S. C. Chang, W. F. Creighton, and B. W. Lee, A comparison of some important surface properties of elemental and tetrahedrally coordinated compound semiconductors, *CRC Crit. Rev. Solid-State Sci.* **5**, 189–229 (1975).
110. P. Mark and W. F. Creighton, The effects of surface index and atomic order on the GaAs–oxygen interaction, *Thin Solid Films* **56**, 19–38 (1979).
111. D. W. Welkie and M. G. Lagally, Correlation of short-range order and sputter dose in GaAs (110) using a vidicon-based LEED system, *J. Vac. Sci. Technol.* **16**, 784–788 (1979).
112. A. Kahn, D. Katnani, and P. Mark, The GaAs (110)–oxygen interaction: A LEED analysis III, *Surf. Sci.* **94**, 547–554 (1980).
113. A. Kahn, D. Katnani, P. Mark, C. Y. Su, I. Lindau, and W. E. Spicer, Order–disorder effects in GaAs (110)–oxygen interaction: A LEED–UPS analysis, *Surf. Sci.* **87**, 325–332 (1979).
114. R. Ludeke, The oxidation of the GaAs (110) surface, *Solid State Commun.* **21**, 815–818 (1977).
115. J. Joannopoulos and E. J. Mele, Extrinsic surface states for oxygen chemisorbed on the GaAs (110) surface, *J. Vac. Sci. Technol.* **15**, 1287–1289 (1978).
116. J. Stohr, R. S. Bauer, J. C. McMenamin, L. I. Johansson, and S. Brennan, Surface EXAFS investigation of oxygen chemisorption on GaAs (110), *J. Vac. Sci. Technol.* **16**, 1195–1199 (1979).
117. C. R. Brundle and D. Seybold, Oxygen interaction with GaAs surfaces: An XPS/UPS study, *J. Vac. Sci. Technol.* **16**, 1186–1190 (1979).
118. I. Lindau, C. Y. Su, P. R. Skeath, and W. E. Spicer, *Proceedings of the Fourth International Conference on Solid Surface*, Cannes (1980), pp. 979–983.
119. W. E. Spicer, I. Lindau, P. Pianetta, P. W. Chye, and C. M. Garner, Fundamental studies of III–V surfaces and the (III–V)-oxide interface, *Thin Solid Films* **56**, 1–13 (1979).
120. R. H. Williams and I. T. McGovern, Surface characterization of indium phosphide, *Surf. Sci.* **51**, 14–28 (1975).
121. V. Montgomery, Ph.D. thesis, The New University of Ulster (1982).
122. W. Monch and R. Enninghorst, Charge Transfer to oxygen chemisorbed on cleaved GaAs (110) surfaces, *J. Vac. Sci. Technol.* **17**, 942–945 (1980).
123. P. W. Chye, I. Lindau, P. Pianetta, C. M. Garner, C. Y. Su, and W. E. Spicer, Photoemission study of Au Schottky-barrier formation on GaSb, GaAs and InP using synchrotron radiation, *Phys. Rev. B* **18**, 5545–5559 (1978).
124. M. Liehr and H. Luth, Gas adsorption on cleaved GaAs (110) surfaces studied by surface photovoltage spectroscopy, *J. Vac. Sci. Technol.* **16**, 1200–1206 (1979).

125. C. D. Thault, G. M. Guichar, and C. A. Sebenne, Effects of low oxygen exposure on the electronic surface properties of GaAs (110), *Surf. Sci. 80*, 273–277 (1979).
126. S. C. Dahlberg, Effect of adsorbed gases and temperature on the photovoltage spectrum of GaAs, *J. Vac. Sci. Technol. 13*, 1056–1059 (1976).
127. H. Iwasaki, Y. Mizokawa, R. Nishitani, and S. Nakamura, X-ray photoemission study of the initial oxidation of the cleaved (110) surfaces of GaP and InSb, *Surf. Sci. 86*, 811–818 (1979).
128. M. G. Dowsett and E. H. C. Parker, Study of low coverage adsorption of cleaved (110) InP surfaces using SIMS, *J. Vac. Sci. Technol. 16*, 1207–1210 (1979).
129. G. M. Guichard, C. A. Sebenne, and C. D. Thualt, Electronic surface states on cleaved GaP (110): Initial steps of the oxygen chemisorption, *J. Vac. Sci. Technol. 16*, 1212–1215 (1979).
130. Y. Ohno, S. Nakamura, and T. Kuroda, FEM studies of oxygen and gold adsorption and field-desorption on GaAs and GaP surfaces, *Surf. Sci. 91*, L7–L16 (1980).
131. E. W. Kreutz, E. Rickus, and N. Sotnik, Oxidation properties of InSb (110) surfaces, *Surf. Sci. 68*, 392–398 (1977).
132. H. Iwasaki, Y. Mizokawa, R. Nishitani, and S. Nakamura, Effects of water vapor and oxygen excitation on oxidation of GaAs, GaP and InSb surfaces studied by X-ray photoemission spectroscopy, *Japan. J. Appl. Phys. 18*, 1525–1529 (1979).
133. D. Norman and D. K. Skinner, *J. Phys. D (Applied Phys.) 10*, L151–154 (1977).
134. P. E. Gregory and W. E. Spicer, Photoemission study of the adsorption of O_2, CO and H_2 on GaAs (110), *Surf. Sci. 54*, 229–258 (1976).
135. D. J. Miller and D. Haneman, Electron–paramagnetic-resonance study of clean and oxygen-exposed surfaces of GaAs, AlSb and other III–V compounds, *Phys. Rev. B 3*, 2918–2928 (1971).
136. A. E. Morgan, Ellipsometric studies of chemisorption on GaP (110) single crystals, *Surf. Sci. 43*, 150–172 (1974).
137. P. K. Larsen, N. V. Smith, M. Schluter, H. H. Farrell, K. M. Ho, and M. L. Cohen, Surface energy bands and atomic position and Cl chemisorbed on cleaved Si (111), *Phys. Rev. B 17*, 2612–2619 (1978).
138. J. E. Rowe, G. Margaritondo, and S. B. Christman, Chlorine chemisorption on silicon and germanium surfaces: Photoemission polarization effects with synchrotron radiation, *Phys. Rev. B 16*, 1581–1589 (1977).
139. G. Margaritondo, J. E. Rowe, C. M. Bertoni, C. Calandra, and F. Mangi, Chemisorption geometry on cleaved III–V surfaces: Cl and GaAs, GaSb, and InSb, *Phys. Rev. B 20*, 1538–1545 (1979).
140. V. Montgomery, R. H. Williams, and R. R. Varma, The interaction of chlorine with indium phosphide surfaces, *J. Phys. C 11*, 1989–2000 (1978).
141. K. Jacobi, G. Steinert, and W. Ranke, Iodine etching of the GaAs(111) As surface studied by LEED, AES, and mass spectroscopy, *Surf. Sci. 57*, 571–579 (1976).
142. D. D. Pretzer and H. D. Hagstrum, Ion neutralization studies of the (111), ($\overline{111}$) and (110) surfaces of GaAs, *Surf. Sci. 4*, 265–285 (1966).
143. A. A. Morgan and W. J. M. Van Velzen, Chemisorption of gallium phosphide surfaces, *Surf. Sci. 39*, 255–259 (1973).
144. See also Ref. 134.
145. V. Montgomery and R. H. Williams, The adsorption of water on InP and its influence on Schottky Barrier formation, *J. Phys. C 15*, 5887–5897 (1982).
146. F. Mangi, C. M. Bertoni, C. Calandra, and E. Molinari, Theoretical investigation of hydrogen chemisorption on Ga-containing III–V compounds, *J. Vac. Sci. Technol. 21*, 371–374 (1982).
147. M. Buchel and H. Luth, Adsorption of water and methanol on GaAs (110) surfaces studied by ultraviolet photoemission, *Surf. Sci. 87*, 285–294 (1979).

148. G. J. Hughes, T. P. Humphreys, V. Montgomery, and R. H. Williams, The influence of adlayers on Schottky barrier formation; The adsorption of H_2S and H_2O on indium phosphide, *Vacuum 31*, 10–12 (1981).
149. V. Montgomery and R. H. Williams, The adsorption of water on InP and its influence on Schottky barrier formation, *J. Phys. C 15*, 5887–5897 (1982).
150. J. Massies, F. Dezaly, and N. T. Linh, Effects of H_2S adsorption on GaAs (100) grown in situ by MBE, *J. Vac. Sci. Technol. 17*, 1134–1140 (1980).
151. V. Montgomery, R. H. Williams, and G. P. Srivastava, The influence of adsorbed layers in controlling Schottky barriers, *J. Phys. C 14*, L191–L194 (1981).
152. G. Ottaviani and J. W. Mayer, in: *Reliability and Degradation* (M. J. Howes and D. V. Morgan, eds.), Chapter 2, Wiley, New York (1981).
153. K. N. Tu and S. S. Lau, in: *Thin Films—Interdiffusion and Reactions* (J. M. Poate, K. N. Tu, and J. W. Mayer, eds.), Chapter 5, Wiley, New York (1978).
154. K. N. Tu and S. S. Lau, Selective growth of metal-rich silicide of near-noble metals, *Appl. Phys. Lett. 27*, 221–224 (1975).
155. H. B. Kim, A. F. Lovas, G. G. Sweeney, and T. M. S. Heng, *Inst. of Phys. Conf. Ser. 336*, 145 (1977).
156. J. W. Steeds, private communcation.
157. A. McKinley, G. J. Hughes, and R. H. Williams, Cleaved surfaces of indium phosphide and their interfaces with metal electrodes, *J. Phys. C 10*, 4545–4557 (1977).
158. A. McKinley, A. W. Parke, and R. H. Williams, Silver overlayers on (110) indium phosphide: Film growth and Schottky barrier formation. *J. Phys. C 13*, 6723–6736 (1980).
159. B. Tuck, *Introduction to Diffusion on Semiconductors*, IEEE Monograph Series No. 16, Peregrinus (1974).
160. D. V. Morgan in: *Reliability and Degradation* (M. J. Howes and D. V. Morgan, eds.), Chapter 3, Wiley, New York (1981).
161. A. K. Sinha and J. M. Poate, in: *Thin films, interdiffusion and reactions* (J. M. Poate, K. N. Tu, and J. W. Mayer, eds.), pp. 407–432, Wiley, New York (1978).
162. J. W. Matthews, *Proceedings of the Fourth International Vacuum Congress* (1968), p. 479.
163. J. C. Phillips, in: *Thin Films, Interdiffusion and Reactions* (J. M. Poate, K. N. Tu, and J. W. Mayer, eds.), Wiley, New York (1978), pp. 57–67.
164. R. Kern, G. Lelay, and J. J. Metois, *Current Topics in Materials Science*, Vol. 3, Chapter 3, North-Holland, Amsterdam (1979).
165. M. Volmer and A. Weber, *Z. Phys. Chem. 119*, 277 (1926).
166. F. C. Frank and J. H. Van der Merwe, The-dimensional dislocations. I. Static Theory, *Proc. R. Soc. London, Ser. A 198*, 205–215 (1949).
167. J. N. Stranski and L. Krastanov, *Ber. Akad. Wiss. Wien. 146*, 797 (1938).
168. L. Buene, Interdiffusion and phase formation at room temperature in evaporated gold–tin films, *Thin Solid Films 47*, 159–166 (1977).
169. V. Simic and Z. Marinkovic, Room temperature interactions in Ag-metal thin film couples, *Thin Solid Films 61*, 149–160 (1979).
170. V. Simic and Z. Marinkovic, Thin film interdiffusion of Au and In at room temperature, *Thin Solid Films 41*, 57–61 (1977).
171. J. A. Venables, J. Derrien, and A. P. Janssen, Direct observation of the nucleation and growth modes of Ag/Si(111), *Surf. Sci. 95*, 411–430 (1980).
172. J. J. Scheer and J. Van Laar, GaAs–Cs: A new type of photoemitter, *Solid State Commun. 3*, 189–193 (1965).
173. R. F. Steinberg, Photoemission from GaAs thin films, *Appl. Phys. Lett. 12*, 63–65 (1968).
174. A. J. Van Bommel and J. E. Crombeen, Leed, Auger electron spectroscopy (AES) and photoemission studies of the cesium covered GaAs(110) surface, *Surf. Sci. 45*, 308–313 (1974).

175. A. J. Van Bommel and J. E. Crombeen, LEED, Auger electron spectroscopy (AES) and photoemission studies of the adsorption of cesium on the epitaxially grown GaAs(110) surface, *Surf. Sci.* 57, 109–117 (1976).
176. A. J. Van Bommel and J. E. Crombeen, The GaP(001) surface and the adsorption of Cs, *Surf. Sci.* 76, 499–408 (1978).
177. A. J. Van Bommel and J. E. Crombeen, The adsorption and desorption of Cs on GaP and GaSb(001), (110), (111) and ($\overline{111}$) surfaces, studied by LEED, AES and photoemission, *Surf. Sci.* 93, 383–397 (1980).
178. A. J. Van Bommel, J. E. Crombeen, and T. G. J. Van Oirschot, LEED, AES and photoemission measurements of epitaxially grown GaAs(001), (111)A and ($\overline{111}$)B surfaces and their behavior upon Cs adsorption, *Surf. Sci.* 72, 95–108 (1978).
179. T. E. Fischer, Photoelectric emission and interband transitions of GaP, *Phys. Rev.* 147, 603–607 (1966).
180. P. E. Viljoen, M. S. Jazzar, and T. E. Fischer, Electronic properties of clean and cesiated (110) surfaces of GaSb, *Surf. Sci.* 32, 506–518 (1972).
181. B. Goldstein and D. Szostak, Different bonding states of Cs and O on highly photoemissive GaAs by flash-desorption experiments, *Appl. Phys. Lett.* 26, 111–113 (1975).
182. J. Derrien, F. Arnaud d'Avitaya, and M. Bienfait, Isobar, Low energy electron diffraction and loss spectroscopy measurements of cesium covered (110) gallium arsenide, *Solid States Commun.* 20, 557–560 (1976).
183. J. Derrien and F. Arnaud d'Avitaya, Adsorption of cesium on gallium arsenide (110), *Surf. Sci.* 65, 668–686 (1977).
184. W. E. Spicer, I. Lindau, P. E. Gregory, C. M. Garner, P. Pianetta, and P. Chye, Synchrotron radiation studies of electronic structure and surface chemistry of GaAs, GaSb and InP, *J. Vac. Sci. Technol.* 13, 780–785 (1976).
185. J. J. Uebbing, R. L. Bell, Cesium–GaAs Schottky barrier height, *Appl. Phys. Lett.* 11, 357–358 (1967).
186. C. A. Mead, Metal–semiconductor surface barriers, *Solid-State Electron.* 9, 1023–1033 (1966).
187. H. Clemens and W. Monch, *CRC Crit. Rev. Solid-State Sci.* 273–280 (October 1975).
188. P. W. Chye, I. A. Babola, T. Skukegawa, and W. E. Spicer, Photoemission studies of surface states and Schottky-barrier formation on InP, *Phys. Rev. B* 13, 4439–4446 (1976).
189. T. C. McGill, Phenomenology of metal–semiconductor electrical barriers, *J. Vac. Technol.* 11, 935–942 (1974).
190. W. Spitzer and C. A. Mead, Barrier heights on metal-semiconductor systems, *J. Appl. Phys.* 34, 3061–3069 (1963).
191. H. J. Clemens, J. Von Wienskowski, and W. Monch, On the interaction of cesium with cleaved GaAs(110) and Ge(111) surfaces: Work function measurements and adsorption site model, *Surf. Sci.* 78, 648–666 (1978).
192. C. B. Duke, A. Paton, R. J. Meyer, L. J. Brillson, A. Kahn, D. Kanani, J. Carelli, J. L. Yeh, G. Margaritondo, and A. D. Katuani, Atomic geometry of GaAs(110)–p (1 × 1)–Al, *Phys. Rev. Lett.* 46, 440–443 (1981).
193. P. Skeath, C. Y. Su, I. Lindau, and W. E. Spicer, Column III and V elements on GaAs(110): Bonding and adatom–adatom interaction, *J. Vac. Sci. Technol.* 17, 874–879 (1980).
194. L. J. Brillson, R. Z. Bachrach, R. S. Bauer, and J. McMenamin, Chemically induced charge redistribution at Al–GaAs interfaces, *Phys. Rev. Lett* 42, 397–401 (1979).
195. R. Z. Bachrach and R. S. Bauer, Surface reactions and interdiffusion, *J. Vac. Sci. Technol.* 16, 1149–1153 (1979).
196. J. R. Chelikowsky, S. G. Louie, and M. L. Cohen, Surface states and metal overlayers on the (110) surface of GaAs, *Solid State Commun.* 20, 641–644 (1976).

197. D. J. Chadi and R. Z. Bachrach, Chemisorption site geometry and interface electronic structure of Ga and Al on GaAs(110), *J. Vac. Sci. Technol.* **16**, 1159–1163 (1979).
198. C. A. Swarts, J. J. Barton, W. A. Goddard, and T. C. McGill, Chemisorption of Al and Ga on the GaAs(110) surface, *J. Vac. Sci. Technol.* **17**, 869–873 (1980).
199. J. R. Chelikowsky, D. J. Chadi, and M. L. Cohen, Electronic structure of the Al–GaAs(110) surface chemisorption system, *Phys. Rev. B* **23**, 4013–4022 (1981).
200. A. Zunger, Al on GaAs(110) interface: Possibility of adatom cluster formation, *Phys. Rev. B* **24**, 4372–4391 (1981).
201. E. Mele and J. D. Joannopoulos, Surface-barrier formation for Al chemisorbed on GaAs(110), *Phys. Rev. Lett.* **42**, 1094–1097 (1979).
202. E. Mele and J. D. Joannopoulos, Electronic structure of Al chemisorbed on GaAs(110), *J. Vac. Sci. Technol..* **16**, 1154–1158 (1979).
203. J. Ihm and J. D. Joannopoulos, Structural energies of Al deposited on the GaAs(110) surface, *Phys. Rev. Lett.* **47**, 679–682 (1981).
204. J. Ihm and J. D. Joannopoulos, First-principles determination of the structure of the Al/GaAs(110) surface, *J. Vac. Sci. Technol.* **21**, 340–343 (1982).
205. A. Kahn, D. Katnani, J. Carelli, J. L. Yeh, C. B Duke, R. J. Meyer, and A. Paton, LEED intensity analysis of the structure of Al on GaAs(110), *J. Vac. Sci. Technol.* **18**, 792–796 (1981).
206. R. Z. Bachrach, Metal–semiconductor surface and interface states on (110) GaAs, *J. Vac. Sci. Technol.* **15**, 1340–1343 (1978).
207. P. Skeath, I. Lindau, P. Pianetta, P. W. Chye, C. Y. Su, and W. E. Spicer, Photoemission study of the interaction of Al with a GaAs(110) surface, *J. Electron. Spectrosc.* **17**, 259–265 (1979).
208. A. Huijser, J. Van Laar, and T. L. Van Rooy, Angular-resolved photoemission from GaAs(110) surfaces with adsorbed Al, *Surf. Sci.* **102**, 264–270 (1981).
209. R. Ludeke and G. Landgren, Interface behavior and crystallographic relationships of aluminum on GaAs(100) surfaces, *J. Vac. Sci. Technol.* **119**, 667–673 (1981).
210. R. H. Williams, A. McKinley, G. J. Hughes, V. Montgomery, and I. T. McGovern, Metal–GaSe and metal–InP interfaces: Schottky barrier formation and interfacial reactions, *J. Vac. Sci. Technol.* **21**, 594–598 (1982).
211. J. Van Laar, A. Huijser, and T. L. Van Rooy, Adsorption of type III and V elements on GaAs(110), *J. Vac. Sci. Technol.* **16**, 1164–1167 (1979).
212. P. W. Chye, I. Lindau, P. Pianetta, C. M. Garner, C. Y. Su, and W. E. Spicer, Photoemission study of Au Schottky-barrier formation on GaSb, GaAs and InP using synchrotron radiation, *Phys. Rev. B* **18**, 5545–5559 (1978).
213. R. S. Bauer, R. Z. Bachrach, G. V. Hansson, and P. Chiaradia, Dissociative surface reactions at Schottky and heterojunction interfaces with AlAs and GaAs, *J. Vac. Sci. Technol.* **19**, 674–680 (1981).
214. G. Apai, S. T. Lee, and M. G. Mason, Valence band formation in small silver clusters, *Solid State Commun.* **37**, 213–217 (1981).
215. J. F. Hamilton and P. Logel, Nucleation and growth of Ag and Pd on amorphous carbon by vapor deposition, *Thin Solid Films* **16**, 49–63 (1973).
216. R. Ludeke, T. C. Chiang, and D. E. Eastman, Crystallographic relationships and interfacial properties of Ag on GaAs(100) surfaces, *J. Vac. Sci. Technol.* **21**, 598–606 (1982).
217. A. Hiraki, S. Kim, W. Kammamura, and M. Iwami, Room-temperature interfacial reaction in Au–semiconductor systems, *Appl. Phys. Lett.* **13**, 611–612 (1977).
218. A. Hiraki, S. Kim, W. Kammanura, and M. Iwami, Chemical effects in (LVV) Auger spectra of third-period elements (Al, Si, P, and S) dissolved in copper, *Appl. Phys. Lett.* **34**, 194–195 (1979).

219. A. Hiraki, S. Kim, W. Kammanura, and M. Iwami, Dynamical observation of room temperature interfacial reaction in metal–semiconductor system by Auger electron spectroscopy, *Surf. Sci.* 86, 706–710 (1979).
220. J. R. Waldrop and R. W. Grant, Interface chemistry of metal–GaAs Schottky-barrier contacts, *Appl. Phys. Lett.* 34, 630–632 (1979).
221. I. Lindau, P. W. Chye, C. M. Garner, P. Pianetta, C. Y. Su, and W. E. Spicer, New Phenomena in Schottky barrier formation on III–V compounds, *J. Vac. Sci. Technol.* 15, 1332–1339 (1978).
222. L. J. Brillson, C. F. Brucker, N. G. Stoffel, A. D. Katnani, and G. Margaritondo, Abruptness of semiconductor–metal interfaces, *Phys. Rev. Lett.* 46, 838–841 (1981).
223. L. J. Brillson, C. F. Brucker, A. D. Katnani, N. G. Stoffel, and G. Margaritondo, Atomic and electronic structure of InP–metal interfaces: A prototypical III–V compound semiconductor, *J. Vac. Sci. Technol.* 19, 661–666 (1981).
224. R. H. Williams, Surface defects on semiconductors, *Surf. Sci.* 132, 122–142 (1983).
225. A. G. Cullis and R. F. C. Farrow, A study of the structure and properties of epitaxial silver deposited by atomic beam techniques on (001) InP, *Thin Solid Films* 58, 197–202 (1974).
226. S. Knight and C. Paola, in: *Ohmic Contacts to Semiconductor* (B. Schwartz, ed.), Electrochemical Society (1969), p. 2.
227. R. H. Williams, G. P. Srivastava, and I. T. McGovern, Photoelectron spectroscopy of solids and their surfaces, *Rept. Progr. Phys.* 43, 1357–1414 (1980).
228. A. Amith and P. Mark, Schottky barriers on ordered and disordered surfaces of GaAs(110), *J. Vac. Sci. Technol.* 15, 1344–1352 (1978).
229. C. M. Mead and W. Spitzer, Fermi level position at metal–semiconductor interfaces, *Phys. Rev.* 134, A713–A716 (1964).
230. J. M. Palau, E. Testemale, A. Ismail, and L. Lassebatere, Silver Schottky diodes on Kelvin, AES and LEED characterized (100) surfaces of GaAs cleaned by ion bombardment, *Solid State Electron.* 25, 285–294 (1982).
231. R. H. Williams, V. Montgomery, and R. R. Varma, Chemical effects in Schottky barrier formation, *J. Phys. C* 11, L735–L742 (1978).
232. V. Montgomery, A. McKinley, and R. H. Williams, The influence of intermediate adsorbed layers on the metal contacts formed to indium phosphide crystals, *Surf. Sci.* 89, 635–652 (1979).
233. E. Hokelek and G. Y. Robinson, Schottky contacts on chemically etched p- and n-type indium phosphide, *Appl. Phys. Lett.* 40, 426–432 (1982).
234. J. O. McCaldin, T. C. McGill, and C. A. Mead, Correlation for III–V and II–VI semiconductors of the Au Schottky barrier energy with anion electronegativity, *Phys. Rev. Lett.* 36, 56–58 (1976).
235. J. R. Chelikowsky, S. G. Louie, and M. L. Cohen, Surface states and metal overlayers on the (110) surface of GaAs, *Solid State Commun.* 20, 641–644 (1976).
236. J. M. Andrews and J. C. Phillips, Chemical bonding and structure of metal–semiconductor interfaces, *Phys. Rev. Lett.* 35, 56–59 (1975).
237. L. J. Brillson, Transition in Schottky barrier formation with chemical reactivity, *Phys. Rev. Lett.* 40, 260–263 (1978).
238. A. D. Muckherjee, *Reliability and Degradation* (M. J. Howes and D. V. Morgan eds.), Chapter 1, Wiley, New York (1981).
239. J. M. Poate, K. N. Tu, and J. W. Mayer (eds.), *Thin Films, Interdiffusion and Reactions*, Wiley, New York (1978).
240. V. L. Rideout, A review of the theory, technology and application of metal–semiconductor rectifiers, *Thin Solid Films* 48, 261–291 (1978).

241. R. H. Williams, A. McKinley, G. J. Hughes, and T. P. Humphreys, Metal contacts on semiconductors: The adsorption of Sb, Sn and Ga on InP (110) cleaved surfaces, *J. Vac. Sci. Technol. B 2(3)*, 561–568 (1984).
242. J. L. Freeouf and J. M. Woodall, Schottky barriers: An effective work function model, *Appl. Phys. Lett. 39*, 727–729 (1981).
243. J. L. Freeouf, T. N. Jackson, S. E. Laux, and J. M. Woodall, Size dependence of "effective" barrier heights of mixed-phase contacts, *J. Vac. Sci. Technol. 21*, 570–573 (1982).
244. J. L. Freeouf, Silicide interface stoichiometry, *J. Vac. Sci. Technol. 18*, 910–916 (1981).
245. K. Kajiama, Y. Mizushima, and S. Sakata, Schottky barrier height of n-In$_x$ Ga$_y$As diodes, *Appl. Phys. Lett. 23*, 458–459 (1973).
246. J. S. Best, The Schottky-barrier height of Au on n-GaAlAs as a function of AlAs content, *Appl. Phys. Lett. 34*, 522–524 (1979).
247. J. D. Levine, Schottky-barrier anomalies and interface states, *J. Appl. Phys. 42*, 3991–3999 (1971).
248. J. D. Levine, Power law reverse current–voltage characteristic in Schottky barrier, *Solid-State Electron. 17*, 1083–1086 (1974).
249. C. R. Crowell, The physical significance of the T_0 anomalies in Schottky barriers, *Solid-State Electron. 20*, 171–175 (1977).
250. W. Monch and H. Gant, Chemisorption-induced defects on GaAs(110) surfaces, *Phys. Rev. Lett. 48*, 512–515 (1982).
251. R. W. Grant, J. R. Waldrop, S. P. Kowalczyk, and E. A. Kraut, Correlation of GaAs surface chemistry and interface Fermi-level position: A simple defect model interpretation, *J. Vac. Sci. Technnol. 19*, 477–480 (1981).
252. J. Massies and N. T. Linh, *J. Cryst. Growth 56*, 25 (1982).

2

Schottky Diodes and Ohmic Contacts for the III–V Semiconductors

Gary Y. Robinson

1. Introduction

In this chapter a review of the electrical properties of metal–semiconductor contacts to the III–V semiconductors is given. Metal–semiconductor structures play an important role in devices based on the III–V compound semiconductors in the form of Schottky-barrier diodes or ohmic contacts. Important III–V devices utilizing Schottky-barrier junctions include solar cells, microwave mixer diodes, and metal semiconductor field-effect transistors (MESFETs) and their associated integrated circuits. Schottky diodes also find widespread use for III–V semiconductor materials characterization, including carrier concentration profiling and deep-level identification. Ohmic contacts with low resistance are necessary for high performance in many III–V devices. For example, the efficiency of light-emitting diodes and lasers is strongly influenced by contact resistance, and the noise behavior and the gain of an FET are significantly affected by the character of ohmic contacts. In all of these cases, the metal–semiconductor interface is formed on a chemically etched, as compared to an atomically clean, semiconductor surface. Thus, in this chapter the properties of Schottky diodes and ohmic contacts prepared by chemically etching the III–V semiconductors are emphasized, while the previous chapter dealt with metal–semiconductor interface formation on atomically clean surfaces. After a brief development of the concepts underlying the parameters used to characterize the electrical properties of Schottky diodes and ohmic contacts, the trends

Gary Y. Robinson • Department of Electrical Engineering, Colorado State University, Fort Collins, CO 80523.

in parameter variation with composition for the III–V semiconductors is given. Recent results in the measurement of the electrical properties for the III–V semiconductors are tabulated and correlations to various models of metal–semiconductor interface formation are discussed.

For a more detailed review of the theory of Schottky diodes, the reader is referred to Rhoderick[1] or Sze,[2] and recent developments in the understanding of the formation of the metal–semiconductor interface can be found in the annual Proceedings of the Conference on the Physics of Compound Semiconductor Interfaces.[3] Sharma and Gupta[4] have recently provided an extensive list of references on Schottky diodes, and Sinha and Poate[5] have discussed in detail interdiffusion and interaction at the metal–semiconductor interface for some III–V compound semiconductors.

2. Electrical Properties of Metal–Semiconductor Contacts

Based on their current–voltage characteristics, metal–semiconductor contacts can be divided into two classifications. Contacts with rectifying characteristics are called Schottky barriers, and those with linear characteristics are called ohmic contacts. The single most important parameter describing a Schottky barrier is the barrier energy ϕ_B. This section briefly reviews current models of Schottky-barrier formation for the III–V semiconductors and presents theoretical expressions for the electrical characteristics of ideal Schottky diodes. Ohmic contacts are best characterized using the specific contact resistance r_c. In this section the theoretical concepts underlying r_c are presented.

2.1. Classical Models of the Interface

2.1.1. Schottky Model

The classical description of the Schottky barrier formed between a metal and a semiconductor starts with the band diagram shown in Fig. 1. The work function of the metal, ϕ_m, is the energy needed to remove an electron from the Fermi level of the metal (E_F^m) to an infinite distance from the metal surface (i.e., the vacuum level). The work function of the semiconductor is ϕ_s, but a more useful quantity is the electron affinity χ_s, such that

$$\phi_s = \chi_s + \xi \tag{1}$$

where ξ is the position of the Fermi level in the semiconductor (E_F^s) relative to the conduction-band edge. Note that we have chosen the

Schottky Diodes and Ohmic Contacts

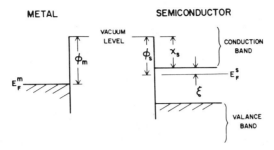

Figure 1. Energy-band diagrams of a metal and a semiconductor separated by a large distance. The work junction for the metal is ϕ_m.

flat-band condition for an n-type semiconductor and thus no electric field exists within the semiconductor. This implies that the semiconductor terminates at the surface without distortion of the electronic energy levels and hence no electronic states exist at the surface. Furthermore, we have chosen the special case of $\phi_s < \phi_m$.

Now if we suddenly bring the metal and the semiconductor together and assume the vacuum level is the same in both materials, electrons flow from the semiconductor to the metal, undercovering the positive donor ions and accumulating at the surface of the metal. The resulting dipole electric field opposes further electron flow, and in equilibrium we must have

$$E_F^m = E_F^s \tag{2}$$

As shown in Fig. 2 if the semiconductor is uniformly doped, the charge density is uniform to a depth w, called the depletion depth, and the field $\mathscr{E}$ is linear with distance. If we define the barrier energy ϕ_B as the energy difference between the Fermi level in the metal and the bottom of the conduction band in the semiconductor at the interface, then

$$\phi_B = \phi_m - \chi_s \tag{3}$$

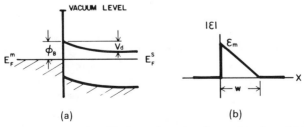

Figure 2. (a) Band diagram of an ideal metal–semiconductor interface. (b) Electric field distribution at the interface for an n-type semiconductor of uniform doping.

This description of a metal–semiconductor contact was first presented by Schottky[6] and Eq. (3) is referred to as the *Schottky limit*. Equation (3) states that the barrier ϕ_B is directly proportional to the metal work function ϕ_m. As we shall see in Section 3, Schottky diodes formed on many of the III–V semiconductors do not show this behavior.

The term "barrier" is chosen because ϕ_B is the energy necessary for electrons in the metal to acquire in order to penetrate the semiconductor. Note that the barrier for electrons in the semiconductor relative to the metal is V_d, the diffusion potential or the band bending in the semiconductor at equilibrium. The fact that V_d is bias dependent and that ϕ_B is not, is responsible for the rectifying I–V characteristics of Schottky diodes.

If we carry the Schottky theory one step further, we obtain the four possible cases of Fig. 3. The term "ohmic" in Fig. 3 refers to an interface without a barrier for majority-carrier flow. Thus the Schottky model should predict that by simply choosing a metal with the appropriate work function, one could obtain a rectifying diode or an ohmic contact. However, experimentally it is found that for most III–V semiconductors, the value of ϕ_B is almost independent of the metal used to form the contact. Thus the Schottky limit of Eq. (3) is not a complete description of all metal–semiconductor interfaces.

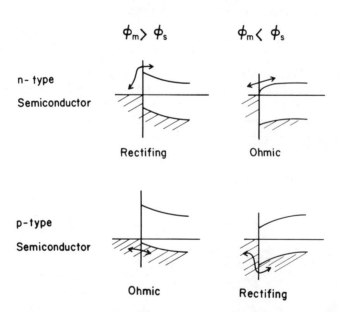

Figure 3. Band diagrams describing the Schottky theory of the metal–semiconductor interface. Four types of conduction are possible depending on the difference $\phi_m - \phi_s$ and the semiconductor type. The arrows indicate majority-carrier current flow.

2.1.2. Bardeen Model

In 1947 Bardeen[7] proposed that if surface states existed at the metal–semiconductor interface in sufficient numbers, then ϕ_B would be independent of ϕ_m. Surface states are electronic states localized at the surface of the semiconductor crystal and are produced by the interruption of the perfect periodicity of the crystal lattice. The states can be occupied or empty depending on their position in energy relative to the Fermi level at the surface. Traditionally, one defines a neutral level ϕ_0 as the energy level, measured relative to the valence band, to which the surface states are filled when the surface is neutral. If the states are filled to an energy greater than ϕ_0, the surface possesses a net negative charge and the states are acceptor-like in behavior; if the states are filled to a level below ϕ_0, the surface has a net positive charge and the states behave in a donor-like manner. The parameter ϕ_0 is used as a convenience in describing surface states and its value depends on the particular surface under consideration.

Assume that a thin insulating layer separates the metal from the semiconductor. The layer is so thin as to be transparent to electron flow yet can withstand a potential difference across it. If the number of surface states is large, then the Fermi level at the surface of the semiconductor will be at ϕ_0. The band bending in the Schottky model was due entirely to the difference between ϕ_s and ϕ_m. However, if an interfacial layer exists and the surface state density is large, the potential difference $\phi_m - \phi_s$ will appear entirely across the interfacial layer since the charge in the surface states will fully accommodate the necessary potential difference. Thus, no change in the charge within the depletion region of the semiconductor is necessary when a metal is brought into contact with the semiconductor. Hence, ϕ_B is independent of ϕ_m. Then E_F^s at the surface of the semiconductor is the same as in the metal E_F^m and we obtain

$$\phi_B = E_g - \phi_0 \tag{4}$$

Thus, the Fermi level is "pinned" or "stabilized" by the surface states to an energy ϕ_0 above the valence band. Equation (4) is known as the *Bardeen limit*.

2.1.3. General Case

In general the value of the barrier energy ϕ_B will be somewhere between the Schottky limit (3) and the Bardeen limit (4). Consider the metal–semiconductor interface shown in Fig. 4, where the semiconductor is n-type and has permittivity ε_s; the interfacial layer has thickness δ and permittivity ε_i; and the surface states are characterized by the density

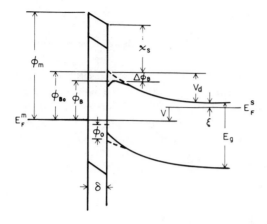

Figure 4. Band diagram for a nonideal metal–semiconductor interface under forward bias V. An insulating layer of thickness δ exists between the metal and semiconductor and surface states are filled to the level ϕ_0.

N_{ss} (per unit area per unit energy) and the neutral level ϕ_0. Cowley and Sze[8] were the first to analyze this general case and it can be shown[1] from their results that ϕ_B^0, the barrier energy with no electric field inside the semiconductor (i.e., the flat-band condition), is given by

$$\phi_B^0 = \gamma(\phi_m - \chi_s) + (1 - \gamma)(E_g - \phi_0) \tag{5}$$

where

$$\gamma \equiv \varepsilon_i/(\varepsilon_i + qN_{ss}\delta) \tag{6}$$

The dimensionless parameter γ varies between zero and unity, depending primarily on the density of surface states. If $qN_{ss}\delta \ll \varepsilon_i$, then $\gamma \approx 1$ and Eq. (5) reduces to the Schottky limit of Eq. (3). If $qN_{ss}\delta \gg \varepsilon_i$, then $\gamma \ll 1$ and Eq. (5) reduces to the Bardeen limit (4). Assuming $\delta \approx 10$ Å and a value for ε_i typical of common insulating films, N_{ss} would have to exceed about 10^{13} states/cm²-eV to pin the Fermi level at ϕ_0. A density of 10^{13} cm^{-2} eV^{-1} corresponds to about one surface state for every 500 atoms of the semiconductor at the metal–semiconductor interface. Such high surface state concentrations have been measured directly on many III–V semiconductors prepared by chemical etching, and thus Schottky diodes formed on such surfaces exhibit values of ϕ_B which are independent of the metal employed as well as the electron affinity of the semiconductor.

Because part of the potential difference $\phi_m - \phi_s$ appears across the interfacial layer and part across the space-charge region of the semiconductor, ϕ_B will not be equal to ϕ_B^0 at zero bias. The effect of the electric field at the metal–semiconductor interface can be written as[1]

$$\phi_B = \phi_B^0 - \alpha \mathscr{E}_m \tag{7}$$

where $\mathscr{E}_m$ is the maximum value of the electric field in the semiconductor, and the parameter α has dimensions of length and is given by

$$\alpha = \delta\varepsilon_s/(\varepsilon_i + qN_{ss}\delta) \tag{8}$$

When $qN_{ss}\delta \gg \varepsilon_i$, $\alpha \approx 0$, $\phi_B \approx \phi_B^0$, and the barrier energy is independent of the electric field and the band bending inside the semiconductor. Thus a large concentration of surface states screens the interior of the semiconductor from the metal. When $N_{ss} = 0$, $\alpha = \delta\varepsilon_s/\varepsilon_i$ and ϕ_B will depend on $\mathscr{E}_m$. This latter case could be of practical importance in a Schottky diode under reverse bias, since $\mathscr{E}_m$ is bias dependent and is largest in reverse bias. The dependence of $\mathscr{E}_m$ on the applied voltage V can be found from the solution of Poisson's equation, assuming depletion of mobile carriers from the space-charge region of a n-type semiconductor uniformly doped with donors in concentration N_D, to be

$$\mathscr{E}_m = [(2qN_D/\varepsilon_s)(V_d - kT/q)]^{1/2} \tag{9}$$

where $V_d = \phi_B^0 - \xi - V$, with V measured positive with respect to the semiconductor as shown in Fig. 4. For example, if $V = -20$ V for a GaAs Schottky diode with $N_D = 10^{15}$ cm^{-3} and $\phi_B^0 = 0.8$ eV and assuming an interfacial layer with $\delta = 10$ Å and $\varepsilon_i = 4\varepsilon_0$, using Eqs. (7)–(9) we obtain $\alpha\mathscr{E}_m \approx 4$ meV (about 0.5% of ϕ_B^0) for $N_{ss} = 0$. Thus the correction to ϕ_B^0 resulting from the field in the interfacial layer is small and is often ignored.

A separate and distinct mechanism, that can also lead to a reduction in ϕ_B^0, is the image-force lowering of a Schottky barrier. An electron just inside the semiconductor induces a sheet of charge on the surface of the metal such that the electrostatic force exerted on the electron is equivalent to that of a positive image charge located an equal distance inside the metal. Thus, it takes less energy to remove the electron from the semiconductor with the image charge present than without. The image-force lowering $\Delta\phi_B$ is found by equating the force due to electric field in the space-charge region to that of the image force; the result is[1]

$$\Delta\phi_B = \left(\frac{q\mathscr{E}_m}{4\pi\varepsilon_s'}\right)^{1/2} \tag{10}$$

where ε_s' is the dynamic permittivity of the semiconductor and is usually smaller than the static permittivity ε_s.[2] As shown by the band diagram of Fig. 4, the image force lowers the barrier by the amount $\Delta\phi_B$.

Equation (10) was derived assuming no interfacial layer present.[1] If we further assume that the effects of the interfacial layer and the image force are independent, then

$$\phi_B = \phi_B^0 - \alpha\mathscr{E}_m - \Delta\phi_B \tag{11}$$

or

$$\phi_B = \phi_{B0} - \Delta\phi_B \tag{12}$$

where $\phi_{B0} = \phi_B^0 - \alpha\mathscr{E}_m$ and is usually referred to in the literature as "the barrier height at zero bias without image-force lowering." This definition is not entirely correct since, as shown above, $\alpha\mathscr{E}_m$ does depend on the bias voltage, even if only very weakly. It has been pointed out by Rhoderick[1] that the image-force lowering $\Delta\phi_B$ arises from an electron in the conduction band near the top of the barrier. If the barrier energy is measured by a technique that involves movement of electrons across the barrier and thus current flow, the measured energy will be ϕ_B. If one determines the barrier energy by techniques that are not dependent on the presence of electrons at the barrier, the measured energy will be ϕ_{B0}.

Finally, it should be noted that the barrier energy at flatband for a metal on a n-type semiconductor ϕ_{Bn}^0 is related to the barrier energy ϕ_{Bp}^0 for the same metal on the same semiconductor but doped p-type, by the equation[1]

$$\phi_{Bn}^0 + \phi_{Bp}^0 = E_g \tag{13}$$

For the above to be true, the character of the interfacial layer and the surface states must be the same on the n- and p-type material; this requires δ, N_{ss}, and ϕ_0 to be identical for both materials. With these assumptions it can be shown that

$$\phi_{Bn0} + \phi_{Bp0} = E_g \tag{14}$$

where ϕ_{Bn0} and ϕ_{Bp0} are the barrier energies at zero bias for n- and p-type materials, respectively. This last equation if very useful, particularly for predicting, for example, the value of ϕ_{Bp0} when only ϕ_{Bn0} is known.

2.2. Mechanisms of Barrier Formation

The fundamental physical mechanisms determining the barrier to electron and hole flow at the metal–semiconductor interface in a practical Schottky diode are not well understood. The formation of a practical Schottky diode consists of depositing a metal film onto a real semiconductor surface, a surface which is not ideal but contaminated by a few monolayers of adsorbed foreign atoms or covered by a thin layer of native oxide. In the case of GaAs and some other III–V semiconductors, the Fermi level E_F^s at the interface is found to be pinned and thus almost independent of the metal work function. If the Fermi level is pinned as a result of surface states, what is the underlying physical mechanism giving rise to the surface states? Alternatively, is the Fermi level pinned as a result of chemical reaction between the metal and the constituents of the semiconductor substrate? What is the role of defects (crystal imperfections, impurities, etc.) at the interface and what role do these defects play in the pinning of the Fermi level? Is the interface truly abrupt on the atomic level and thus is the picture of Fig. 4 an accurate representation of a real metal–semiconductor interface? A simple physical model of a real metal–semiconductor interface does not yet exist and thus unequivocal answers to the above questions have yet to be found. However, in the last few years extensive experimental and theoretical research is beginning to provide us with an accurate picture of the atomic energy levels and the chemical bonding at an ideal metal–semiconductor interface in the III–V semiconductors. The formation of a clean metal–semiconductor interface was discussed in the previous chapter; in this section we will briefly review several recent interface models which are relevant to understanding the formation of practical Schottky diodes in the III–V semiconductors.

Early models of the metal–semiconductor interface were phenomenological in nature and were based on the empirical fact that the ϕ_B for some semiconductors obeyed the Schottky limit (3) and for other semiconductors, principally the III–V compound semiconductors and the elemental semiconductors, followed more closely the Bardeen limit (4). To fit this wide range of observed behavior, Kurtin et al.[9] proposed a "linear interface" model where the barrier energy ϕ_{Bn} was assumed to be linearly dependent on the work function of the metal. Furthermore, they found it more convenient to relate the barrier energy ϕ_{Bn} to the electronegativity χ_m of the metal rather than to the work function ϕ_m. Equation (3) was then modified and written in the form

$$\phi_B = S\chi_m + C$$

where S is the dimensionless parameter that measures the sensitivity of ϕ_B to the metal ($S = \partial\phi_{Bn}/\partial\chi_m$) and C is a constant. McGill and Mead[10] took S to be equal to unity for the Schottky limit and $S = 0$ for the Bardeen limit. By examining $\phi_B - \chi_m$ data for a wide variety of semiconductors, Mead[11] and others[10] found that S depended strongly on the type of chemical bonding in the semiconductor, with ionic materials exhibiting values of S of approximately unity and covalently bonded materials (Ge, Si, GaAs, and InP) exhibiting values of S of about 0.1. Recently, however, the data on which the linear interface model of the covalent–ionic transition is based have been reexamined as a result of more recent experiments[12] and the validity of the parameter S is now in question.

Another empirical rule, developed by Mead and Spitzer,[13] stated that for covalent semiconductors the Schottky-barrier energy on n-type semiconductors ϕ_{Bn} was approximately $2E_g/3$. This rule has been found to be approximately correct for GaAs, GaP, and AlAs, but does not correctly predict the barrier energy on InP, InAs, and GaSb. This empirical rule was further developed by McCaldin et al.,[14,15] using gold as a reference metal to a variety of p-type compound semiconductors, to show that ϕ_{Bp} varied inversely with the electronegativity of the anion of the semiconductor. This evolved into the "common anion rule" which states that $E_F - E_v$ is determined by the anion of the semiconductor and is fixed and independent of the cation or the metal.[15] Thus, those semiconductors with a common anion should exhibit the same ϕ_{Bp}. The common anion rule implies that the Schottky-barrier energy is a property of the bulk semiconductor and not the interface. The common anion rule has correctly predicted the Schottky-barrier energies in studies of the ternary alloys of the III–V compounds GaAs–InAs and GaP–InP, but has failed in describing the GaAs–AlAs system. The ϕ_B data for these materials and related ternary alloys are discussed in Section 3.2.4.

More recently, Spicer and co-workers[16,17] have introduced a defect model to explain Fermi-level pinning at the metal–semiconductor interface. Using X-ray photoemission spectroscopy, Spicer et al. have measured directly E_F^s on clean (vacuum cleaved) GaAs, InP, and GaSb surfaces while metal atoms are deposited under ultrahigh vacuum conditions. Less than a monolayer of metal or oxygen was found to perturb the semiconductor, producing lattice defects at or near the surface. The defects in turn produce surface states which pin the Fermi level in the resulting Schottky diodes. The measured distribution of interface states for Si, GaAs, InP, and GaSb are shown in Fig. 5, where the acceptor-like states have been identified with anion vacancies and the donor-like states have been identified as being produced by cation vacancies. Since it was found to require less than 1% surface coverage of oxygen to pin E_F^s in GaAs, InP, and GaSb, small amounts of oxygen contamination during diode fabrication may be responsible for the observed values of ϕ_B in practical Schottky diodes in these materials. Theoretical studies tend to support this defect model of the interface; Daw and Smith[18] calculated the energy levels of anion vacancies on the surfaces of GaAs and InP and found reasonable agreement with the photoemission data of Spicer et al.[17]

Early X-ray photoemission spectroscopy measurements showed that ideal clean surfaces of III–V semiconductors exhibited no intrinsic surface states in the energy gap. Thus the Bardeen model of Section 2.1.2 does not apply to clean surfaces on the III–V semiconductor since no intrinsic surface states exist to pin the Fermi level. However, the defect model predicts that Fermi-level pinning will occur on a real surface as a result of extrinsic surface states, those induced on the surface as a result of perturbation by a metal or an oxide layer.

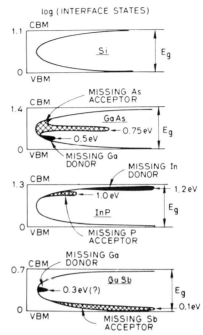

Figure 5. Qualitative distributions of interface states for the semiconductors Si, GaAs, InP, and GaSb as measured by X-ray photoemission spectroscopy. (Reprinted with permission from Ref. 17, copyright 1980, American Institute of Physics.)

The data of Williams et al.[19] supports the defect model of Spicer. Williams measured ϕ_B as well as E_F^s and found that coverage of GaAs by a monolayer of metal produced Fermi-level pinning at the energy levels measured by Spicer. However, on InP Williams observed a more complicated behavior: the Fermi level was pinned at two distinctly different positions depending on the metal used. Furthermore, no simple relationship was found to exist between ϕ_B and χ_m. The data indicated that for those metals which reacted with InP the Fermi level was pinned near the conduction-band minimum (i.e., at low ϕ_{Bn}), while those metals that did not react with the InP produced Fermi-level pinning about 0.5 eV below the conduction-band minimum. Thus, in the case of InP the chemical reactivity of the metal with the semiconductor is important as is the creation of defects during metal–semiconductor interface formation.

For real semiconductor surfaces, the role of chemical reactions between the metal and native oxide has been examined to determine if reduction of the oxide layer will lead to unpinning of E_F^s. The experiments of Kowalczyk et al.[20–22] have demonstrated that the surface of GaAs covered with approximately 10 Å of As_2O_3 and Ga_2O_3 or pure Ga_2O_3 can be altered by chemical reaction with certain metals during

Schottky-barrier formation. For example, the reactive metals Al, Mg, Ti, and Cr were found to reduce the oxides on GaAs and, in some cases, new oxides (Al_2O_3, Cr_2O_3) were formed even during metal deposition at room temperature. However, little change in the position of the Fermi level at the interface occurred as a result of these chemical reactions and thus the defect model of Spicer still applied. Similar experiments are needed on other III–V semiconductors to determine the role of metal–oxide reactions in practical Schottky diodes.

The importance of microscopic chemical effects at the metal–semiconductor interface was first considered by Phillips[23] and later by Andrews and Phillips[24] and Brillson.[25] Brillson showed that the chemical reactivity of the semiconductor constituents with the metal film can play a key role in determining ϕ_B in several II–VI semiconductors and in the III–V compound GaP.[25] Williams also found similar behavior for InP.[19] The reactivity of the metal with a given semiconductor depends on the stability of the semiconductor. For example, semiconductors with small heats of formation will react readily with the metal and ϕ_B will be determined by the composition of the resulting interface. Those semiconductors exhibiting large heats of formation are relatively chemically stable and do not react readily with a deposited metal; the resulting ϕ_B should then depend on the choice of metal (i.e., the Schottky limit should apply).

In addition to the chemical reaction between the metal and the semiconductor, interdiffusion of the elements at the interface can also affect ϕ_B. Brillson et al.[26] have examined the deposition of several metals onto atomically clean InP using X-ray photoemission spectroscopy and observed the outdiffusion of P into an overlaying Au film and the outdiffusion of In into Al and Ni films.

The importance of interdiffusion, as well as chemical reactivity, in determination of ϕ_B for metals deposited on chemically etched semiconductors is not yet fully understood. Williams first examined the chemically etched $\langle 110 \rangle$ surface of n-type InP and found little dependence of ϕ_B on the heat of reaction ΔH_r of the metal with phosphorus.[19] A later study[27] found that ϕ_B varied with ΔH_r in chemically etched n- and p-type InP Scottky diodes. Furthermore, outdiffusion of In into Au and Ag overlayers was observed but not the type of interdiffusion reported by Brillson on atomically clean InP. In summary, interdiffusion and chemical reactions between the metal and the semiconductor have been demonstrated to be important for two III–V semiconductors, GaP and InP, when both are atomically clean. For practical Schottky diodes, the only report available is for InP, where chemical effects remain important in determining ϕ_B but are less pronounced than for atomically clean surfaces.

Another approach to understanding the metal–semiconductor interface is the effective-work-function model of Woodall and Freeouf.[28,29] In this model the value of ϕ_B is not determined by surface states but rather is related to the work functions of several different metallic-like phases at the metal–semiconductor interface. The phases result from oxygen contamination or metal–semiconductor reactions occurring during metalization. Since each phase will have its own work function, Eq. (3) can be written in the modified form $\phi_{Bn} = \phi_{eff} - \chi_S$, where ϕ_{eff} is the weighted average of the work functions of the different phases present. Woodall and Freeouf believe that for the III–V semiconductors the anion is the primary contributor to ϕ_{eff}. Thus, for metal–GaAs Schottky diodes, ϕ_{eff} is the work function of elemental As, not the work function of the metal; the GaAs serves as its own reference and ϕ_B is independent of the metal. The Fermi level is pinned, not by surface states, but rather by the presence of excess As at the metal–semiconductor interface. The effective-work-function model is a reasonable description of practical Schottky diodes since the existence of oxides on the semiconductor surface or reactions of the metal and semiconductor could easily produce mixed phases at the interface. Furthermore, the effective-work-function model has been used to explain both the successes and failures of the common anion rule.[29]

In summary, several distinctly different physical and chemical models have been seen to explain Fermi-level pinning. Defect formation, chemical reaction, and interdiffusion can all play a role in the formation of a metal–semiconductor interface and thus no simple, self-consistent, comprehensive model for the formation of Schottky barriers on real semiconductor surfaces yet exist. Fortunately, in order to accurately predict the electrical behavior of real Schottky diodes, it is sufficient to use the Bardeen model assuming that ϕ_B, N_{ss}, and δ are known. However, the ability to engineer the metal–semiconductor interface and thus precisely control ϕ_B prior to or during metal deposition is not yet possible.

2.3. Current Transport

The current–voltage characteristics of a metal–semiconductor contact are governed by the transport of the charge carriers, electrons or holes, across the metal–semiconductor interface and its associated space-charge region. One of several different physical mechanisms can control the transport of the carriers depending on the semiconductor material, doping, temperature, and energy ϕ_B. For the III–V semiconductors under forward bias, current flows by thermionic emission, field emission, or a combination of the two. In thermionic emission, the majority carriers with kinetic energies in excess of V_d (see Fig. 4) are emitted from the semiconductor to the metal, and any collisions within the space-charge region are relatively unimportant. In field emission, electrons tunnel quantum mechanically through the barrier at energies near the Fermi level; tunneling at energies above the Fermi level is called thermionic-field emission. Other mechanisms of carrier flow may be important in reverse bias, such as recombination inside the space-charge region or in the bulk of the semiconductor, but they will not be discussed here.

2.3.1. Thermionic Emission: Rectification

The current–voltage relationship for thermionic emission in metal–semiconductor contacts was first derived by Bethe[30] and a detailed description is given in Sze.[2] In thermionic emission, the current is assumed to be controlled only by the transfer of carriers across the top of the barrier, and the drift and diffusion that occur as a result of collisions within the space-charge region are assumed unimportant. For an n-type semiconductor under forward bias assuming $q\phi_B \gg kT$, the electrons emitted over the barrier into the metal will be in equilibrium with the electron population in the semiconductor and thus will have a Maxwellian energy distribution. The resulting current density J for an applied bias V (measured positive with respect to the n-type semiconductor) is given by[2]

$$J = A^* T^2 \exp(-q\phi_B/kT)[\exp(qV/kT) - 1] \qquad (15)$$

where $A^* = 4\pi q m_e^* k^2 / h^3$. Here m_e^* is the effective mass of the majority carriers (electrons), k is the Boltzmann's constant, and h is Planck's constant. Equation (15) correctly predicts the often observed J–V behavior of rectifying Schottky diodes on lightly doped III–V semiconductors: an exponentially increasing current under forward bias and an almost constant current under reverse bias.

The parameter A^* is called Richardson's constant and is equal to 120 A/cm^2-K^2 for $m^* = m_0$, the free electron mass. In the case of a real semiconductor, the correct value of m^* and thus A^* depends on the band structure of the semiconductor in question. For a single energy valley with a spherical energy surface, the isotropic effective mass m^* is used. For multiple valleys and ellipsoidal constant-energy surfaces, Crowell[31] has provided a method to calculate m^*. Furthermore, an effective Richardson constant A^{**} has been defined by Crowell and Sze[32] to account for two additional effects which can occur near the top of the metal–semiconductor barrier: (1) emission of optical phonons by electrons as a result of collisions and (2) quantum-mechanical reflection by the barrier and tunneling of electrons through the barrier. Both effects are dependent on the electric field at the barrier and thus the applied voltage V and the doping N_D. Crowell and Sze[32] calculated values of A^{**} for Si and GaAs. As pointed out by Rhoderick,[1] A^{**} is within a factor of 2 of A^* for the range of V and N_D for which Eq. (15) is valid; hence the difference between A^* and A^{**} is probably not important in the only III–V semiconductor for which A^{**} has been calculated, namely GaAs.

In Table 1 a listing of theoretical and experimental values of A^* for GaAs and InP is given. The value of A^* has been calculated using $A^* = 120(m^*/m_0) \text{ A/cm}^2\text{K}^2$. The experimental values of Richardson's constant were obtained at low doping and at high temperatures in order to ensure current flow by thermionic emission, and by measuring the temperature dependence of J at a fixed bias V and fitting the resulting data to Eq. (15). The values in Table 2 are the experimental values after correction for the dependence of ϕ_B on T, and the large spread in A^* for GaAs is believed to result from an interfacial layer[33,34] or deep levels at the metal–semiconductor interface.[35] Unfortunately, experimental values of A^* have only been reported for GaAs and InP.

Equation (15) was derived[2] assuming that the barrier energy ϕ_B was independent of bias. In the previous section we showed that as a result of image-force lowering and an interfacial layer, ϕ_B depends on the electric field in the semiconductor $\mathscr{E}_m$ which in turn depends on the applied potential V. If we assume that the dependence of ϕ_B on V can be represented by a Taylor series and we retain only the terms linear in V, then

$$\phi_B = \phi_{B1} + \beta V \tag{16}$$

where β is a constant to be determined and ϕ_{B1} is the barrier energy at

Table 1. Richardson Constant for GaAs and InP

Material	m^*/m_0	A^* (A/cm^2 – K^2)		References
		Theory	Experiment	
n-GaAs	0.072	8.64 (A^{**} = 4.4)		32
			8.6 ± 1.0	33
			0.8–2.8	34
			3	35
			0.1–19.1	36
p-GaAs	0.62	74.4		
n-InP	0.82	9.8		
			1.8 ± 0.5	37
			0.30	38
			3–12	39
p-InP	0.5	60	~30	40

Table 2. Schottky-Barrier Energies on n-Type GaAs at 300 K

Metal	Orientation	Surface preparation	ϕ_B (eV) I–V	ϕ_B (eV) C–V	ϕ_B (eV) Photo-response	References
Au	(110)	AC[a]		0.95	0.90	11
	(111)	CE[b]	0.99			62
	(100)	CE	0.95			62
	(110)	CE		0.89		65
	(111)	CE		1.03		61
	($\overline{111}$)	CE		0.97		61
	(110)	CE		0.91		61
	(111)	CE	0.90			66
	($\overline{111}$)	CE	0.93			66
	(110)	CE	0.98			66
	?	CE	0.85–0.92[c]			35
	($\overline{111}$)	CE	0.89 ± 0.05	0.94		67
	?	CE	0.91	0.92 ± 0.03		34
	(100)	CE	0.82			52
	(110)	AC		0.95		1
	(110)	CE		0.94		1
	(100)	CE	0.90	0.97		68
	?	CE	0.95	0.95		50
	(100)	CE	0.80	0.81		36
	(100)	CE	0.89	0.89	0.89	33
	(100)	CE	0.94	0.97		69
	(100)	AC		0.94		70
	?	CE	0.86			71
Ag	(110)	AC		0.93	0.88	11
	(111)	CE	0.94	0.94		62
	(100)	CE	0.85	0.85		62
	(110)	AC		0.88		1
	(100)	AC	0.60[a], 0.67[c]			72
	(100)	CE	0.82	1.05		73
	(110)	CE	0.91			66
Pt	(110)	AC		0.94	0.86	11
	?	CE	0.9	0.93		74
	(100)	CE	0.84	0.93		75
	(100)	CE	0.92 (n = 1.12)	1.02		76
Al	(110)	AC		0.80	0.80	11
	(111)	CE	0.80			62
	(100)	CE	0.71			62
	?	CE	0.73–0.77[c]			35
	(100)	AC	0.66[d], 0.72[e]			64
	?	CE	0.80 ± 0.03	0.84 ± 0.03		34
	(100)	CE	0.73	0.83	0.90	77

(continued)

Table 2. (*Continued*)

Metal	Orientation	Surface preparation	ϕ_B (eV) I–V	ϕ_B (eV) C–V	ϕ_B (eV) Photo-response	References
Cu	(110)	AC		0.87	0.82	11
	(111)	CE	0.80			62
	(100)	CE	0.73			62
	(110)	AC		0.86		1
	(110)	CE		0.86		1
W	(110)	AC	0.71	0.77	0.80	11
	(100)	CE	0.73			62
	?	CE	0.64	0.7		74
	(100)	CE	0.65	0.69		75
	(111)	CE	0.81	0.81		78
Au/W	(100)	CE	0.66			79
Pt/W	(100)	CE	0.68			70
Cr	?	CE	0.73 0.77			80
Au/Cr	?	CE	0.78			80
Pt/Cr	?	CE	0.7			81
Mo	?	CE	0.90	0.90		82
Au/Mo	(100)	CE	0.54 ($n = 1.15$)	0.86^f		69
Sn	(111)	CE	0.70			62
	(100)	CE	0.68			62
	(110)	AC		0.79		1
	(110)	CE		0.67		1
	(110)	AC		0.77	0.79	70
	(111)	CE	0.65			66
	($\overline{111}$)	CE	0.68			66
	(110)	CE	0.77			66
Ni	(111)	CE	0.83			62
	(100)	CE	0.70			1
	($\overline{111}$)	CE	0.81			1
	(100)	CE	0.83			49
	(111)	CE	0.67			66
	($\overline{111}$)	CE	0.90			66
	(110)	CE	0.79			66
Be	(110)	AC		0.82	0.81	11
In	(110)	AC		0.76		1
	(110)	CE		0.64		1
	(110)	AC		0.91	0.91	70
Pd	(100)	CE	0.87	0.93	0.82	47
Au/Mo/Pd	(100)	CE		0.94		85
Hf	?	CE			0.72	84

(*continued*)

Table 2. (Continued)

Metal	Orientation	Surface preparation	ϕ_B (eV)			References
			I–V	C–V	Photo-response	
Pb	(110)	CE		0.76		1
	(110)	AC		0.83	0.84	70
Mg	(110)	CE	0.68			66
Ca	(110)	CE	0.57			66

[a] AC = atomically clean.
[b] CE = chemically etched.
[c] Determined from Richardson plot.
[d] Arsenic-stabilized surface in MBE growth.
[e] Gallium-stabilized surface in MBE growth.
[f] Estimated from voltage intercept of $1/C^2$–V data.

zero bias given by Eq. (11) to be

$$\phi_{B1} = \phi_B^0 - \alpha \mathscr{E}_m (V=0) - \Delta \phi_B (V=0) \tag{17}$$

Following Rhoderick,[1] substitution of Eq. (16) into (15) leads to

$$J = A^* T^2 \exp(-q\phi_{B1}/kT)[\exp(qV/nkT)] \tag{18}$$

for $V \geq 3kT/q$. This equation is in the form of the classical rectifier equation for forward bias

$$J = J_{0T} \exp(qV/nkT) \tag{19}$$

where

$$J_{0T} = A^* T^2 \exp(-q\phi_{B1}/kT) \tag{20}$$

$$n = (1 - \beta)^{-1} \tag{21}$$

The parameter n is called the "ideality factor" (i.e., if $n = 1.0$ the J–V characteristic is that of an ideal diode). Since for both image-force lowering and the interfacial layer, $\beta = \partial \phi_B / \partial V > 0$, n is greater than unity. At moderate doping where thermionic emission is dominant, the calculated values of n vary from 1.010 to 1.025 in n-type GaAs. Furthermore, most of the contribution to β is due to image-force lowering. As we shall see in Section 3, the measured values of n are often found to exceed 1.03 in GaAs as well as other III–V Schottky diodes, and thus there must be other physical mechanisms contributing to nonideal J–V behavior besides the voltage dependence of the barrier energy.

2.3.2. Field Emission and Thermionic-Field Emission: Ohmic Behavior

As the concentration of the dopant is increased in the semiconductor, another mechanism for current flow across a metal–semiconductor interface becomes important. As the width of the depletion layer decreases with increasing doping, quantum-mechanical tunneling of electrons through the barrier can occur. At very high doping, the barrier can be thin enough to permit appreciable field emission. At moderately high doping, the barrier is somewhat wider and only electrons with sufficient thermal energy to tunnel near the top of the barrier produce an appreciable current. This latter process is known as thermionic-field emission. Field emission is independent of temperature, while thermionic-field emission is temperature dependent. Both types of quantum-mechanical tunneling can lead to ohmic behavior in contacts to the III–V semiconductors, we shall see below. In this section we review the theoretical description of thermionic-field emission first given by Padovani and Stratton[41] and later expanded by Crowell and Rideout,[42] and show how the specific contact resistance r_c depends on doping, temperature, and ϕ_B.

The transmission probability P that an electron of energy E can successfully tunnel through a triangular-shaped potential energy barrier with diffusion potential V_d is given by[1]

$$P = \exp\left(-\frac{2(qV_d - E)^{3/2}}{3E_{00}(qV_d)^{1/2}}\right) \qquad (22)$$

where

$$E_{00} = \frac{hq}{4\pi}\left(\frac{N_D}{m^*\varepsilon_s}\right)^{1/2} \qquad (23)$$

The parameter E_{00} has dimensions of energy and is very useful in determining the range of doping and temperature for which field emission, thermionic-field emission, or thermionic emission is valid. Note that for a given applied potential (i.e., qV_d) and electron energy E, Eq. (22) predicts an exponential dependence of P, and thus carrier flow, on $N_D^{-1/2}$.

Assuming uniform doping of concentration N_D, complete depletion of the space-charge region, and ignoring image-force lowering, Padovani and Stratton[41] solved for the flux of electrons tunneling from the semiconductor to the metal under forward bias V using a probability function similar to Eq. (22). Their analytical expressions for the J–V characteristics can be summarized as follows:

1. At very high doping or at low temperatures, where $E_{00} \gg kT$, electrons are field emitted directly from states at the Fermi level E_F^s in

the semiconductor to vacant states in the metal, and

$$J \approx J_{0F} \exp(qV/E_{00}) \tag{24}$$

where

$$J_{0F} = \frac{\pi A^* T}{kC_1 \sin(\pi kTC_1)} \exp(-q\phi_B/E_{00})$$

$$C_1 = (2E_{00})^{-1} \ln[-4(\phi_B - V)/\xi] \tag{25}$$

Since $\xi \equiv E_c - E_F^s$, ξ will be negative for a degenerately doped semiconductor. Note that the J–V characteristics of (24) exhibit only a weak temperature dependence through J_{0F}, and the exponential factor E_{00}^{-1} is independent of temperature.

2. At moderate temperatures and doping levels where $E_{00} \sim kT$, electrons tunnel at energy E_m relative to the lower edge of the conduction band in the bulk, and

$$J \approx J_{0TF} \exp(qV/E_0) \tag{26}$$

where

$$J_{0TF} = \frac{A^* T\sqrt{[\pi E_{00} q(\phi_B - V - \xi)]}}{k \cosh(E_{00}/kT)} \exp\left(-\frac{q\xi}{kT} - \frac{q}{E_0}(\phi_B - \xi)\right) \tag{27}$$

$$E_0 = E_{00} \coth(E_{00}/kT) \tag{28}$$

For thermionic-field emission of Eq. (26), Padovani and Stratton[41] found the energy at which tunneling occurred to be

$$E_m = q(\phi_B - V - \xi) \cosh^{-2}(E_{00}/kT) \tag{29}$$

3. At high temperatures or low doping where $E_{00} \ll kT$, the carriers are thermionically emitted over the barrier and the result of Section 2.3.1, namely

$$J = J_{0T} \exp(qV/nkT) \tag{19}$$

applies.

It is useful to examine the effects of field emission and thermionic-field emission on the J–V characteristics in terms of the ideality factor n. By comparing Eq. (19) with (28), we obtain

$$n = \frac{E_0}{kT} = \frac{E_{00}}{kT} \coth(E_{00}/kT) \tag{30}$$

which predicts a rapid increase in n as the doping N_D is increased. Figure 6 shows a plot of Eq. (30) and the regions of thermionic, thermionic-field, and field emission. For n-type GaAs at room temperature, the crossover

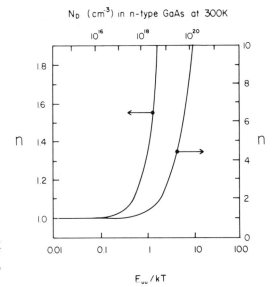

Figure 6. Ideality factor n for a Schottky diode as a function of the characteristic energy E_{00} using Eq. (30).

from thermionic emission to thermionic-field emission occurs at the relatively low doping of about 10^{17}–10^{18} cm^{-3}; this is primarily because of the low value of the electron effective mass in GaAs.

Tunneling at the metal–semiconductor interface can be used to produce an ohmic contact. An ohmic contact is defined as a metal–semiconductor contact which exhibits a linear J–V characteristic over a useful range of V, is nonrectifying, and contributes a voltage drop which is insignificnt compared to the voltage appearing across the device to be contacted. As we have seen, all metal–semiconductor contacts have a J–V characteristic which is nonlinear and of the form

$$J \approx J_0[\exp(V/V_0)-1] \qquad (31)$$

where V_0 and J_0 depend on the mechanism of carrier flow. At sufficiently small voltages (i.e., $V \ll V_0$)

$$J \approx J_0(V/V_0) \qquad (32)$$

and thus the contact can exhibit a linear J–V behavior, if we ignore the dependence of J_0 on V. For a practical ohmic contact, J_0 must be large as possible so that the resistance of the contact is small, ensuring that $V \ll V_0$ under all operating conditions. Using Eqs. (20), (25), and (27), it can be shown that the largest value of J_0 occurs for the case of field emission when the semiconductor doping is high. Thus contacts which employ highly doped regions directly under the metal electrode produce the lowest-resistance ohmic contacts.

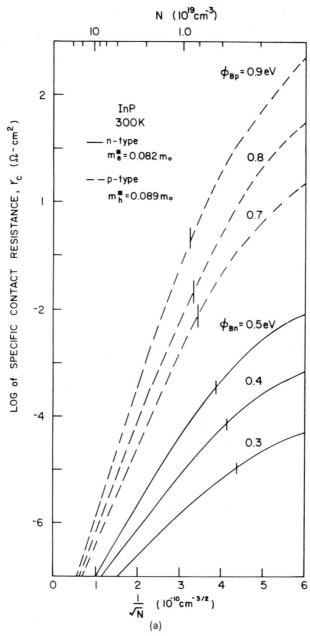

Figure 7. Specific contact resistance as a function of net doping $N = |N_D - N_A|$ for (a) InP and (b) GaAs at 300 K based on Eq. (34) and (35). The vertical dash marks indicate the transition from field emission to thermionic-field emission. For the p-type material in both (a) and (b), the effective mass for light holes was used.

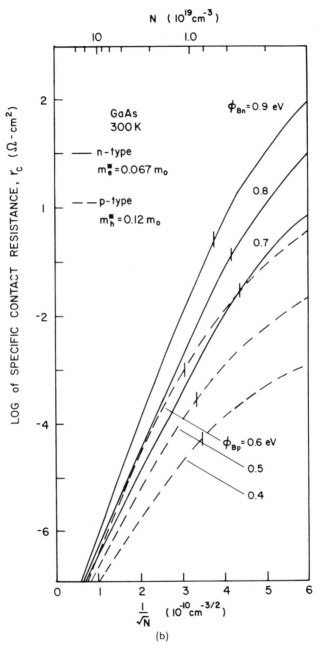

Figure 7. (*Continued*)

The single parameter most often used to characterize an ohmic contact is the specific contact resistance r_c, defined by

$$r_c \equiv \left(\frac{\partial V}{\partial J}\right)_{V=0} \tag{33}$$

Applying this definition to the above three cases, we obtain:
1. For field emission with $E_{00} \gg kT$,

$$r_c = \left(\frac{A^* T \pi q}{k \sin(\pi C_1 kT)} \exp(-q\phi_B/E_{00}) - \frac{A^* q}{C_1 k^2} \exp[-(q\phi_B/E_{00}) + C_1 q\xi]\right)^{-1} \tag{34}$$

2. For thermionic-field emission with $E_{00} \sim kT$,

$$r_c = \frac{k^2 \cosh(E_{00}/kT)}{qA^*[\pi(q\phi_B + q\xi)E_{00}]^{1/2}} [\coth(E_{00}/kT)]^{1/2}$$
$$\cdot \exp\{[q(\phi_B - \xi)/E_0] + q\xi/kT\} \tag{35}$$

3. For thermionic emission with $E_{00} \ll kT$,

$$r_c = \frac{k}{qA^*T} \exp(q\phi_B/kT) \tag{36}$$

These expressions have been evaluated for InP and GaAs at 300 K and the results are plotted in Fig. 7 for a range of dopant concentrations $N = |N_D - N_A|$. At very high concentrations, r_c is due to field emission and is dependent on ϕ_B, N, and m^* through the factor $\exp[\phi_B(m^*/N)^{1/2}]$; thus a plot of log r_c vs. $N^{-1/2}$ is a straight line. At lower concentrations r_c is controlled by thermionic-field emission and the dependence on ϕ_B, N, and m^* is more complicated and given by Eq. (35).

For the semiconductors InP and GaAs, it is interesting to compare the roles of m^* and ϕ_B in determining the value of r_c for low-resistance ohmic contacts. For InP, $m_h^* > m_e^*$ and $\phi_{Bp} > \phi_{Bn}$ and thus r_c for p-type material is much larger than for n-type material for all values of N_A and N_D. This can be clearly seen in Fig. 7(a). For GaAs, again $m_h^* > m_e^*$; however, $\phi_{Bp} < \phi_{Bn}$ and r_c is lowest on p-type material when $N_A < 5 \times 10^{18}$ cm^{-3}, where the effect of the lower value of ϕ_B dominates, r_c is lowest for n-type material when $N_D > 2 \times 10^{19}$ cm^{-3}, where the effect of the lower value of m^* dominates [see Fig. 6(b)]. Thus, it is necessary to consider *both* the metal–semiconductor barrier energy and the effective mass of the majority carriers for a given semiconductor in order to accurately calculate r_c and correctly predict the difficulty in forming ohmic contacts to p-type material relative to n-type material. It should be noted that we have used the effective mass for light holes in Fig. 7 since the tunnel current due to the heavy holes will be small compared to that for the light holes.

Schottky Diodes and Ohmic Contacts

From Eqs. (34)–(36), one finds r_c insensitive to temperature for high doping and exponentially dependent on temperature for low doping. Since many semiconductor-device applications require an $r_c \leqslant 10^{-6}\,\Omega\text{-cm}^2$ (see Section 4), it is evident from Fig. 6 that, theoretically, N_D must be greater than about $6 \times 10^{19}\,\text{cm}^{-3}$ in n-type GaAs and about $2 \times 10^{19}\,\text{cm}^{-3}$ in n-type InP. Current flow in such metal–semiconductor contacts is by field emission.

2.4. Capacitance of a Schottky Diode

The capacitance of the space-charge region of a Schottky diode can provide a variety of useful information about the nature of the metal–semiconductor interface. The barrier energy ϕ_B can be obtained from the voltage dependence of the capacitance. If an interfacial layer exists, the thickness of the layer can be extracted from C–V data. Furthermore, the concentration of shallow dopants and deep traps in the semiconductor can be determined from the Schottky-barrier capacitance. A detailed review of the theory of the capacitance of a Schottky barrier can be found in Rhoderick[1]; a brief outline of the concepts relevant to interpretation of Schottky diode C–V data is given here.

Consider the Schottky model of a metal–semiconductor diode (i.e., no interfacial layer) for a semiconductor with spatially uniform doping N. With reverse bias $V = -V_r$ applied, the charge per unit area, Q_s, in the space-charge region is related to the incremental capacitance C per unit area of the diode by

$$C = \frac{dQ_s}{dV_r} \tag{37}$$

From Gauss's law, the electric field $\mathscr{E}$ inside the space-charge region is $Q_s = \varepsilon_s \mathscr{E}$, and thus

$$C = \varepsilon_s \frac{d\mathscr{E}}{dV_r} \tag{38}$$

Assuming that Q_s is due only to the uncompensated ionized donors and that the minority carriers can be ignored, and then using Eq. (9) for the electric field at the interface in Eq. (38) the classical expression for the capacitance of the Schottky diode can be obtained:

$$C = \left(\frac{q\varepsilon_s N}{2}\left(\phi_{B0} - \xi + V_r - \frac{kT}{q}\right)\right)^{1/2} \tag{39}$$

Without an interfacial layer present, $\phi_{B0} = \phi_B^0$. The use of this equation to measure ϕ_{B0} and N will be discussed in Section 3.

For a metal–semiconductor diode characterized by the Bardeen model, the existence of the interfacial layer and associated surface states make ϕ_B dependent on the field at the interface and the charge distribution is altered at the interface. The capacitance–voltage relationship $(C-V_r)$ is different from that of Eq. (39). Cowley[43] has examined theoretically the effect of an interfacial layer on the capacitance of a Schottky diode and found that the form of Eq. (39) is preserved and only the apparent value of ϕ_B need be modified if the layer thickness δ is small ($\delta < 30$ Å in GaAs diodes). Very thick interfacial layers (i.e., $\delta > 200$ Å) lead to low currents at forward-biased voltages comparable to ϕ_{B0} and to $C-V$ behavior typical of classical metal–insulator–semiconductor (MIS) diodes rather than conventional Schottky diodes. Intermediate layer thickness (30–200 Å) leads to so-called "MIS Schottky diodes" which exhibit rectification similar to conventional Schottky diodes but $C-V_r$ characteristics greatly different from that predicted by Eq. (39).[36] Recently, GaAs MIS Schottky diodes used as solar cells have exhibited high conversion efficiencies and large open-circuit voltages,[44] and thus there is considerable interest in developing a practical MIS Schottky-diode technology.

The existence of traps (i.e., energy levels located more than a few kT from the band edges) in the space-charge region can significantly alter the $C-V_r$ relationship from that of the simple form of Eq. (39). The occupation of the traps vary with bias voltage V_r, which must be properly accounted in deriving the capacitance of the diode. In 1963 Goodman[45] treated the case of a single-level trap of density N_T and the results of his analysis can be summarized as follows:

1. Under certain measurement conditions, the occupation of the traps can vary with time and thus the frequency dependence of C or the transient behavior of C can be used to extract the density, energy level, and capture cross section of the traps.
2. For a large concentration of traps ($N_T \geq |N_D - N_A|$), the $C-V_r$ relationship is more complicated than Eq. (3).

Thus, in order to study only the metal–semiconductor interface experimentally and to avoid the complication introduced by bulk traps in the semiconductor, the semiconductor material is selected to have $|N_D - N_A| \gg N_T$. This condition can be easily achieved in many III–V semiconductors.

3. Schottky-Diode Technology

In this section values of ϕ_B for Schottky diodes on the binary III–V compound semiconductors and their alloys are tabulated and correlated to the various models of metal–semiconductor interface formation.

Schottky Diodes and Ohmic Contacts

3.1. Measurement of ϕ_B

Before summarizing the barrier energies measured on the III–V semiconductors, it is useful to review the methods of experimentally obtaining ϕ_B in practical metal–semiconductor structures. The values of ϕ_B obtained by different techniques often do not agree and thus an understanding of the inherent assumptions in each technique, as well as the practical limitations of each measurement, can be useful in interpreting ϕ_B data.

3.1.1. Photoresponse Measurements

One of the earliest and most direct techniques for measurement of ϕ_B consists of photoexcitation of electrons from the Fermi level in the metal to the conduction band of the semiconductor by illuminating the Schottky diode with monochromatic light, as shown in Fig. 8.[1,11] As the photon energy $h\nu$ is varied, the diode photocurrent will increase sharply when $h\nu > q\phi_B$ (process a in Fig. 8) and when $h\nu > E_g$ the photocurrent will increase even more rapidly as a result of band-to-band excitation (process b in Fig. 8). The photocurrent per absorbed photon I_p for photon energies more than about $3kT$ larger than $q\phi_B$ but less than E_g is given by Fowler's theory[46] for classical photoemission as

$$I_p \propto (h\nu - q\phi_B)^2$$

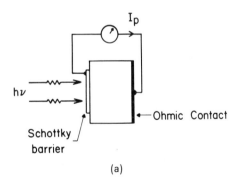

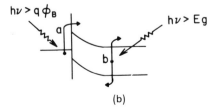

Figure 8. Measurement of ϕ_B using internal photoemission. (a) Experimental arrangement for determination of the short-circuit photocurrent I_p. (b) Two processes for carrier excitation.

or

$$\sqrt{I_p} \propto h\nu - q\phi_B \qquad (40)$$

Thus by plotting the square root of I_p normalized to the photon flux, a straight line results with an extrapolated intercept which is $q\phi_B$. An example of the photoresponse data for Pd Schottky diodes[47] on GaAs and InP is shown in Fig. 9.

In practice, the incident light is chopped to avoid edge effects and other contributions to the diode leakage current, and if the diode is illuminated from the front side, the metal is made thin enough to allow adequate transmission of the light to the metal–semiconductor interface. The photoresponse technique has been used not only to determine ϕ_B directly, but also to measure the voltage dependence of image-force lowering $\Delta\phi$, the temperature dependence of ϕ_B, and the direct- and indirect-band-gap energies in several ternary alloy semiconductors.[2]

Recently, Anderson et al.[48] have examined theoretically the effects of thermal excitation and quantum-mechanical tunneling on the determination of Schottky-barrier energies by extrapolation of the square-root photoresponse. They calculated the difference between the true barrier energy ϕ_B and the apparent barrier energy ϕ_A and found that for n-type GaAs, $q(\phi_B - \phi_A)$ can exceed 50 meV in many cases. Hence precise determinaton of ϕ_B by the photoresponse technique requires that these effects be taken into account; however, the only III–V semiconductor for which the necessary calculations have been performed is n-type GaAs.[48]

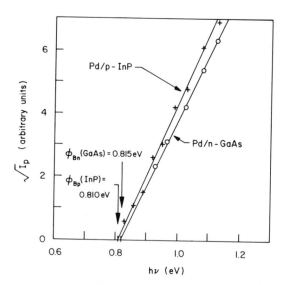

Figure 9. Photoresponse of Pd Schottky barriers on p-type InP and n-type GaAs.

3.1.2. Current–Voltage Measurements

The current–voltage characteristic of a rectifying Schottky diode can be used to determine ϕ_B by several different experimental methods. The simplest method is to measure the diode current density J under forward bias V at a fixed temperature and fit the resulting data to Eq. (19). In a plot of log J vs. V, the intercept on the log J axis yields the saturation current J_0 and the slope provides the ideality factor n. If n is close to unity (from Fig. 5, less than 1.10 for n-type GaAs), then the current flows by thermionic emission, $J_0 = J_{0T}$, and the barrier energy at zero bias ϕ_{B1} can be found from

$$\phi_{B1} = \frac{kT}{q} \ln(J_{0T}/A^*T^2)^{-1} \qquad (41)$$

if A^* is known. In practice the log J vs. V curve should be linear over at least three orders of magnitude for reliable results. At large forward currents, the voltage drop across any series resistance due to the semiconductor bulk or diode connections can produce significant deviations from Eq. (19).

Typical forward-biased J–V characteristics for InP Schottky diodes measured at 300 K are shown in Fig. 10. For the p-type diode the data are linear over five decades of current, $n < 1.1$, and ϕ_{Bp} can be easily

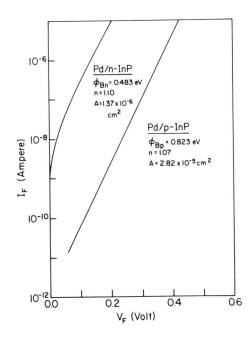

Figure 10. Forward-biased current–voltage characteristics of Pd Schottky diodes on p-type and n-type InP at 300 K. The current I_F is related to the current density J by $I_F = JA$. The values of ϕ_B were calculated using Eqs. (20) and (41) and the theoretical values of A^* are from Table 1.

obtained to within ±0.5%. While for the n-type diode, the data cannot be fitted as accurately with a single straight line and thus ϕ_{Bn} could be determined only to within about ±10%. For the III–V semiconductors near room temperature, barrier energies of less than about 0.5 eV are difficult to measure accurately using only forward-biased J–V data.

An alternative method to determine ϕ_B experimentally is to measure J_0 as a function of temperature T at a fixed forward bias V. If thermionic emission is dominant (i.e., $E_{00} \ll kT$), $J_0 = J_{0T}$ and a plot of $\log(J_{0T}/T^2)$ vs $1/T$ (i.e., the so-called "Richardson plot") should yield ϕ_{B1} and A^* by fitting the data to

$$J_{0T} = A^* T^2 \exp(-q\phi_{B1}/nkT) \tag{42}$$

Note that Eq. (42) is similar to the traditional form of the saturation current given by Eq. (20) except that n is included in the exponential factor. This is because the experimental data for Au/GaAs[36] and Ni/GaAs[49] diodes cannot be explained by the simpler and more widely used form of Eq. (20). The Richardson plot reported by Ashok et al.[36] for Au/n-type GaAs diodes, prepared by the chemical etching of the GaAs prior to Au metalization, is shown in Fig. 11, where the plot of $1/nT$ yields a straight line and the plot of $1/T$ shows significant deviation from linearity at low temperatures. Since n was found to be 1.08 at 300 K

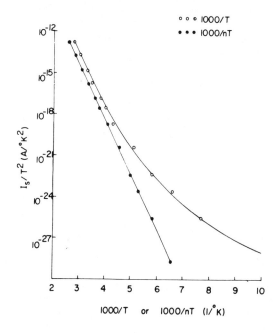

Figure 11. The saturation current $I_s = J_{0T} A$ as a function of inverse temperature in Au/n-GaAs Schottky diodes. (Reprinted with permission from Ref. 36, copyright 1979, Pergamon Press, Ltd.)

and the doping in the GaAs was 10^{17} cm^{-3}, conduction by thermionic-field emission must be ruled out and thus cannot account for the dependence of n on T and J_0 on n. Hackam and Harrop[49] argued that an interfacial layer strongly affects n and that the effects are still present at $V = 0$; thus n should be included in the expression for the saturation current. Whatever the physical basis for modification of Eq. (20), there exists substantial experimental evidence indicating that the classical expression for the thermionic emission saturation current does not always hold for n-type GaAs Schottky diodes.

If the diode J–V characteristics are controlled by thermionic-field emission, then n will be strongly temperature dependent. In this case, the critical parameter E_{00} is found by plotting E_0 vs. kT/q, measuring E_0 as a function of T using $E_0 = kT[n(T)]$, and fitting the results to Eq. (28). Typical results for Au/GaAs diodes[50] which have been heat treated to alter their $n(T)$ behavior are shown in Fig. 12(a). Note that E_{00} is found from the E_0-axis intercept. With E_{00} known, the effective doping under the metal contact is

$$N_D = m^* \varepsilon_s (4\pi E_{00}/hq)^2 \qquad (43)$$

For thermionic-field emission, $J_0 = J_{0TF}$ and the barrier energy can be found by plotting $\log[J_{0TF} \cosh(E_{00}/kT)]$ vs. $1/E_0$ and comparing the resulting data to Eq. (27). For example, it was found that heat treatment of Au/n-GaAs Schottky diodes decreased the apparent barrier energy ϕ_{Bn}, measured from forward-biased J–V data assuming thermionic emission,[51,52] as shown in Fig. 12(b). However, measurement of the temperature dependence of the saturation current for heat-treated Au/GaAs diodes revealed that the actual barrier energy remained unchanged with heat treatment while the net donor density increased significantly, resulting in the observed increase in saturation current.[50] The change in the electrical characteristics of the Au/GaAs diodes was due to dissociation of the GaAs during heat treatment, with the loss of either Ga or As resulting in an increase in uncompensated donors at the metal–semiconductor interface.[52]

Accurate measurement of low barrier energies can be difficult, and Tantraporn[53] has shown that barrier energies as low as 0.26 eV can be measured by observing the temperature dependence of the voltage applied to a metal–semiconductor diode for two polarities for a constant current I. As the diode is cooled to a sufficiently low temperature, rectification appears at temperature T. The barrier energy is then obtained from a Richardson plot of $\log(I^2/T)$ vs. $1/T$. Tantraporn demonstrated that this technique can be used to measure ϕ_B for each of two back-to-back Schottky diodes on the same piece of GaAs.

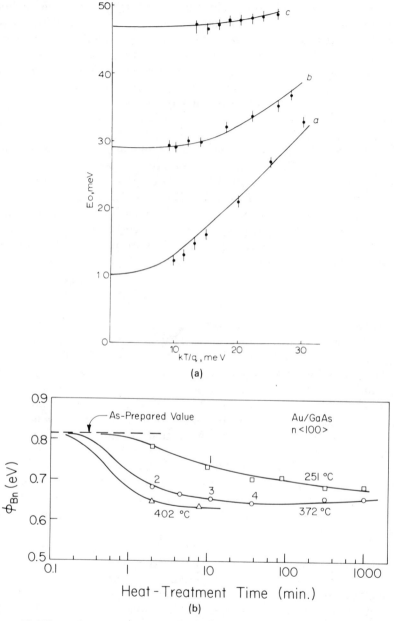

Figure 12. Effects of heat treatment on Au/n-GaAs Schottky diodes. (a) E_0-T data fitted using thermionic-field emission theory. Prior to measurement, diode a was heat treated at 260°C for 30 min, diode b at 300°C for 60 min, and diode c at 450°C for 30 min. (Reprinted with permission from Ref. 50, copyright 1975, IEE.) (b) Measured values of ϕ_{Bn} obtained from forward-biased current–voltage data assuming thermionic emission. The actual value of ϕ_{Bn} does not change with heat treatment (see text).

3.1.3. Capacitance–Voltage Measurements

The capacitance of an ideal Schottky diode was derived in Section 2.4 and the result, Eq. (39), can be written in the following form:

$$C^{-2} = (2/q\varepsilon_s N)(V_r + V_I) \tag{44}$$

where $V_I = \phi_{B0} - \xi - kT/q$. A plot of C^{-2} vs. the reverse bias voltage V_r is then a straight line of slope inversely proportional to the net doping $N = |N_D - N_A|$ and of intercept V_I from which ϕ_{B0} can be found. Note that ϕ_{B0} does not include the image-force lowering $\Delta\phi_B$ since no carriers are transported across the interface during the capacitance measurement.[1]

An interfacial layer at the metal–semiconductor interface can alter the C–V relationship. Cowley[43] has shown that an interface layer of thickness δ (see Fig. 4) on an n-type semiconductor with surface state density N_{ss} produces an apparent intercept voltage V'_I in a C^{-2} vs. V_r plot that is larger than the actual V_I. If the occupation of the interface states is determined by the semiconductor Fermi level (i.e., thick interfacial layer), Cowley found[43]

$$V'_I = V_I + \frac{qN_D\alpha^2}{2\varepsilon_s} \tag{45}$$

where α is defined in Eq. (8). For reasonable values of N_{ss} and δ, the last term in Eq. (45) is less than kT/q and thus can be ignored. If the layer is much thinner, such that the charge in the surface states is determined by the Fermi level in the metal, Cowley found[43]

$$V'_I = V_I + \frac{\delta}{\varepsilon}(2qN_D\varepsilon_s V_I)^{1/2} + q\varepsilon_s N_D \delta^2/2\varepsilon_i^2 \tag{46}$$

The last term in Eq. (46) can be neglected except at very high doping concentrations. Therefore, if an insulating film exists at the interface, we expect to obtain $V'_I > V_I$ and the apparent barrier energy will be larger than the actual barrier energy ϕ_{B0}. Furthermore, V'_I should increase linearly with δ.

The data of Pruniaux and Adams[54] for Au/GaAs diodes tend to support Cowley's theory. Figure 13 shows the ϕ_{B0} measured by the C–V technique for Au/GaAs diodes as a function of the thickness of an oxide layer formed on the GaAs surface in air prior to Au metalization. The oxide thickness was determined with ellipsometry. Samples numbered 2 and 5 were prepared by a slightly different method than the other samples. The linear dependence of the measured barrier energy on δ in Fig. 3.6 agrees with the form of Eq. (46), but the lack of reproducibility in the experiments of Pruniaux and Adams prevents quantitative comparison

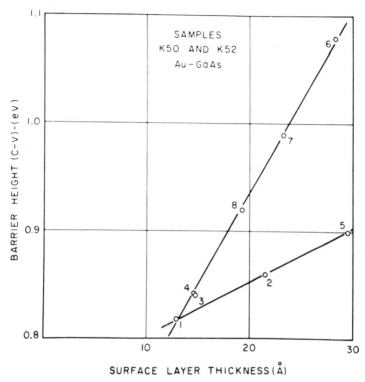

Figure 13. Schottky-barrier energy ϕ_{B0} for Au/n-GaAs diodes, measured by the C–V method, as a function of interface oxide layer thickness. (Reprinted with permission from Ref. 54, copyright 1972, American Institute of Physics.)

with Cowley's theory. Smith and Rhoderick[55] have also examined experimentally the effect on the C–V characteristics of purposely introducing an interfacial layer in Au/GaAs diodes. They not only found that the observed V'_I was greater than V_I, but that the slope of the C^{-2}–V curve is increased, leading to an underestimate of N as well as an overestimate of ϕ_{B0}.

Even after accounting for the influence of an interfacial layer on C–V measurements, it is not uncommon to find that the barrier energy obtained from C–V measurements, $\phi_B(C$–$V)$, is larger than the barrier energy found from I–V measurements, $\phi_B(I$–$V)$, in practical Schottky diodes for both the elemental semiconductors and the III–V semiconductors.[1] One explanation for $\phi_B(C$–$V) > \phi_B(I$–$V)$ may be that during barrier formation a thin compensated layer of thickness W_I forms immediately adjacent to the metal and the potential energy barrier is reduced at the interface in the manner proposed in Fig. 14. The parabolic extension of the energy bands from the edge of the depletion region

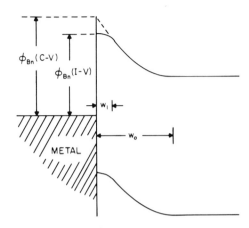

Figure 14. Compensation of the surface of an n-type semiconductor can lead to $\phi(C-V) > \phi_B(I-V)$.

back to the surface shown in Fig. 14 would project a value of $\phi_B(C-V) > \phi(I-V)$, the actual barrier energy as determined by the net doping density at the metal–semiconductor interface. This model has been used to explain the behavior of n-type silicon Schottky diodes partially compensated by aluminum[56] as a result of heat treatment and the behavior of GaAs diodes with sintered Pd/Ge contacts.[57]

Another mechanism that may account for the observed differences between $C-V$ and $I-V$ data is the possibility of a lateral nonuniformity of ϕ_B across the metal–semiconductor interface. As discussed in Section 2.2, Woodall and Freeouf[28] suggest that for the III–V semiconductors the interface is made up of clusters of different phase materials and depending on the relative amounts and the barrier energy of each phase, the value of the barrier energy would depend on the technique of measurement. $I-V$ measurements would emphasize the lower value of ϕ_B, while $C-V$ measurements would give a value of ϕ_B averaged over the interface.

Recently, nonideal $C^{-2}-V$ plots have been observed for GaAs Schottky diodes which can be explained in terms of an excess capacitance C_0 which is independent of bias.[58,59] The measured $C-V_r$ data can be transformed to a linear plot of $1/(C - C_0)^2$ vs. V_r, which gives the correct value of N and ϕ_B. The physical interpretation of C_0 is not yet established but C_0 may be due to surface states and an interfacial layer, in agreement with an extension of Cowley's theory.[58] However, this simple explanation may not be adequate since other researchers have found C_0 to also be dependent on the doping and deep levels in the semiconductor.[59]

Finally, the effects of series resistance, deep traps, and a rough surface on determination of ϕ_B by the $C-V$ technique have been discussed by Goodman.[45] The effect of the depletion region along the edges of the diode has been examined by Copeland.[60]

3.2. Barrier Energies

In this section, a summary is given of the values of ϕ_B measured for Schottky diodes on the III–V semiconductors and reported in the scientific literature up through mid-1981. Mead[11] first compiled ϕ_B data for a variety of semiconductors in 1966, and the inclusion of Mead's summary in a book by Sze[2] has made values of ϕ_B readily available to a large number of users. In the tables presented in this chapter, the values of ϕ_B taken from Mead[11] are given as the first entry for comparison to more recent data where appropriate. Furthermore, data are listed for both atomically clean (AC), and chemically etched (CE) semiconductor surfaces. The category AC refers to semiconductors cleaved under vacuum conditions in a beam of metal atoms, semiconductor surfaces freshly grown by molecular-beam epitaxy and metalized *in situ*, and semiconductor surfaces which were sputter etched and thermally annealed in vacuum prior to *in situ* metalization. The category CE refers to Schottky diodes fabricated using conventional device processing steps with the semiconductor surface prepared by polishing, chemical etching, and exposure to air prior to metalization; such diodes usually exhibit an oxide layer and other contamination at the metal–semiconductor interface. The values of ϕ_B are classified by measurement technique and in many cases the error in ϕ_B was ±0.03 eV or less, and for the I–V data, n was less than 1.10 unless otherwise noted. However, since the care exercised in measurement of ϕ_B varies greatly, the reader should consult the references cited for more experimental details. Finally, since the measurement error was usually greater than the barrier lowering $\Delta\phi$, no distinction is made in the tables between ϕ_B and ϕ_{B0} [see Eq. (12)].

3.2.1. GaAs

The most widely studied III–V semiconductor is GaAs, primarily because of its commercial importance. Schottky-barrier energies have been measured for a wide variety of metals on n-type GaAs, but there have been fewer reports for Schottky diodes on p-type material. Since the electron mobility is much higher than the hole mobility, n-type materials finds wider device application than p-type GaAs, and ϕ_{Bn} is larger than ϕ_{Bp}, making it easier to measure ϕ_{Bn} than ϕ_{Bp}. The values of ϕ_B reported for n-type GaAs are given in Table 2 and for p-type GaAs in Table 3.

For both AC and CE GaAs surfaces, an examination of Table 2 shows that ϕ_{Bn} depends on crystal orientation with the $(\overline{1}\overline{1}\overline{1})$ and (110) faces exhibiting a larger ϕ_{Bn} than the (100) face. This orientation dependence was first reported by Kahng[61] and substantiated by the

Table 3. Schottky-Barrier Energies on p-Type GaAs at 300 K

Metal	Orientation	Surface preparation	ϕ_B (eV) I–V	C–V	Photo-response	References
Al	(110)	AC		0.63	0.50	11
		AC (77 K)		0.61		11
	(100)	CE	0.65			77
Au	(110)	AC		0.48	0.42	11
		AC (77 K)		0.46		11
Ag	(110)	AC (77 K)		0.44		11
	(110)	CE	0.50			66
Pt	(110)	AC (77 K)		0.48		11
Ag	(110)	AC (77 K)		0.44		11
Cu	(110)	AC (77 K)		0.52		11
Pd	(100)	CE	0.49			47
Hf	?	CE			0.68	84
Mg	(110)	CE	0.72			66
Ni	(110)	CE	0.58			66

experiments of Genzabella and Howell[62] who fabricated a series of GaAs mixer diodes with seven different metals on (111) and (100) material, and by the experiments of Smith[63] who carefully measured ϕ_{Bn} for the $(\bar{1}\bar{1}\bar{1})$, (111), (110), and (100) orientations in chemically etched Au/GaAs diodes. If the Schottky limit of Eq. (3) applied, the dependence of ϕ_B on crystal orientation could be due to the dependence of the semiconductor electron affinity χ_s on crystal structure. However, the lack of dependence of ϕ_B on the metal work function ϕ_m indicates the Schottky limit does not apply for GaAs. If ϕ_B was given by the Bardeen limit, $\phi_B = E_g - \phi_0$, and if the surface state distribution and thus the neutral level ϕ_0 was orientation dependent, the experimental results would be explainable. The number of surface states may be related to the number of atomic bonds on a given crystal surface and thus alteration of the surface structure could alter ϕ_B. Experiments using molecular-beam epitaxy (MBE) showed that ϕ_{Bn} for Al/GaAs diodes formed *in situ* depend on the atomic structure of the GaAs surface prior to Al metalization.[64] Since the (100) surface is used most frequently for device fabrication, ϕ_{Bn} for GaAs devices tends to the lowest possible value for a given metal.

The method used to prepare the GaAs surface prior to metallization can significantly affect ϕ_B. Rhoderick[1] has discussed methods of preparing chemically etched GaAs surfaces in order to obtain reproducible Schottky diodes with ideal-like J–V and $1/C^2$–V behavior. The large spread in ϕ_{Bn} for Au/n-GaAs diodes of 0.80–1.03 eV given in Table 2 is most probably the result of the wide range of GaAs preparation methods used. Well-controlled experiments where any differences in ϕ_B for AC and CE surfaces could be detected are few in number. For n-type GaAs on the surface that corresponds to the direction of easy cleavage, (110), Smith[63] observed approximately the same value of ϕ_{Bn} for Au and Cu on CE and AC surfaces, but noticeably higher ϕ_{Bn} values on AC compared to CE surfaces for Sn and In.

As is evident from Table 3, little data are available from independent measurements of ϕ_B on p-type GaAs. In those cases where ϕ_{Bn} and ϕ_{Bp} were measured by the same researcher using a single technique, the sum of $\phi_{Bn} + \phi_{Bp}$ was reasonably close to the band gap E_g of GaAs. This is illustrated in Fig. 15 where ϕ_{Bn} and ϕ_{Bp} are plotted as functions of the work function ϕ_m for the metals Hf, Al, Ni, Ag, Pd, and Au. Assuming a simple linear fit to the ϕ_{Bn} data yields a straight line that fits the ϕ_{Bp} data reasonably well if $\phi_{Bp} = E_g - \phi_{Bn}$. Thus Eq. (14) is seen to apply to GaAs.

If ϕ_B data are examined for a larger number of metals than shown in Fig. 15, the dependence of ϕ_B on ϕ_m is not as readily apparent. Figure

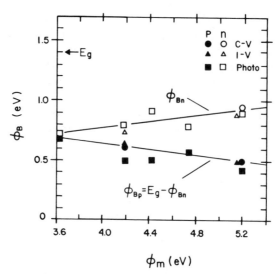

Figure 15. Schottky-barrier energy ϕ_B for chemically etched n-type and p-type GaAs as a function of the metal work function ϕ_m. The values of ϕ_B are from Refs. 11, 47, 66, 77 and 84, and the values of ϕ_m are taken from Table 2.1 of Ref. 1.

16 shows ϕ_{Bn} vs. ϕ_m for chemically etched n-type GaAs using much of the data from Table 2. If we assume that

$$\phi_{Bn} = S'\phi_m + C' \tag{47}$$

where S' and C' are constants determined solely by the semiconductor material, then there appears to be no simple linear dependence of ϕ_{Bn} on ϕ_m, and thus the linear interface theory of Mead and co-workers, discussed in Section 2.2, does not appear to apply for chemically etched GaAs surfaces. For comparison two limiting cases are shown in Fig. 16: $S' = 0$ for the Bardeen limit and $S' = 1$ for the Schottky limit, where S is defined by $S' = \partial\phi_B/\partial\phi_m$. Schlüter[12] has pointed out that the linear interface model does not apply to AC semiconductors as well. Thus there seems to be little correlation between ϕ_B and ϕ_m for GaAs, and for most metals, the Fermi level E_F^s appears to be pinned between 0.7 and 1.0 eV below the conduction-band minimum at the surface. This does correspond roughly to the position of E_F^s in GaAs for the defect model of Spicer et al.[16] The defect model was based on experiments with Schottky

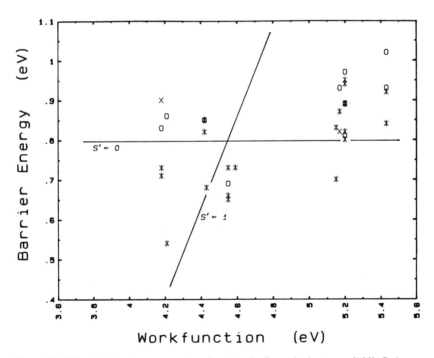

Figure 16. Schottky-barrier energy ϕ_{Bn} for chemically etched n-type ⟨100⟩ GaAs as a function of the metal work function ϕ_m. The values of ϕ_B were taken from Table 2: (*) for $I-V$, (0) for $C-V$, and (x) for photo data. The values of ϕ_m are taken from Table 2.1 of Ref. 1.

diodes on AC surfaces, and thus does not appear to account for the differences in ϕ_B from metal to metal for CE surfaces. Finally, in comparing the data of Fig. 16 and Tables 2 and 3 to the predicted behavior based on the models of metal–semiconductor interface formation discussed in Section 2.2, the effects of chemical reaction between the metal and the GaAs and the role of interdiffusion at the interface are not easily discerned.

3.2.2. InP

InP has been considered as an alternative to GaAs for a variety of device applications and thus the Schottky-barrier energy of metal contacts to InP have been examined by a number of investigators. Mead and Spitzer[13] were the first to measure ϕ_B on InP and their results are

Table 4. Schottky-Barrier Energies on *n*-type InP at 300 K

Metal	Orientation	Surface preparation	ϕ_B (eV)			References
			I–V	C–V	Photo-response	
Au	(110)	AC		0.49	0.52	11
	(110)	CE	0.40	0.40		37
	(110)	AC	0.43	0.50		85
	(110)	CE	0.49	0.57		85
	(111)	CE		0.45		90
	(110)	CE		0.47		86
	(100)	CE		0.50		86
	(110)	CE	0.47	0.50		39
	(110)	CE	0.62^a			39
	(100)	CE	0.43	0.53		91
	(100)	CE^b	0.5	0.56	0.53	88
	(100)	CE^b	0.4^a			88
	(100)	CE	0.49			87
	(100)	CE	0.46	0.45		92
	(100)	CE	0.55 (n = 1.34)			93
Ag	(110)	AC		0.54	0.57	11
	(110)	AC		0.42	0.47	85
	(110)	CE		0.48	0.55	85
	(110)	CE	0.49	0.59		39
	(110)	CE	0.69^a			39
	(100)	AC	Ohmic^c			72
	(100)	AC	0.43^d			72
	(100)	CE	0.41			27

(continued)

Table 4. (*Continued*)

Metal	Orientation	Surface preparation	ϕ_B (eV) I–V	ϕ_B (eV) C–V	Photo-response	References
Al	(110)	CE	0.47	0.52		85
	(110)	AC	Ohmic			19
	(100)	CE	0.50	0.53		91
	(100)	CE^b	0.3			89
	(100)	CE	0.37			27
Ni	(110)	CE	0.50	0.50		85
	(110)	AC	Ohmic			19
	(110)	CE (183 K)	0.36	0.41		39
	(100)	CE	0.36			27
Cu	(110)	AC		0.49		19
	(110)	CE	0.46	0.44		39
	(110)	CE	0.50^a			39
Cr	(100)	CE	Ohmic			91
	(110)	CE (80 K)	0.21 ($n = 2.1$)			39
In	(110)	AC, CE	Ohmic			19
	(110)	CE (183 K)	0.32 ($n = 1.29$)			39
Au/Ti	(100)	CE	0.42			91
	(100)	CE	0.51	0.53		38
Fe	(110)	CE	0.55	0.55		19
	(110)	AC	Ohmic			19
Co	(100)	CE	0.46			27
Pd	(100)	CE	0.45			27
Pt	(110)	CE	0.52	0.54		39
Sn	(110)	CE (172 K)	0.27 ($n = 1.13$)			39

^{a} Determined from Richardson plot.
^{b} Heated in vacuum prior to metalization.
^{c} Phosphorus-stabilized surface.
^{d} Indium-stabilized surface.

summarized in Tables 4 and 5 along with more recent data. Smith[37] examined Au/n-InP diodes using a bromine methanol solution to etch the InP surface and found a low barrier energy of 0.40 eV. Careful studies by Williams and co-workers[85] of both vacuum cleaved and chemically etched n-type InP have verified that the Au/n-InP barrier energy is between 0.43 and 0.50 eV. Furthermore, it is now apparent from these early measurements, as well as from the more recent data summarized

Table 5. Schottky-Barrier Energies on p-Type InP at 300 K

Metal	Orientation	Surface preparation	ϕ_B (eV) I–V	ϕ_B (eV) C–V	ϕ_B (eV) Photoresponse	Referenced
Au	(110)	AC		0.76		11
	($\overline{1}\overline{1}\overline{1}$)	CE			0.77	94
	(100)	CE			0.77	94
	(110)	CE	0.82			39
	(100)	CE	0.79	0.93	0.81	27
Ag	(111)	CE			0.74	95
	(100)	CE			0.78	95
	(100)	CE	0.79	0.86	0.81	27
Al	(100)	CE	0.89 ($n = 1.30$)			96
	(100)	CE	0.89	1.12	0.92	27
Ni	(100)	CE	0.90	1.14	0.90	39
	(110)	CE	0.96			39
Cu	(110)	CE	0.85			39
Pd	(100)	CE	0.82	0.90	0.81	27
Co	(100)	CE	0.81	0.87	0.78	27

in Tables 4 and 5, that the barrier energy is larger on p-type InP than on n-type InP for both CE and AC surfaces, which is just the opposite from that of GaAs.

In early work on InP n-channel MESFETs it was found that the Schottky-gate electrode was leaky under reverse bias and large gate currents flowed for small forward-biased voltages (see Fig. 10) as a result of the low ϕ_{Bn} for InP.[86] In order to raise the effective barrier energy on n-type material, Wada and Majerfeld[87] introduced a thin (<100 Å) oxide film between a Au electrode and the InP by oxidizing the InP in HNO_3 under illumination. The resulting ϕ_{Bn} increased from 0.49 eV for an unoxidized surface to 0.94 eV with the oxide present. Morgan and Frey[88] have extended this technique to the Au/n-InGaAs Schottky barrier and Christou and Anderson[89] have shown that P_2O_5 at the barrier can also substantially increase ϕ_{Bn}.

In order to determine which of the mechanisms of interface formation discussed in Section 2.2 applies to InP, we have used the data of Tables 4 and 5 to plot ϕ_B as a function of the metal work function ϕ_m in Fig. 17. If the linear interface theory of Kurtin et al.[9] applied to InP, ϕ_{Bn} would be given by Eq. (47) with a value of S' between 0 and 1. However, it is evident from Fig. 17 that there is no obvious correlation

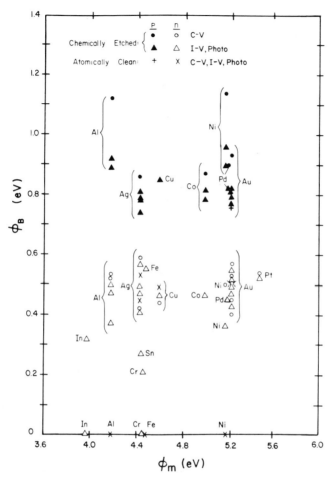

Figure 17. Schottky-barrier energy ϕ_B for both chemically etched and atomically cleaned InP as a function of the metal work function ϕ_m. The values of ϕ_B are from Tables 4 and 5, and the values of ϕ_m are from Table 2.1 of Ref. 1.

between ϕ_{Bn} (or ϕ_{Bp}) and ϕ_m, and a single value of S' cannot accurately represent all of the ϕ_B data on InP. Thus, the metal work function is not the dominant controlling parameter in metal/InP interface formation.

The work of Williams et al.[19] on the (110) face of n-type InP has demonstrated that chemical effects play an important role in determining ϕ_B. Shown in Fig. 18 are the barrier energies measured by Williams et al. for several different metals on atomically clean n-type InP as a function of the heat of reaction per formula weight ΔH_r for the most stable metal–phosphorus compound formed between the contact metal and InP. It is apparent from this plot that there is a systematic dependence

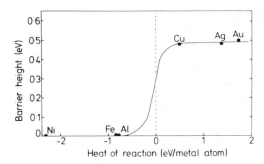

Figure 18. Schottky-barrier energy on atomically clean n-type InP as a function of the heat of reaction for the most stable metal–phosphorus compound which can form between the contact metal and InP. (Reprinted with permission from Ref. 19, copyright 1978, American Institute of Physics.)

of ϕ_B on the reactivity of the metal with the InP. Metals (i.e., Ni, Fe, and Al) which form compounds with phosphorus that are significantly more stable than InP yield contacts with low barrier energies and ohmic behavior. Metals (i.e., Au, Ag, and Cu) which form phosphides that are less stable than InP yield Schottky diodes with $\phi_{Bn} \sim 0.5$ eV. Recent experiments by Brillson et al.[26] have confirmed the observations of Williams et al. and have also shown that reactive metal–InP diodes exhibit an abrupt interface and In-rich outdiffusion, while unreactive metal–InP

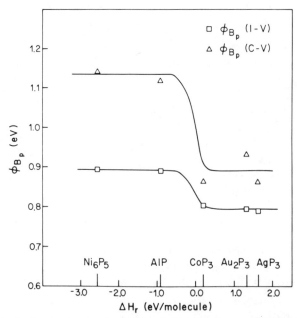

Figure 19. Schottky-barrier energy ϕ_{Bp} for chemically etched p-type $\langle 100 \rangle$ InP as a function of the heat of reaction ΔH_r for the most stable metal–phosphorus compounds. For $\Delta H_r > 0$, the metal–phosphorus compounds are less stable than InP and for $\Delta H_r < 0$, more stable than InP. Data from Ref. 27.

diodes exhibit diffuse interfaces and P-rich outdiffusion. Thus the data for Schottky diodes formed on AC n-type InP indicate that the work function of the metal is unimportant and microscopic chemical effects are dominant during metal/InP interface formation.

The chemical effects demonstrated in Fig. 18 are less pronounced on etched surfaces than on atomically clean surfaces. Hökelek and Robinson[27] have measured ϕ_{Bp} for chemically etched InP Schottky diodes, where the large value of ϕ_{Bp} is relatively easy to measure with high accuracy at room temperature, and found a dependence of ϕ_{Bp} on ΔH_r similar to that of Williams et al. The InP was etched in a solution of 10 wt% HIO_3 in H_2O and exposed to air prior to metallization. As shown in Fig. 19, the reactive metals Ni and Al exhibited higher values of ϕ_{Bp} than the nonreactive metals Au, Ag, and Co. Although the difference between the two limiting cases is not as great as that observed by Williams and co-workers for the vacuum cleaved InP, the chemical effects are none the less apparent. Finally, the sum of the independently measured values of ϕ_{Bp} and ϕ_{Bn} was found to agree with the accepted value of E_g (1.34 eV) for InP.[47]

3.2.3. Other Binary Compounds

The barrier energies measured for the III–V semiconductors other than GaAs and InP are given in Table 6. The wide-band-gap semiconductor GaP has been examined by numerous investigators since the original work of White and Logan[97] and ϕ_B was found to depend on the crystal orientation and cleanliness of the GaP surface. Cowley[43] first reported that ϕ_{Bn} for Au/GaP diodes where the Au was deposited in an oil diffusion pump system was larger than ϕ_{Bn} for Au/GaP diodes prepared in an ion pump system. Apparently the oil vapors contaminated the GaP surface and an interfacial layer was formed between the Au and the GaP. Smith[98] and Smith and Abbott[99] later showed that by heating the GaP to 120°C in an oil diffusion pump vacuum system prior to metalization, ϕ_{Bn} values similar to the ion pump values of Cowley's could be reproducibly obtained. Goldberg et al.[100] were also able to obtain Au/n-GaP diodes with ideal-like diode characteristics by using an electrodeless plating technique. Furthermore, Smith and Abbott[99] showed that ϕ_{Bn} and ϕ_{Bp} in GaP depended on crystal orientation, with the $(\overline{111})$ face exhibiting a larger ϕ_{Bn} than the (100) face for both Au and Al Schottky diodes—the same dependence as was found for GaAs (see Fig. 16).

The ϕ_B data available to date are not adequate to completely explain the mechanism of metal–semiconductor interface formation on GaP. Early work on AC surfaces[11] indicated that ϕ_{Bn} was approximately $2E_g/3$, independent of the metal; however, the data summarized in Table 6 and

Table 6. Schottky-Barrier Energies at 300 K for the III-V Compounds Other Than GaAs and InP

Semiconductor	Metal	Surface preparation	ϕ_B (eV)			References
			I–V	C–V	Photoresponse	
n-AlAs (110)	Au	AC			1.2	11
(100)		AC[a]	0.77			101
(110)	Pt	AC			1.0	11
n-AlSb	Au	CE	1.03			105
p-AlSb (110)	Au	AC		0.53	0.55	11
n-GaP (110)	Au	AC	1.1	1.3	1.3	11
		CE		1.34	1.28	11
		CE[b]		1.17		98
(100)		CE[b]		1.31		99
($\overline{1}\overline{1}\overline{1}$)		CE	1.17	1.18		100
		CE			1.34–1.45	106
		CE	1.30	1.30		107
	Pt	CE		1.52	1.45	11
		CE	1.35–1.75	1.5–1.9	1.5	108
		CE		1.45		107
	Ag	CE			1.20	11
		CE	1.6–1.75	1.61–1.88	1.6–1.85	108
		CE	1.14	1.14		107
	Al	CE		1.14	1.05	11
(100)		CE[b]		0.96		99
($\overline{1}\overline{1}\overline{1}$)		CE[b]		1.06		99
		CE		1.16		107
	Cu	CE		1.34	1.20	11
		CE		1.32	1.30	107

Schottky Diodes and Ohmic Contacts

(111)	Cr	CE	1.25[c]	1.19	109
		CE	1.20	1.18	107
	Ni	CE	1.17	1.18	100
		CE	1.30	1.27	107
	Mg	CE		1.09	11
	Mo	CE		1.13	107
p-GaP (110)	Au	AC	0.68	0.75	11
(100)		CE			11
$(\overline{111})$		CE		1.00	99
		CE		0.87	99
(100)	Al	CE		1.25	99
$(\overline{111})$		CE		1.15	99
n-GaSb (110)	Au	AC		0.61	11
		CE			110
(110)		AC			111
		CE			105
	Pd	CE			110
p-GaSb (111)	Au	CE	Ohmic (77 K)		11
n-InAs (110)	Au	AC, CE	Ohmic		11
		?	Ohmic		103
p-InAs (110)	Au	AC		0.47 (77 K)	11
(110)		AC, CE		0.51 (80 K)	112
		?		≥0.42 (77 K)	103
(110)	Al	AC, CE		0.38 (80 K)	112
(110)	Cu	AC, CE		0.38 (80 K)	112

(continued)

Table 6. (*Continued*)

Semiconductor	Metal	Surface preparation	ϕ_B (eV)			References
			$I-V$	$C-V$	Photoresponse	
n-InSb (110)	Au	AC, CE		0.17 (77 K)		11
(211)		CE	0.076 (77 K)			104
(110)	Ag	AC, CE		0.18 (77 K)		11
(211)		CE	0.076 (77 K)			104
p-BN	Au	CE		3.1		11
p-BP	Au	AC			0.87	11

[a] AlAs surface grown by MBE.
[b] Heated to 120°C *in vacuo* prior to metalization.
[c] Determined from Richardson plot.
[d] From determination of Fermi level at surface using X-ray photoemission.

particularly the work of Smith and Abbott,[99] demonstrate that ϕ_B depends on the metal, and in the case of Al on (100) GaP, ϕ_{Bp} can even exceed ϕ_{Bn}. If the linear interface model of Kurtin et al.[9] is assumed to apply to GaP, then the interface index is found to be $S' = 0.27 \pm 0.05$ from a fit of ϕ_{Bn} vs. ϕ_m as shown in Fig. 20. On the other hand, chemical effects appear to be as important as the effect of the metal work function.

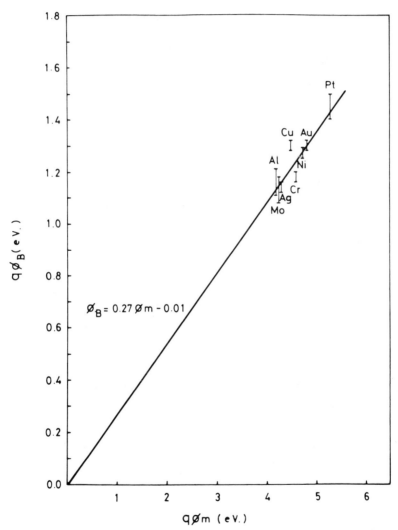

Figure 20. Schottky–barrier energy ϕ_B for chemically etched n-type GaP as a function of the metal work function ϕ_m. (Reprinted with permission from Ref. 107, copyright 1979, Pergamon Press, Ltd.)

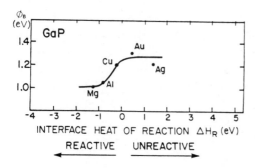

Figure 21. Schottky-barrier energy ϕ_B for n-type GaP as a function of the heat of reaction ΔH_r of metal–phosphorus compounds. (Reprinted with permission from Ref. 25, copyright 1978, American Institute of Physics.)

Taking only the ϕ_{Bn} data summarized by Mead,[11] which were measured by photoemission, Brillson[25] has recently shown that ϕ_{Bn} abruptly increases above a critical value of the heat of reaction ΔH_r of metal with GaP. Figure 21 shows the correlation of ϕ_{Bn} with ΔH_r for GaP; note the similarity to Fig. 18 for InP. Thus, at this time, the metal work function as well as microscopic chemical effects appear to influence the barrier energy in GaP diodes.

For the remaining III–V compound semiconductors, only a limited amount of experimental facts are known. For the wide-band-gap material AlAs, there is a large disparity in the ϕ_{Bn} values yet measured.[11,101] For the smaller-bandgap materials InAs and InSb, capacitance–voltage measurements of ϕ_B are usually made at low temperatures in order to reduce the shunt conductance produced by the large intrinsic carrier concentration at room temperature. In the case of InAs, the Fermi level at the semiconductor surface E_F^s is found to be above the conduction-band minimum, as shown in Fig. 22. Thus the InAs surface is strongly inverted in p-type material and accumulated in n-type material, leading to a negative resistance in the J–V characteristics of Au/p-InAs diodes[102] and ohmic behavior in Au/n-InAs diodes.[103]

The present data on InSb Schottky diodes indicate that the method of surface preparation greatly influences ϕ_B. The early work of Mead and Spitzer[13] demonstrated that E_F^s was at or above the conduction-band minimum, similar to InAs, in vacuum cleaved samples. However, more recent work[104] indicated that if the InSb is chemically etched by the proper procedure, the barrier energy will be $\phi_{Bn} \simeq 0.076$ eV, implying

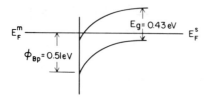

Figure 22. Band diagram of a p-type InAs Schottky diode at equilibrium.

that E_F^s is about midgap for InSb. More measurements, particularly on p-type InSb, are needed.

3.2.4. III–V Alloys

In order to combine the desirable optical or electronic properties of several different III–V compounds and thus achieve optimum device performance for a particular application, it is possible to form alloys of two or more binary compound semiconductors. For example, the alloy of AlAs and GaAs is $Al_xGa_{1-x}As$, where x is the mole fraction of AlAs present. Such alloys are now widely used in light-emitting diodes ($GaAs_{1-x}P_x$), lasers ($Al_xGa_{1-x}As$, $Ga_yIn_{1-y}P_x$), and heterojunction transistors ($Al_xGa_{1-x}As$). The recent availability of high-quality epitaxial layers of many III–V alloys has permitted experimental investigations of the role of the cation (group III element) and anion (group V element) in the formation of the metal–semiconductor interface and thus in the determination of ϕ_B. The ternary and quaternary alloys which have been employed as Schottky diodes are given in Table 7, along with the appropriate references. For the cations Al, Ga, and In and the anions P, As, and Sb, there are 9 binary compounds and 18 ternary alloys possible, yet investigations of the compositional dependence of ϕ_B in Schottky diodes have been reported to date in only 7 ternary alloys.

The Schottky barrier ϕ_B has been measured on n-type $Al_xGa_{1-x}As$ for the full range of x in both chemically etched and atomically clean

Table 7. Summary of III–V Alloy Semiconductors for Which Schottky-Barrier Energies Have Been Measured

Alloy	E_g (eV)	ϕ_{Bn} (eV)	References
(A) Ternaries:			
1. Common Anion:			
$Al_xGa_{1-x}As^a$	1.44–2.16	0.7–1.1	101, 113, 115
$Al_xGa_{1-x}Sb^a$	0.72–1.44	0.6–1.0	105
$Al_xIn_{1-x}As$	0.36–2.16	$\sim 0.8^b$	118
$Ga_xIn_{1-x}P$	1.35–2.26	0.5–1.0	120
$Ga_xIn_{1-x}As$	0.36–1.44	0–0.9	88, 119, 122
$Ga_xIn_{1-x}Sb$	0.17–0.72	0.1–0.6	121
2. Common cation:			
$GaAs_{1-x}P_x$	1.44–2.26	0.9–1.4	114
(B) Quaternaries:			
$Ga_yIn_{1-y}As_{1-x}P_x$	0.36–2.26	0.4–0.8	94, 123

[a] Does not obey the common anion rule.
[b] For $x = 0.48$.

material. In 1972 Goldberg et al.[113] reported ϕ_B values for $Al_xGa_{1-x}As$ alloys in the composition range $0 < x < 0.4$, corresponding to the direct-gap band structure, using Au on chemically etched (100) surfaces. It was later pointed out by Rideout[114] that the data of Ref. 113 could be fitted by the equation

$$\phi_{Bn} \simeq E_g - 0.55 \text{ eV}$$

This implied that the barrier energy for holes $E_g - \phi_{Bn}$ was independent of composition and equal to 0.55 eV, as shown in Fig. 23. In 1979 Best[115] examined chemically etched (100) n-type $Al_xGa_{1-x}As$ diodes using a Au metalization and found that $E_g - \phi_{Bn}$ increased linearly with x, also shown in Fig. 23. Even more recently, Okamoto et al.[101] have grown Al epitaxial films *in situ* on (100) n-type AlGaAs using MBE under ultrahigh vacuum conditions. AlAs and its alloys are very reactive semiconductors and an oxide-free interface in a Schottky diode must be formed under ultrahigh vacuum conditions. Okamoto's results, shown in Fig. 23, indi-

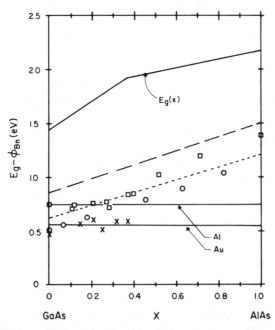

Figure 23. Comparison of theory and experiment for $Al_xGa_{1-x}As$ Schottky diodes as a function of composition x. For reference, the variation of the energy gap with x is shown as the solid curve marked $E_g(x)$. The results of measurement of ϕ_{Bn} are noted as follows: (×) Au/AlGaAs CE,[113] (○) Au/AlGaAs CE,[115] and (□) Al/AlGaAs AC.[101] Three theoretical models are shown: (——) the common anion rule[114] for Al and Au metalization, (— —) the antisite defect model,[116] and (- - -) the anion vacancy model.[117]

cate that $E_g - \phi_{Bn}$ is independent of x for $x < 0.4$ but increases with x for $x > 0.4$, the compositional range corresponding to the indirect-gap band structure. In Fig. 23, $E_g - \phi_{Bn}$ for Al/AlGaAs diodes is larger than for Au/AlGaAs diodes because of the difference in the work functions of Al and Au (see Fig. 15).

If the alloy system $Al_xGa_{1-x}As$ obeyed the common anion rule of McCaldin et al.,[14,15] then $E_g - \phi_{Bn}$ should be independent of x. As seen by the solid lines in Fig. 23, the common anion rule agrees with the data for $x < 0.4$ but fails to describe the compositional dependence of ϕ_B on x for $x > 0.4$.

As discussed in Section 2.2, Freeouf and Woodall[29] have proposed an effective-work-function model where the hole barrier is given by

$$E_g - \phi_{Bn} = (E_g + \chi_s) - \phi_{\text{eff}}$$

where ϕ_{eff} is the effective work function of a mixed phase assumed to be in contact with the semiconductor. If ϕ_{eff} is the work function of As for the AlGaAs system as Freeouf and Woodall postulate,[29] then ϕ_{eff} does not vary with x. Hence the observed variation in $E_g - \phi_{Bn}$ of Fig. 23 must result from a variation in $E_g + \chi_s$ with composition. The variation of $E_g + \chi_s$ tabulated in Ref. 29 for the end points $x = 0$ (GaAs) and $x = 1$ (AlAs) could account for the overall variation of $E_g - \phi_{Bn}$ observed in the data of Fig. 23, but definitive values of χ_s as a function of x are as yet lacking for the $Al_xGa_{1-x}As$ system.

Calculations of the energy levels introduced by point defects on the surface of AlGaAs give reasonable agreement with the position of the Fermi level at the semiconductor surface E_F^s determined experimentally in AlGaAs Schottky diodes. Allen and Dow[116] calculated E_F^s as a function of x for $Al_xGa_{1-x}As$ assuming an antisite defect mechanism to be responsible for the Fermi-level pinning in GaAs–AlAs alloys; their results for a cation on an As vacancy site are shown by the dashed line in Fig. 23. Also, Daw et al.[18,117] have calculated the energy levels of the highest neutral As vacancy on the surface of AlGaAs alloys and their results are shown as the dotted line in Fig. 23. Both defect models are for the (110) orientation, while all of the experimental data is for (100) surfaces; but both models do predict the geneal trend in the experimental data: an increase in $E_g - \phi_{Bn}$ with an increase in x.

The dependence of barrier energy ϕ_{Bn}, for Schottky diodes of Au on n-type $Al_xGa_{1-x}Sb$, on x has been measured by Chin et al.,[105] and their results are shown in Fig. 24. The hole barrier $E_g - \phi_{Bn}$ is not constant, but increases with increasing x, similar to the trend observed in the AlGaAs Schottky diodes. For the remaining ternary alloy containing Al, $Al_xIn_{1-x}As$, the barrier energy has not been reported except for $x = 0.48$ where ϕ_{Bn} was found to be 0.8 eV.[118]

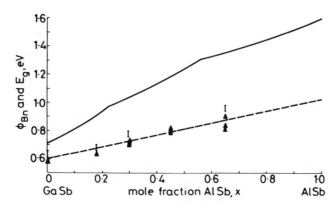

Figure 24. The energy gap E_g (solid line) and the Schottky-barrier energy ϕ_{Bn} (dashed line) for n-type alloys of $Al_xGa_{1-x}Sb$. Note that the hole barrier $E_g - \phi_{Bn}$ is not independent of composition. (Reprinted with permission from Ref. 105, copyright 1980, IEE.)

Of the Schottky diodes of the common anion ternary alloys containing Ga and In, all have been found to follow the common anion rule. The $Ga_xIn_{1-x}As$ system has been studied by Kajiyama et al.[119] and their results are shown in Fig. 25(a). The dependence of ϕ_{Bn} on x for the $Ga_xIn_{1-x}P$ system has been measured by Kuech and McCaldin[120] and their results are shown in Fig. 25(b). Note that in both cases the difference $E_g - \phi_{Bn}$ is approximately constant for the range of composition measured. In the case of $Ga_xIn_{1-x}As$, the variation of E_g, and thus ϕ_{Bn}, with x is monotonic since both GaAs and InAs are direct-band-gap semiconductors. On the other hand, for $Ga_xIn_{1-x}P$ there is an abrupt change in the variation of E_g and ϕ_{Bn} with x at $x = 0.75$, the crossover point from the direct-band-gap structure of InAs to the indirect-band-gap structure of GaP. In both cases the common anion rule holds. Furthermore, calculations using the antisite defect model of Allen and Dow[116] and the anion vacancy model of Daw et al.[117] both agree reasonably well with ϕ_{Bn} dependence on x for the $Ga_xIn_{1-x}As$ system.

Keeler et al.[121] have measured ϕ_B as a function of x for the $Ga_xIn_{1-x}Sb$ system using Au Schottky diodes on p-type material for $x > 0.62$ and for n-type material for $x < 0.42$. For their chemically etched polycrystalline GaInSb diodes, they found $\phi_{Bp} \approx 0.04 E_g$ in agreement with the common anion rule; however, for the n-type material they found $\phi_{Bn} = 0.8 E_g$, in contradiction to the common anion rule.

The compositional dependence of the Schottky-barrier energies for $Au/GaAs_{1-x}P_x$ diodes was first reported by Rideout[114] where he found, within experimental error, $\phi_{Bn} = E_g = 0.55$ eV, implying that $\phi_{Bp} = 0.55$ eV, independent of the As/P anion ratio. From Table 6, $\phi_{Bp} = 0.5$ eV for Au/p-GaAs diodes and $\phi_{Bp} = 0.7 - 0.8$ eV for Au/p-GaP diodes. If

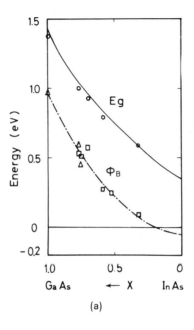

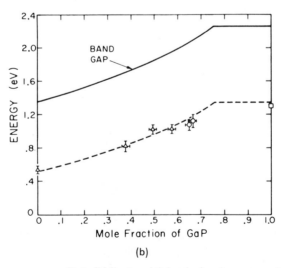

Figure 25. The energy gap E_g (solid line) and Schottky-barrier energy ϕ_B (dashed line) for n-type ternary alloys containing Ga and In. (a) $Ga_xIn_{1-x}As$ system from Ref. 119, copyright 1973. (b) $Ga_xIn_{1-x}P$ system from Ref. 120, copyright 1980. (Reprinted with permission, American Institute of Physics.)

Fermi-level pinning is determined primarily by the anion, ϕ_{Bp} should vary smoothly between these limits as the composition x varies in $GaAs_{1-x}P_x$ alloys. This concept has been tested by Escher et al.[94] using Au Schottky diodes on the quaternary system $Ga_yIn_{1-y}As_{1-x}P_x$. The compositional dependence of the barrier energy on p-type material is shown in Fig. 26. As x is varied, and y adjusted for lattice matching to the InP substrate, the barrier energy ϕ_{Bp} increases monotonically from that of $Ga_yIn_{1-y}As$ to that of $Ga_yIn_{1-y}P$ with the ratio ϕ_{Bp}/E_g remaining approximately constant and independent of x.

Thus, all experimental evidence to date indicates that the ternary and quaternary III–V alloys containing the cations Ga and In obey the common anion rule, implying that the anion alone determined the position of the Fermi level in Schottky diodes made from these materials. However, for the III–V alloys containing the cation Al, the common anion rule fails to correctly predict the observed dependence of Schottky-barrier energy on composition. Models of the metal–semiconductor interface based on point defects at the semiconductor surface agree reasonably well with experiment in the two alloy systems yet examined theoretically. More precise calculations for a wider variety of alloys are necessary to fully test the defect models.

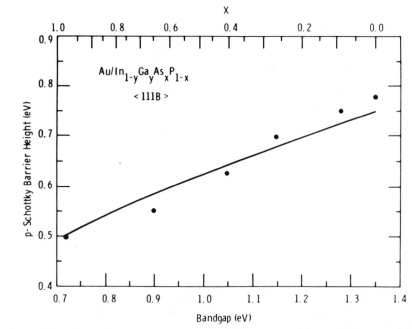

Figure 26. Schottky-barrier energy of Au/p-InGaAsP alloys lattice matched to InP. (Reprinted with permission from Ref. 94, copyright 1976, American Institute of Physics.)

4. Ohmic-Contact Technology

Low-resistance ohmic contacts are necessary in most III–V semiconductor devices for optimum performance and reliable operation. For light-emitting diodes, lasers, and Gunn diodes, values of the specific contact resistance r_c in the range of 10^{-2}–10^{-5} Ω-cm^2 have been found to be adequate, since the ohmic contacts employed in these devices are relatively large in area and the resulting total contact resistance can easily be less than a few ohms. This case is illustrated in Fig. 27(a) where the device current is distributed uniformly over the area A. For example, in GaAs and InP r_c values in this range can be easily achieved using n^+ and p^+ layers, directly in contact with the metal, with dopant concentrations of about 5×10^{18}–2×10^{19} cm^{-3} (see Fig. 7). In most applications, the conventional device processing schemes for forming thin n^+ or p^+ layers (i.e., alloying, diffusion, epitaxy) have been adequate to achieve the

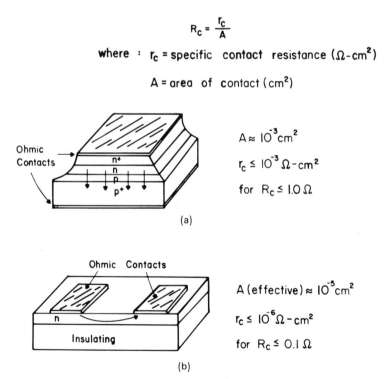

$$R_c = \frac{r_c}{A}$$

where : r_c = specific contact resistance (Ω-cm^2)

A = area of contact (cm^2)

$A \approx 10^{-3}$ cm^2

$r_c \leq 10^{-3}$ Ω-cm^2

for $R_c \leq 1.0$ Ω

(a)

A (effective) $\approx 10^{-5}$ cm^2

$r_c \leq 10^{-6}$ Ω-cm^2

for $R_c \leq 0.1$ Ω

(b)

Figure 27. Specific contact resistance of ohmic contacts for several device structures. (a) Current flow perpendicular to large-area contacts. (b) Current flow parallel to small-area contacts.

necessary doping levels. However, as the materials technology of the III–V semiconductors improves, new devices are being developed which demand much smaller values of r_c. For example, the GaAs MESFETs used in integrated circuits have small effective areas for the ohmic contacts at the source and drain electrodes, while device performance requires very small parasitic resistance at these contacts. As shown in Fig. 27(b), the combination of these conditions necessitates r_c values of 10^{-6} Ω-cm^2 or smaller for ohmic contacts to n-type GaAs. Conventional contacting methods have proven inadequate to meet these requirements and new methods of forming ohmic contacts, such as epitaxial heterojunction structures, are now being explored. In this section the technology of forming ohmic contacts to the III–V semiconductors is reviewed, along with the various methods of measuring r_c. The electrical properties of alloyed metal contacts recently measured on several III–V semiconductors are also summarized.

4.1. Methods of Forming Ohmic Contacts

In low-resistance ohmic contacts current flow across the metal–semiconductor interface is by field emission; this is true for all the III–V semiconductors except n-type InAs, where a surface barrier does not exist (see Section 3.2.3). According to the theoretical expression (34) for field emission, the specific contact resistance r_c should be proportional to the factor $\exp(\phi_B/N^{1/2})$, where N is the net concentration of the dopant at the semiconductor surface. As seen in Section 3 for a given III–V semiconductor, ϕ_B depends only weakly on the choice of metal and on the method of surface preparation in chemically etched material. Thus, in practice, most low-resistance ohmic contacts use the maximum surface concentration of dopant that is technologically feasible. This approach is widely used for GaAs, InP, and GaP, where a thin highly doped n^+ or p^+ layer is introduced directly under the metal electrode, as shown by the band diagrams of Fig. 28(a). The thickness t of the heavily doped layer is chosen to be larger than the depletion width w of the metal–semiconductor barrier. Current flow for the structure is governed by field emission through the metal–semiconductor barrier if the contact potential of the n^+/n (or p^+/p) homojunction is approximately equal to or less than kT/q.

An alternative approach to achieving low-resistance contacts to wide-band-gap III–V semiconductors is to introduce a narrow-band-gap semiconductor at the metal–semiconductor interface in the form of a thin heavily doped layer.[124] The band diagrams for $n^+/n^+/n$ and $p^+/p^+/p$ heterojunction contacts are shown in Fig. 28(b), where the space-charge regions and the effect of electronic states at the heterojunction have been

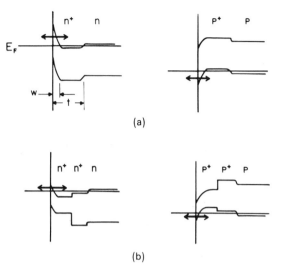

Figure 28. Band diagrams for ohmic contacts at zero bias. Current flow by majority carriers is shown by the arrows. (a) Homojunction $n+/n$ or $p+/p$ contacts. (b) Heterojunction $n+/n+/n$ or $p+/p+/p$ contacts.

ignored. The heterojunction ohmic contact thus combines a low ϕ_B with a large N to achieve very small values of r_c.

4.1.1. Diffusion and Ion Implantation

Diffusion and ion implantation can be used to introduce n^+ and p^+ regions locally in the surface of III–V semiconductor wafers. Diffusion must be carried out at elevated temperatures and thus care must be exercised to prevent the loss of the volatile group V constituent. Diffusion of a dopant is usually carried out in an evacuated, sealed ampoule or in an open-tube furnace with an overpressure of As or P. Selective doping by diffusion through windows in chemically vapor deposited SiO_2 masks is possible, but in the case of GaAs, excessive lateral diffusion along the SiO_2/GaAs interface can result in loss of pattern definition. The maximum dopant concentration is limited by the solid solubility of the impurity; in GaAs donor concentrations as high as 5×10^{19} cm^{-3} and acceptor concentrations of about 1×10^{20} cm^{-3} at 800°C can be achieved.[125] A novel approach to utilizing diffusion in the formation of ohmic contacts to n-GaAs has recently been reported by Nissim et al.[126] where laser radiation was used to assist the thermal diffusion of the donor Sn from a spin-on film of SnO_2/SiO_2. A contact resistance of 1×10^{-6} Ω-cm^2 was achieved using this technique. Laser-assisted diffusion has also been employed to form p^+ ohmic contacts to p-type InP; Cd was released from the metal alkyl $Cd(CH_3)_2$ by photolysis and diffused into the InP by laser heating.[127]

Doping concentrations in excess of the solid solubility limit can be obtained locally on the surface of III–V semiconductors using ion implantation. For example, donor concentrations greater than 4×10^{19} cm^{-3} have been achieved in GaAs using low-energy implantation of Se.[128] Such high surface concentrations can only be produced with very high ion doses (10^{15}–10^{16} cm^{-2}), and thus annealing of the semiconductor after implantation is necessary to restore crystallinity and activate the dopant. Loss of As during thermal annealing of GaAs in a furnace can be prevented by encapsulating the GaAs with a Si-N film deposited by plasma CVD[129] or using a proximity technique where the GaAs wafer is to be annealed is placed face down on a bed of powered GaAs.[130] In InP, Si-implanted n^+ layers have been furnace annealed with a CVD SiO$_2$ cap to produce ohmic contacts[131] with $r_c = 2 \times 10^{-5}$ Ω-cm^2 for ion doses higher than 2×10^{14} cm^{-2}.

An alternative to furnace annealing of ion-implanted layers is laser or electron-beam annealing. In GaAs, a Q-switched Nd YAG laser was used to anneal high-dose Te-implanted material and the electrically active Te concentration was found to be 10 times the solubility limit.[132] However, loss of As from the surface during laser irradiation produced a Ga-rich layer that had to be removed prior to metalization. Ohmic contacts with $r_c = 2 \times 10^{-5}$ Ω-cm^2 were formed after the additional removal of 50 Å of GaAs. A pulsed electron beam has been used to anneal Se-implanted GaAs in order to form ohmic contacts[128,133]; electron concentrations of 4×10^{19} cm^{-3} have been obtained[128] and used to produce contacts with $r_c < 6 \times 10^{-6}$ Ω-cm^2. Reproducible, low-resistance ohmic contacts to p-InP have been formed using Zn$^+$ and Cd$^+$ implanted layers followed by pulsed-laser annealing; for a dose of 3×10^{15} cm^{-2} at 30 keV, contact resistances of $(0.5-2) \times 10^{-4}$ Ω-cm^2 have been reported.[134]

4.1.2. Epitaxy

Thin highly doped layers of the III–V semiconductors can be formed under well-controlled conditions using liquid-phase epitaxy (LPE), vapor-phase epitaxy (VPE), or molecular-beam epitaxy (MBE). For LPE and VPE the maximum obtainable dopant concentrations are determined by the solid solubility of the dopant and thermodynamic considerations during epitaxial growth. For example, in GaAs, the maximum electron concentration is approximately 8×10^{18} cm^{-3} for Sn as a donor using LPE.[135] Higher impurity concentrations can be obtained by MBE since dopant incorporation is governed by surface kinetics during film growth and not by thermodynamic considerations.[136] Typical values of the maximum achievable carrier concentrations in GaAs are $n = 3 \times 10^{19}$ cm^{-3} for Sn and $p = 2 \times 10^{20}$ cm^{-3} for Ge.[137] Another advantage of MBE for ohmic-contact formation is that oxide contamination at the metal–semiconductor interface can be avoided by *in situ* metalization.

Ohmic contacts to GaAs using MBE n^+ and p^+ layers have been reported by several workers. Barnes and Cho[138] reported $r_c = 2 \times 10^{-6}$ Ω-cm^2 using a Sn-doped layer with $|N_D - N_A| = 6 \times 10^{19}$ cm^{-3}. Similar contacts were used to form the source and drain contacts in a microwave power MESFET.[139] Tsang[140] achieved comparable results using Sn-doped n-GaAs and Be-doped p-GaAs with a graded transition region of less than 100-Å thickness between the heavily doped GaAs and the metallic film. The MBE ohmic contacts did not require annealing or alloying to achieve $r_c < 10^{-5}$ Ω-cm; thus the surface of the contacts was smooth and featureless.

4.1.3. Alloying

The simplest and most widely used method of forming ohmic contacts to the III–V semiconductors is the alloy regrowth technique. A thin metallic film containing a suitable dopant is deposited by thermal evaporation, sputtering, electroplating, or other means onto the surface to be contacted. The contact structure is then heated to a temperature above the melting point of the metal film, and a thin layer of the semiconductor is partially dissolved by the molten metal, forming an alloy containing the dopant in high concentration. Upon cooling, the dopant is incorporated in the semiconductor during epitaxial regrowth. The melting point of the metal film is chosen to be well below the melting point of the semiconductor; metallic alloys of eutectic composition are often used to achieve ohmic-contact formation at relatively low alloying temperatures. Since the critical step is the formation of a liquid on the semiconductor surface, laterally uniform alloying requires uniform etching of the semiconductor. This is difficult to achieve in practice because the molten alloy may not wet the surface uniformly and nucleation at surface irregularities and segregation of the elements present can lead to a nonuniform surface morphology. Thus, additional components are often included in the metal film to improve wettability or a nonreactive layer can be deposited over the metallic layer prior to alloying in order to improve surface uniformity.

Alloyed ohmic contacts can be formed thermally in an open-tube furnace in a flowing gas of H_2 or N_2. For the III–V semiconductors this can result in the loss of the more volatile component(s) during alloying. Sebestyen *et al.*[141] have heated alloy contacts to GaAs and GaP in vacuum and observed the evaporation of As_2 and P_2, respectively, using a quadrupole mass spectrometer. To avoid the loss of the volatile component, Hartnagel and co-workers have used a sealed ampoule to obtain an overpressure of As_2 during alloying ohmic contacts to GaAs[142] and P_2 for ohmic contacts to InP.[143]

In order to avoid loss of the column V element and to improve surface uniformity, alloying of ohmic contacts using laser and electron beams have been used. With a pulsed electron beam or a Q-switched laser, the surface can be heated rapidly (<1 μs) to temperatures higher than used in conventional furnace alloying without extensive heating of the bulk of the semiconductor. Thus, previously established doping profiles remain undisturbed. Furthermore, smooth surface are produced because nucleation of surface irregularities and phase segregation do not have time to occur. Pulsed-laser alloying of contacts to GaAs has been reported,[144-147] and a comparison of furnace-alloyed and pulsed-electron-beam alloyed contacts to n-type GaAs can be found in Ref. 148. In all cases, r_c is comparable to the furnace-alloyed results but the surface morphology can be superior.

An interesting alternative to pulsed-beam alloying is spark alloying, where a high-frequency spark is used to directly heat the contact surface. Low-resistance ohmic contacts to n-type GaAs, with surfaces uniformly comparable to that found with pulsed-beam techniques, has been reported by D'Angelo et al.[149]

As an alternative to alloyed contacts, sintered metal films have been used for ohmic contacts to III–IV semiconductors. Sintering is the formation of low-resistance contacts by low-temperature solid-phase reaction and offers the advantage of smoother contact surfaces and more reproducible electrical characteristics than can be obtained with the conventional alloy regrowth technique. Examples of sintered ohmic contacts to n-type GaAs include Pd/Ge films[57,150] and Ni/Ge films.[146]

4.1.4. Heterojunctions

Since r_c depends exponentially on ϕ_B, another method to obtain low-resistance ohmic contacts is to lower the barrier at the metal–semiconductor interface. Reduction of ϕ_B by chemical treatment of the semiconductor surface has been demonstrated in the case of n-type GaAs, where Massies et al.[151] found that the adsorption of sulfur from H_2S sufficiently lowered ϕ_B for an Al/n-GaAs diode to produce ohmic-contact behavior. However, more often the barrier energy is lowered by introducing a thin layer of dissimilar semiconductor, with a characteristically low ϕ_B, between the semiconductor to be contacted and the metal electrode. In order that the resistance of the heterojunction interface created in the structure does not adversely affect the overall contact resistance, the semiconductor materials are chosen with nearly equal lattice constants and an abrupt interface is formed by epitaxy. As illustrated in Fig. 28(b), the semiconductor materials are chosen to avoid the formation of a discontinuity in the band edge such that a potential energy barrier exists at the heterojunction to impede the flow of majority carriers. To further reduce r_c, the narrow-band-gap semiconductor is heavily doped resulting in a $n^+/n^+/n$ or a $p^+/p^+/p$ structure.

The heterojunction contact was first proposed as an explanation for the ohmic behavior of In contacts alloyed to n-GaAs,[124] since InAs has a small band gap and is likely to form by epitaxial regrowth during low-temperature alloying. However, direct experimental evidence that alloyed contacts are in fact heterojunction contacts is at present lacking.

A heterojunction contact using a n^+ Ge layer grown epitaxially on n-type GaAs has been reported by Stall et al.[152,153] The conduction-band diagram for the Ge/GaAs ohmic contact is shown in Fig. 29; note that the discontinuity ΔE_c in the conduction band at the heterojunction is below the Fermi level for the donor concentrations given. The barrier energy $\phi_{Bn} \approx 0.5$ eV for Ge as compared to 0.8 eV for GaAs and, by using MBE to deposit the Ge at low temperatures with As doping, the donor concentration exceeds 10^{20} cm^{-3} in the Ge layer as compared to a maximum donor concentration of about 5×10^{19} cm^{-3} obtainable by MBE in GaAs.[138] Using Au metalization without alloying, Stall et al.[153] measured $r_c = 5 \times 10^{-8}$ Ω-cm^2 for N_D (GaAs) = 1.5×10^{-18} cm^{-3} and $r_c = 1.5 \times 10^{-7}$ Ω-cm^2 for N_D (GaAs) = 1.0×10^{17} cm^{-3}.

Another approach to epitaxial heterojunction ohmic contacts to n-type GaAs has been reported by Woodall et al.[154] By utilizing the

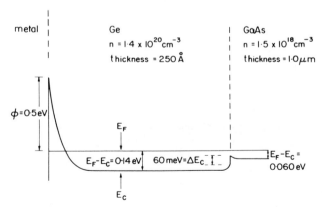

Figure 29. Conduction-band edge of a n^+-Ge/n^+-GaAs heterojunction contact grown by MBE. (Reprinted with permission from Ref. 152, © 1979, IEE.)

Fermi-level pinning that occurs in the conduction band of the narrow-band-gap semiconductor InAs (see Table 6), low-resistance ohmic contacts were formed using MBE InAs/GaAs structures. As shown in Fig. 30, in order to avoid the large conduction-band discontinuity interface, a layer of $Ga_xIn_{1-x}As$ graded in composition was grown between the GaAs substrate and the InAs adjacent to the metal contact. The resulting structure has no abrupt discontinuities in the conduction band and $\phi_{Bn} < 0$ for the metal/n-InAs interface. Using a Ag contact to a graded GaInAs layer of approximately 3×10^{18} cm^{-3} doping grown by MBE, Woodall et al.[154] measured $r_c < 5 \times 10^{-6}$ Ω-cm^2 for contact to a GaAs layer of $N_D = 2 \times 10^{17}$ cm^{-3}. Finally, from the Schottky-barrier energy measurements of Kajiyama et al.[119] to n-type alloys of $GaAs_xIn_{1-x}$, as shown in Fig. 25(a), $\phi_{Bn} \leq 0$ for $0.8 \leq x \leq 1.0$, and thus it is only necessary

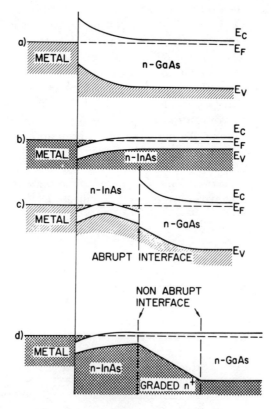

Figure 30. Band diagram for InAs–GaAs heterojunction ohmic contacts: (a) n-GaAs only, (b) n-InAs only, (c) InAs–GaAs abrupt interface, and (d) InAs–$Ga_xIn_{1-x}As$–GaAs graded interface. (Reprinted with permission from Ref. 154, copyright 1981, American Institute of Physics.)

to terminate the graded layer at the metal–$Ga_xIn_{1-x}As$ interface with $x > 0.8$ to achieve a barrierless contact.

4.2. Measurement of r_c

There are several methods that can be used to measure the specific contact resistance r_c. For a homogeneous contact of area A having uniform current density, the contact resistance R_c is simply

$$R_c = r_c/A \tag{48}$$

The measured resistance R will be approximately equal to R_c for most sample geometries when $r_c \geq 10^{-2}$ Ω-cm^2. However, for smaller values of r_c, the spreading resistance of the semiconductor R_b and the series resistance R_0 of the connecting wires and semiconductor substrate must be taken into account. Then in general

$$R = R_c + R_b + R_0 \tag{49}$$

where R_b and R_0 depend on the particular geometry of the metal–semiconductor contact being characterized. For the III–V semiconductors accurate determination of r_c is most often carried out using one of four methods: (1) the technique of Cox and Strack, (2) the four-point method, (3) the Shockley extrapolation technique, or (4) the transmission-line model. We will discuss only these four methods of determining r_c; other methods can be found in the literature and are usually variations of one of the above four techniques.

4.2.1. Cox–Strack Method

The method of Cox and Strack[155] utilizes the structure of Fig. 31, where the resistance of a circular contact of radius a on an n-type film of resistivity ρ and thickness t is to be determined. The current flow pattern is axial, through the layer to the heavily doped n^+ substrate. This structure requires metalization of both the back and the front surfaces of the semiconductor wafer, and can be used for either p- or n-type material and for epitaxial or bulk layers. In later case, t would be the thickness of the wafer. The spreading resistance for the layer is

$$R_b = \frac{\rho}{a}F \tag{50}$$

where F is a function of the ratio a/t and was found experimentally by Cox and Strack to have the approximate form

$$F(a/t) \approx \frac{1}{\pi}\arctan\left(\frac{2t}{a}\right) \tag{51}$$

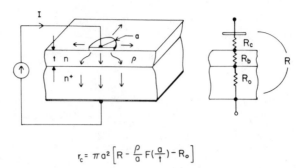

$$r_c = \pi a^2 \left[R - \frac{\rho}{a} F\left(\frac{a}{t}\right) - R_0 \right]$$

Figure 31. Measurement of the specific contact resistance r_c using the Cox–Strack method. The circular ohmic contact has radius a.

In many cases, more accurate values of F are needed than can be found using Eq. (51), and Brooks and Mathes[156] have evaluated numerically an integral expression for $F(a/t)$ in the form of a universal curve. Then with $F(a/t)$ known, we have

$$R = \frac{r_c}{\pi a^2} + \frac{\rho}{a} F + R_0 \tag{52}$$

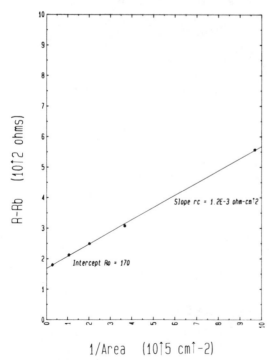

Figure 32. Determination of r_c for a Au–Be alloyed ohmic contact to p-type InP using the Cox–Strack method.[195]

In practice, the resistance of an array of contacts with differing areas are measured, the spreading resistance is calculated for each contact using Eq. (50), and a plot of $R - R_b$ vs. $1/a^2$ is made. As shown in Fig. 32, a straight line fitted to the data points yields the values of r_c and R_0. The minimum value of r_c that can be accurately determined by this technique depends primarily on two factors: (1) the accuracy in measurement of the dimension a, and (2) the error in taking the difference between the two almost equal quantities R and R_b. For n-type GaAs epitaxial layers with $a \geqslant 5\mu$, it is possible to measure r_c values down to approximately 1×10^{-6} Ω-cm^2 with an error of about $\pm 25\%$; at higher values of r_c, the error can be much smaller.

4.2.2. Four-Point Method

The four-point method$^{(157,158)}$ for measurement of r_c requires metalization of only one surface of the wafer as shown in Fig. 33. The layer being contacted is of thickness t and can be an epitaxial layer on a nonconducting substrate or a uniformly doped bulk wafer. Again, the spreading resistance must be first calculated and separated from the total

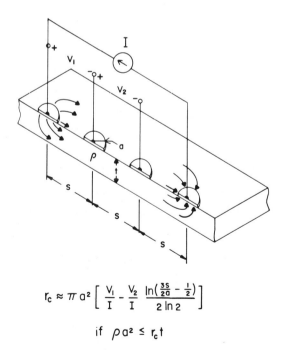

$$r_c \approx \pi a^2 \left[\frac{V_1}{I} - \frac{V_2}{I} \frac{\ln\left(\frac{3s}{2a} - \frac{1}{2}\right)}{2 \ln 2} \right]$$

$$\text{if } \rho a^2 \leqslant r_c t$$

Figure 33. Measurement of the specific contact resistance r_c by the four-point method.

resistance measured. The spreading resistance R_b for radial current flow from a circular contact of radius a is given by Fang et al.[159] in the form of an infinite series. R_b is found to depend on r_c as well as ρ because the current is nonuniformly distributed across the contact area. Fang et al.[159] show that R_b is less than $0.12 R_c$, and thus the contribution of R_b to R can be ignored, if

$$\rho a^2 \leq r_c t \qquad (53)$$

In Section 4.3 we shall see that the ratio ρ/r_c is almost independent of the doping concentration for ohmic contacts to GaAs and InP, and thus one can choose a and t so that Eq. (53) is satisfied for most values of $|N_D - N_A|$ of interest in these materials.

In the four-point measurement of Fig. 33, the voltage V_1 and V_2 are measured for a known current I and assuming that the resistance of the semiconductor film between the contacts is the same everywhere, then[157]

$$V_1 - V_2 = I(R_c + R_b)$$

or

$$r_c = \pi a^2 \left(\frac{V_1}{I} - \frac{V_2}{I} - R_b \right) \qquad (54)$$

Kuphal[158] has recently pointed out that the potential distribution in the plane of the semiconductor layer is logarithmic rather than linear, and the correct expression for the contact resistance is

$$r_c = \pi a^2 \left(\frac{V_1}{I} - \frac{V_2}{I} \frac{\ln(3a/2a - \frac{1}{2})}{2 \ln 2} - R_b \right) \qquad (55)$$

if $a \ll s$ and $t \ll s$. Furthermore, if Eq. (53) is satisfied, then R_b can be neglected in Eq. (55). Kuphal[158] has used the four-point technique to measure values of r_c as small as 1.2×10^{-6} Ω-cm^2 with about $\pm 10\%$ accuracy.

4.2.3. Shockley Technique

For evaluation of ohmic contacts to thin semiconductor layers on nonconducting substrates, a technique which was first proposed by Shockley can be used.[160,161] As shown in Fig. 34 the technique consists of measuring the voltage drop $V(x)$ along the surface of the semiconductor film with coplanar ohmic contacts and using the extrapolated voltage V_0 appearing across the contact, then r_c can be found. Because the sheet resistance R_s (with dimensions of $\Omega/\square$) of the epitaxial film, as well as the specific contact resistance r_c, are nonzero, the current is nonuniformly

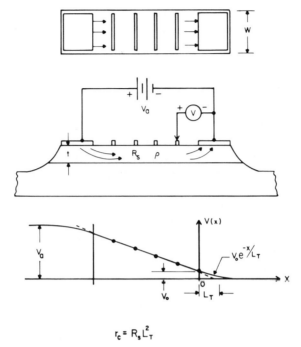

Figure 34. Measurement of the specific contact resistance r_c by the Shockley technique. The voltage applied to the coplanar ohmic contacts is V_a. The linear voltage distribution between the contacts is extrapolated to obtain the transfer length L_T.

distributed and current crowding occurs under each contact. Assuming the sheet resistance is laterally uniform and the semiconductor layer is infinitely thin, Hower et al.[161] have shown that the potential distribution under the contact is

$$V(x) = V_0 \exp(-x/L_T) \tag{56}$$

where $L_T = (r_c/R_s)^{1/2}$ is called the "transfer length." The potential distribution measured under a planar contact to a diffused Si resistor has been found experimentally to be in close agreement with Eq. [56].[162]

The value of r_c is found by extrapolating the linear voltage drop measured between the two contacts to obtain L_T, and r_c is then found from

$$r_c = R_s L_T^2 \tag{57}$$

The minimum value of r_c that can be accurately measured using the Shockley extrapolation method is limited by the error in determination of L_T. Assuming L_T could be measured to within ± 1 μ, it should be possible to measure r_c down to about 5×10^{-7} Ω-cm² with ±25% accuracy for ohmic contacts to n-type GaAs; r_c values as small as 3×10^{-6} Ω-cm² have been reported.[124]

4.2.4. Transmission-Line Model

Another method to determine the resistance of ohmic contacts applied to a thin III–V semiconductor layer on a nonconducting substrate utilized the transmission-line model (TLM). The theory of the TLM was developed independently by Murrmann and Widman[163] and by Berger[164–166] for ohmic contacts to diffused layers in Si integrated circuits. However, the theory can also be used to analyze the source and drain ohmic contacts for field-effect transistors fabricated with GaAs and other III–V semiconductors. As shown in Fig. 35, in the TLM the planar contact is treated as a resistive transmission line with uniform sheet resistance R_s and specific contact resistance r_c. The total resistance R_e of the contact and the epitaxial layer under the contact is found to be[164]

$$R_e = \frac{(r_c R_s)^{1/2}}{W} \coth\left[d \left(\frac{R_s}{r_c}\right)^{1/2} \right] \qquad (58)$$

where $(r_c R_c)^{1/2}/W$ is the characteristic resistance of the transmission line of width W. The current in the semiconductor film under the contact decays with distance as $\exp(-\alpha_l x)$, where the attenuation factor α_l is

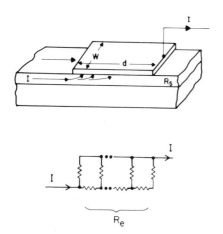

$r_c \approx \dfrac{R_e^2 W^2}{R_s}$ if $d \geq 2\sqrt{\dfrac{r_c}{R_s}}$

Figure 35. Measurement of the specific contact resistance r_c by the transmission-line model (TLM) technique. R_e is the total resistance of the metal–semiconductor interface and the epitaxial layer under the contact.

found to be

$$a_l = \left(\frac{R_s}{r_c}\right)^{1/2} \quad (59)$$

Note that α_l is related to the transfer length L_T of the Shockley method, since Eqs. (57) and (59) yield $\alpha_l = \bar{L}_T^1$.

In order to measure r_c using the TLM, one must first find R_e experimentally. This is most easily accomplished using the arrangement of Fig. 36, where three identical ohmic contacts are spaced at unequal distances l_1 and l_2 along the surface of the layer. If R_1 and R_2 are the resistances measured, one can easily show that

$$R_e = \frac{-R_2 l_1 + R_1 l_2}{2(l_2 - l_1)} \quad (60)$$

One further simplification can be made. Usually $\alpha_l d \geq 2$ and thus Eq. (60) becomes

$$R_e \sim \frac{(r_c R_s)^{1/2}}{W}$$

or

$$r_c = R_e^2 W^2 / R_s \quad (61)$$

R_e is found using Eq. (60) and substituted into Eq. (61) to find r_c. The ultimate accuracy of the TLM method is determined by the errors in measurement of l_1 and l_2 of Fig. 35. Values of r_c of less than 10^{-7} Ω-cm^2 have been measured for contacts to Si[165] and 3×10^{-7} Ω-cm^2 to GaAs (see Section 4.3.1). Finally, the TLM has been extended to transmission lines of circular geometries[167] and to arbitrary shapes.[168]

The use of the TLM avoids the necessity of measuring the $V(x)$ distribution of the Shockley method and hence is somewhat simpler to implement. But both methods assume that the sheet resistance between and under the contacts is identical and that the semi-conducting layer is infinitely thin. In the alloyed ohmic contacts used in FETs, this is not generally true.[169] During contact formation, interdiffusion of the metal and semiconductor layers results in significant alteration of the sheet

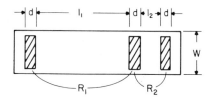

Figure 36. A simple method of determining R_e using a linear array of unequally spaced ohmic contacts. The shaded regions are the contact areas.

resistance under the contacts. Stall et al.[153] have developed a numerical method to take into account finite thickness of the semiconducting layer, and Reeves and Harrison[170] show how to separately determine the sheet resistance between the contacts and under the contacts. In many cases it is not possible to make the separation and thus it is common to use R_e rather than r_c as a measure of the quality of the ohmic contacts in III–V FET structures.

4.3. Alloyed Ohmic Contacts

Since alloyed contacts are the most widely used method of forming ohmic contacts to the III–V semiconductors, a review of the electrical properties of these contacts is given in this section. The status of the technology of alloyed ohmic contacts prior to 1968 can be found in several articles in a book by Schwartz.[171] Rideout[172] has reviewed methods of forming ohmic contacts to the III–V semiconductors and the associated state-of-the-art technology through 1974. In this section, a review of the current literature describing alloyed contacts is given, with emphasis on those contact systems for which r_c has been measured. The values of r_c are summarized in Table 8 for the binary compound semiconductors and in Table 9 for the alloy semiconductors.

4.3.1. GaAs

Alloyed ohmic contacts to GaAs have been reported by many workers. The status of ohmic contacts to GaAs prior to 1964 has been reviewed by Libov et al.[173] with the conclusion that pure In forms a low-resistance contact to n-type GaAs and an alloy of In–Zn is best for contacting p-type GaAs. However, In surface migrates and diffuses rapidly along crystal dislocations, making it impractical for device applications. More often, alloys based on Au or Ag are used to form alloyed contacts in GaAs devices, and usually contain Zn as the dopant for contacting p-type material and Ge or Sn for contacting n-type GaAs.

As shown in Table 8, several multilayered, alloyed thin-film systems have been developed as ohmic contacts to n-type GaAs. The contact exhibiting the lowest r_c and highest reliability is the Au–Ge–Ni system, which was first introduced by Braslau et al.[174] for making ohmic contacts to Gunn diodes and consisted of a Au–Ge film of eutectic composition covered by a Ni film. It was found that when the contact was heat treated above the Au–Ge eutectic temperature (360°C), an ohmic contact was formed to n-type GaAs for a wide range of GaAs resistivity. During alloying the Ge apparently formed an n^+ layer[175] sufficiently heavily doped to produce a linear current–voltage characteristic as a result of

Table 8. Alloyed Ohmic Contacts to the III-V Binary Compounds

Semi-conductor	Type	Contact material	Minimum r_c (Ω-cm^2)	Majority carrier concentration n or p (cm^{-3})	References
AlP	n	Ga–Ag			172
AlAs	n, p	In–Te			172
	n, p	Au			172
	n, p	Au–Ge			172
	n	Au–Sn			172
GaP	p	Au–Be–Ni	7.5×10^{-5}	2×10^{18}	199
	p	Au–Zn–Sb	1.5×10^{-2}	6×10^{17}	198
	p	Au–Be		5×10^{17}	201
	p	Au–Zn	6.5×10^{-4}	2×10^{19}	197
	n	Au–Ge–Ni	8×10^{-5}	3×10^{18}	197, 200
	n	Au–Ge–Ni–Sb	3×10^{-3}	5×10^{16}	198
GaAs	p	Au–Zn	$r_c \sim (1.8 \times 10^{18})/p^{1.3}$	10^{17}–10^{19}	187
	p	Ag–Zn	2×10^{-5}	2×10^{17}	203
	p	Ag–In–Zn	$<10^{-4}$	10^{18}	155
	n	Au–Ge–Ni	$r_c \approx (1.8 \times 10^{12})/n$	10^{15}–10^{19}	180
	n	Ag–In–Ge	6×10^{-4}	5×10^{15}	155
	n	Ni–Ge[a]	3×10^{-5}	1×10^{17}	204
	n	Pd–Ge[a]	3.5×10^{-4}	1×10^{16}	57, 150
GaSb					
	p	Au–Zn			205
	p, n	In			172
	n	Ag–In			206
	n	Au–Ge–Ni			207
InP	p	Au–Be	$r_c \approx (1 \times 10^{14})/p$	10^{16}–10^{19}	195
	p	Au–Mg	10^{-4}	6×10^{17}	192
	p	Au–Zn	5×10^{-3}–1.1×10^{-4}	10^{16}–10^{18}	158, 193, 190
	p	In–Zn	1×10^{-2}	5×10^{16}	191
	n	Au–Ag–Sn	8×10^{-5}	5×10^{17}–2×10^{18}	208
	n	Ag–Sn–In	$<10^{-4}$	3×10^{15}	
	n	Au–Ge–Ni	3×10^{-5}–8×10^{-7}	3×10^{16}–8×10^{17}	209, 210, 189
	n	Au–Sn	1.8×10^{-6}	3×10^{18}	211
InAs	n	In			172
	n	Sn–Te			
InSb	n	In			172
	n	Sn–Te			

[a] Sintered contact.

Table 9. Alloyed Ohmic Contacts to III–V Alloy Semiconductors

Alloy	Type	Contact material	Minimum r_c (Ω-cm^2)	Majority carrier concentration n or p (cm^{-3})	References
$Al_xGa_{1-x}As$					
$x = 0.4$	p	Al	2×10^{-5}	2×10^{19}	197
	p	Au–Zn	8×10^{-6}	2×10^{19}	197
	n	Au–Ge–Ni	2×10^{-4}	1×10^{18}	197
$Ga_xIn_{1-x}As$					
$x = 0.47$	p	Au–Zn	2×10^{-5}	5×10^{18}	212
	n	Au–Ge–Ni	5×10^{-7}	1×10^{17}	189
	n	Au–Sn			213
$GaAs_{1-x}P_x$					
$x = 0.4$	p	Au–Zn	6.5×10^{-4}	2×10^{19}	197
	n	Au–Ge–Ni	2×10^{-4}	4.5×10^{17}	197
	n	Au–Ge–NiCr	10^{-4}	10^{16}–10^{17}	214
$Ga_yIn_{1-y}As_{1-x}P_x$					
$x = 0.4, y = 0.3$	p	Au–Mg			215
vary x, y	p	Au–Zn	10^{-4}–10^{-5}	2×10^{18}–5×10^{18}	212
$x = 0.4, y = 0.3$	n	Au–Ge–Ni	5.8×10^{-6}	1×10^{17}	189
$Al_yGa_{1-y}As_{1-x}Sb_x$					
$x = 0.9, y = 0.5$	p	Au–Zn			207
	n	Au–Ge–Ni			205

field emission at the contact interface. The Au provided a low eutectic temperature and was compatible with existing microelectronic processing and packaging techniques. It was found that the Au–Ge alone did not wet the GaAs surface well during alloying, but the presence of a small amount of Ni (2–11 wt%) greatly improved surface uniformity of the alloyed Au–Ge contact. Braslau et al. proposed that the Ni film did not melt during heat treatment, and thus provided a cover to hold the liquid Au–Ge in uniform contact with the GaAs surface. Later measurements[176] have shown that the Ni improves the wetting of the molten Au–Ge film and increases the incorporation of the Ge in the GaAs. The barrier energy was found to be about 0.68 eV before alloying[176] and about 0.3–0.4 eV after alloying.[176,177] Using high-dose implants to form the n^+ source and drain region under Au–Ge–Ni alloyed contacts in GaAs MESFETs, r_c's as low as 3×10^{-7} and 5×10^{-7} Ω-cm^2 have been reported for Se[178] and Si[179] implementations, respectively.

The characteristics of the Au–Ge–Ni contact to GaAs have been recently reviewed by Braslau[180] and by Heiblum et al.[181] Braslau[180] noted that the r_c data for the Au–Ge–Ni contact varied inversely with the substrate doping N_D as shown in Fig. 37, which summarizes a large number of contact resistance measurements. The $1/N_D$ dependence for r_c had been previously noted by Goldberg and Isanemkou[182] and Edwards et al.[183] and is unexpected if during alloying, it is assumed that the Ge is incorporated as a donor in concentration N^+ to form a n^+ layer of thickness t as shown in Fig. 28(a). If $t > w$, where w is width of the space-charge region at the metal–semiconductor interface, r_c would be controlled by field emission and from Eq. (34) would have the form $\exp(\phi_B/N^{1/2})$, independent of N_D. This case is shown in Fig. 37 as the horizontal lines, where it was assumed $N^+ = 6 \times 10^{19}$ cm^{-3} and $\phi_B = 0.3$ and 0.8 eV. If $t < w$, then r_c would depend on N_D and in the limit of $t \rightarrow 0$, $r_c \alpha \exp(\phi_B/N_D^{1/2})$. This latter case is also shown with Fig. 37 with $\phi_B = 0.8$ and 0.3 eV. It is evident that neither case explains the experimental measurements for the entire range of N_D explored.

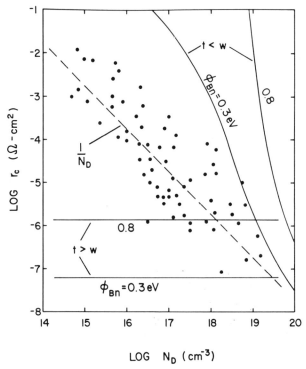

Figure 37. Specific contact resistance r_c vs. substrate doping N_D for Au–Ge–Ni alloyed contacts to n-type GaAs. The data points, from Braslau and references therein, follow a N_D^{-1} dependence (dashed line). Assuming uniform contact area, two theoretical cases (solid lines) are shown: for $t < w$, r_c is controlled by the substrate doping and varies with N_D, and for $t > w$, r_c is controlled by the concentration of the Ge (assumed 6×10^{19} cm^{-3}) and independent of N_D.

Popovic[184] has extended the field emission model to explain the $1/N_D$ dependence of r_c by assuming that t is small enough to permit the hot electrons tunneling through the metal–semiconductor barrier to be transported without collision (i.e., ballistically) across the n^+ region. Thus the current is controlled not by the metal–semiconductor interface but by the potential energy barrier at the n^+–n interface. Since the n^+–n barrier height depends on substrate doping N_D, Popovic[184] showed that in the limit $N^+ \to \infty$, $r_c \propto 1/N_D$. For Popovic's model to be correct, t should be small compared to the carrier mean free path and no larger than w (i.e, about 40 Å at 6×10^{19} cm^{-3} in n-GaAs). As Braslau[180] has pointed out, there is ample experimental evidence that t is about 1000–3000 Å for the alloyed Au–Ge–Ni contact, and thus it unlikely that ballistic transport through the n^+ region is responsible for the observed $r_c - N_D$ behavior.

Braslau[180] has proposed an alternative model to explain the $1/N_D$ dependence of r_c for the Au–Ge–Ni contact. He postulated that the measured r_c is limited by the spreading resistance of the substrate because the current is constrained to flow through small localized regions as shown by the device cross-section of Fig. 38. The current is nonuniformly distributed across the contact surface, with most of the current assumed to flow through Ge-rich protrusions of diameter d and separation a. The measured specific contact resistance is then given by [180]

$$(r_c)_{\text{meas}} \approx \langle a \rangle^2 \left(\frac{\rho}{\pi \langle d/2 \rangle} + \frac{r_c}{2\pi f \langle d/2 \rangle^2} \right) \qquad (62)$$

where $\langle \ \rangle$ indicates mean values, ρ is the resistivity of the substrate with doping N_D, and f is a factor to account for field enhancement of the current at the penetrating points. The first term in Eq. (62) is the substrate spreading resistance and the second term is the actual contact resistance of the conducting protrusions. Braslau[180] shows that the second term may be neglected for $\rho > 10^{-3}$ Ω-cm ($N_D < 4 \times 10^{18}$ cm^{-3} in n-GaAs),

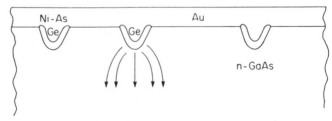

Figure 38. Structural model proposed by Braslau to explain the N_D^{-1} dependence of r_c in Au–Ge–Ni alloyed ohmic contacts. (Reprinted with permission from Ref. 180), copyright 1981, American Institute of Physics.)

and thus the observed contact resistance is proportional to ρ, or $1/N_D$ since $\rho = (q\mu_n N_D)^{-1}$ where μ_n is the electron mobility. The dashed line with negative slope in Fig. 37, which corresponds to Eq. (62) and can be written $r_c \approx (1.8 \times 10^{12})/N_D$ Ω-cm^2 with N_D in cm^{-3}, is taken from Braslau using values of $\langle a \rangle$ and $\langle d \rangle$ obtained from metallurgical analysis of the Au–Ge–Ni contacts.[180] Allowing for the spread in the r_c data, Braslau's model agrees well with the experimental results and if correct, the measured contact resistance of the Au–Ge–Ni contact is determined by localized n^+ regions formed by nonuniform alloying of the contact.

All of the data in Fig. 37 were taken for Au–Ge–Ni contacts heated to above 360°C, the Au–Ge eutectic temperature, since an abrupt drop in r_c is known to occur above the temperature.[176] However, Werthen and Scifres (185) have recently reported the formation of Au–Ge ohmic contacts to n-type GaAs with $r_c = 2.5 \times 10^{-5}$ Ω-cm^2 after heating the contacts to only 274°C for 2 min. No explanation was given for why the low anneal temperature produces ohmic behavior.

Ohmic contacts to p-type GaAs have been examined by a number of researchers and their results are summarized in Table 8. From Fig. 7(b), calculations based on field emission theory predict the contact resistance to be higher on p-type GaAs than on comparably doped n-type GaAs, primarily because the effective mass for holes is much larger than for electrons in GaAs. In practice, it is possible to obtain higher concentrations of electrically active acceptors (i.e., about 10^{20} cm^{-3}) than donors (about 10^{19} cm^{-3}) in GaAs and thus the effects of the difference in effective masses is somewhat offset. Furthermore, in IMPATT diodes, light-emitting diodes, and other devices, ohmic contacts to p-GaAs are usually of large area and thus do not require values of r_c below about 5×10^{-5} Ω-cm^2.

Au–Zn[186,187] and Ag–Zn[188] alloy contacts are most often used to contact p-type GaAs. The Zn concentration is usually 5–15 wt% and alloying takes place at 400–500°C for a few minutes. Evaporated Zn does not adhere well to GaAs, and Gopen and Yu[186] prepared Au–Zn ohmic contacts using sputtering to improve adhesion. They found that for substrate acceptor concentrations N_A greater than about 4×10^{18} cm^{-3}, which is approximately the solid solubility of Zn in GaAs at the alloying temperature, the measured contact resistance agreed with that predicted using the field emission dependence $r_c \propto \exp(\phi/N_A^{1/2})$. Below 4×10^{18} cm^{-3}, Gopen and Yu claimed that r_c should be determined primarily by the concentration of Zn incorporated in the GaAs during alloying and thus should be independent of N_A. More recently, Sanada and Wada [187] added an additional Au layer under the traditional Au–Zn, again to improve adherance, and found that r_c varies with substrate carrier concentration p (assumed to equal N_A) as shown in Fig. 39. Their results can be fitted to the expression $r_c = (1.8 \times 10^{18})/p^{1.3}$ in Ω-cm^2 for p in

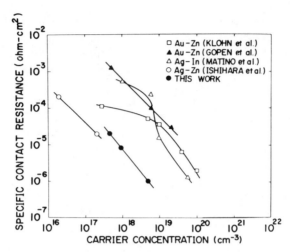

Figure 39. Summary of alloyed ohmic contacts to p-type GaAs. Note the inverse dependence on carrier concentration. The data symbols correspond to the following metalizations: (□) Au–Zn,[217] (▲) Au–Zn,[186] (△) Ag–In,[188] (○) Ag–Zn,[216] and (●) Au–Zn–Au.[187] (Reprinted with permission from Ref. 187, copyright 1980, *Japanese Journal of Applied Physics*.)

cm^{-3}. Note that all of the data on Au-based and Ag-based alloyed contacts to p-GaAs shown in Fig. 39 follow approximately a $1/N_A$ dependence, implying that nonuniform alloying may be producing localized p^+ regions and thus the measured r_c is dominated by the spreading resistance of the substrate. Braslau's model for the Au–Ge–Ni/n-GaAs contact may apply equally well to the Au–Zn/p-GaAs and Ag–Zn/p-GaAs contacts.

4.3.2. InP

For InP, it is easiest to form low-resistance ohmic contacts to n-type material as compared to p-type. As is evident from Fig. 7(a), r_c is much larger for p-type than for n-type material since $\phi_{Bp} > \phi_{Bn}$ and $m_h^* > m_e^*$, making it much more difficult for holes to tunnel through a metal–InP barrier than for electrons in comparably doped material.

The most commonly used alloy contact to n-type InP is Au–Ge–Ni, the same contact as used for n-type GaAs. Contact resistance values as low as 8×10^{-7} Ω-cm^2 have been reported for n-InP of 1×10^{17} cm^{-3} doping using Au–Ge–Ni alloyed contacts.[189]

For ohmic contact to p-type InP, previous studies have shown that Zn, Cd, and Mg can be used as acceptors in Au-based or In-based alloyed contacts and values of r_c as low as 10^{-2} Ω-cm^2 can be easily achieved.[190,191] However, lower values of r_c are difficult to obtain, especially in InP with a substrate acceptor concentration N_A of less than 10^{18} cm^{-3}. Evaporated metal layers of Au–Mg,[192] Au–Zn[158,193] and Au–Be[194,195] have all been used, with the Au–Zn and Au–Be contacts exhibiting the smoothest contact surfaces after alloying. The contact resistance measured for the Au-based alloyed contacts to p-type InP is shown in Fig. 40. The value r_c was found to be weakly dependent on the amount of Zn or Be used, but strongly dependent on N_A. Assuming a $1/N_A$ dependence (dashed line in Fig. 40), the data can be fitted with the equation $r_c \sim (1 \times 10^{14})/N_A$ Ω-cm^2 for N_A in cm^{-3}. Using $\rho = (q\mu_p N_A)^{-1}$ with the hole mobility μ_p for InP, Eq. (62) was fitted to the r_c–N_A data of Fig. 40 (solid lines), allowing for variation of μ_p with the total ionized impurity density $N_A + N_D$, where N_D is the concentration of compensating donor impurities. As is evident from Fig. 40, Braslau's model can also be used to explain the r_c–N_A dependence measured to date for Au-based alloyed contacts to p-type InP.

Although the metallurgy of Au-based InP alloy contacts is very complicated[192] and not yet fully understood, the electrical characteristics may be the result of the structure shown in Fig. 41. This structure is based on the careful experimental work of Piotrowska et al.,[196] who investigated the formation of binary compounds in the Au–InP contact for the temperature range of 320–360°C. They found that Au dissociates the InP substrate, with the Au taking up the In to the solid solubility

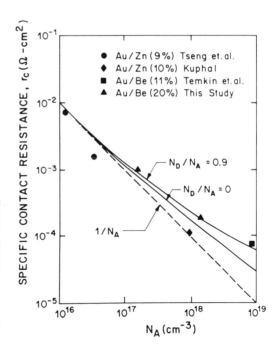

Figure 40. Specific contact resistance for Au-based alloyed ohmic contacts to p-type InP, plotted against substrate doping N_A. The dashed line corresponds to a $N^{-1}/_A$ dependence and the solid lines represent two different levels of compensation by donors N_D. The data symbols correspond to the following references: (●) Ref. 193, (◆) Ref. 158, (■) Ref. 194), and (◆) Ref. 195).

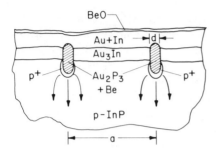

Figure 41. Proposed structural model for a Au–Be alloyed contact to p-type InP.[195] The vertical dimensions are expanded for clarity.

limit. Further heat treatment led to the formation of the compounds Au_3In and Au_2P_3. The Au_2P_3 nucleates at the InP interface to form clusters of about 1 μm in diameter which penetrate into the InP substrate. The reactions continue until all of the Au is transformed to Au_3In and Au_2P_3. Figure 41 is similar to Fig. 6 of Ref. 196 except that it is assumed that during alloying the Be (or Zn) segregates into the Au_2P_3 clusters or diffuses to the surface of the contact where it is oxidized.

4.3.3. Other Binary Compounds

Very little detailed information is available on ohmic contacts to the III–V compound semiconductors other than GaAs and InP. Typical values of r_c for alloyed contacts to GaP are given in Table 8; Au–Zn[197,198] and Au–Be[199] are often used to contact p-type GaP and Au–Ge–Ni[107,200] has been used with modest success to contact n-type GaP. It has been observed that Ga and P migrate through Au-based films during galvanic oxidation of alloyed contacts[201] and subsequent deposition of a Ti/Au or Ti/Pt/Au layer was found to prevent oxidation of the contact surface.[202]

4.2.4. III–V Alloys

Some typical alloyed ohmic contacts to the ternary and quaternary III–V alloys are listed in Table 9. Again the Au–Ge–Ni system is used to contact n-type alloys and Au–Zn is most often used to contact p-type alloys. Additional information on ohmic-contact technology to the mixed III–V semiconductors can be found in a review article by Rideout.[172]

The only detailed study of contact resistance to a III–V quaternary has been reported by Nakano et al.,[212] where Au–Zn was alloyed to p-type InGaAsP alloys of differing compositions. Their results are shown in Fig. 42(a), where r_c is plotted as a function of the wavelength of band-gap emission and thus composition of the InGaAsP. Note that the contact resistance decreases as the band-gap energy decreases. For r_c to

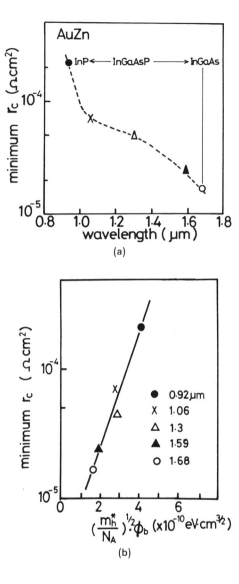

Figure 42. Au–Zn alloyed ohmic contacts to the quaternary semiconductor InGaAsP. (a) Specific contact resistance r_c as a function of the wavelength of band-gap emission (i.e., composition). (b) Specific contact resistance r_c as a function of the tunneling parameter $\phi_B m_n^{*1/2}/N_A^{1/2}$. (Reprinted with permission from Ref. 212, copyright 1980, *Japanese Journal of Applied Physics.*)

be governed by field emission, the dependence of r_c on ϕ_{Bp} should be given by $\exp(\phi_{Bp} m_h^{*1/2}/N_A^{1/2})$. Nakano *et al.* calculated m_h^* as a function of composition, measured N_A for each sample, and used the dependence of ϕ_{Bp} on composition shown in Fig. 26 to obtain the plot in Fig. 42(b). This last plot is perhaps the best experimental evidence yet obtained illustrating that quantum-mechanical tunneling is the mechanism of carrier flow in ohmic contacts.

Acknowledgments

The author would like to acknowledge past and present graduate students, D. Fertig, H. Grinolds, E. Hökelek, L. P. Erickson, A. Waseem, A. Valois, and E. Selvin, who have made many valuable contributions to our research on metal–semiconductor contacts. Preparation of this manuscript was partially supported with funds from the Army Research Office, and the assistance of Barb Dahle, Kim Schave, and MaryLee Nyberg in manuscript preparation was greatly appreciated.

References

1. E. H. Rhoderick, *Metal–Semiconductor Contacts*, Clarendon Press, Oxford (1978).
2. S. M. Sze, *Physics of Semiconductor Devices*, Wiley, New York, Chapter 8 (1969).
3. Proceedings of conference on the physics of compound semiconductors, *J. Vac. Sci. Technol. 13* (1976); *14* (1977); *15* (1978); *16* (1979); *17* (1980); *19* (1981).
4. B. L. Sharma and S. C. Gupta, Metal–semiconductor barrier junctions, *Solid-State Technol. 23*, 97–101 (1980); *23*, 90–95 (1980).
5. A. K. Sinha and J. M. Poate, Metal–compound semiconductor reactions, in: *Thin Films—Interdiffusion and Reactions* (J. M. Poate, K. N. Tu, and J. W. Mayer, eds.) pp. 407–432, Wiley, New York (1978).
6. W. Schottky, *Naturwissenschaften 26*, 843 (1938).
7. J. Bardeen, Surface states and rectification at a metal–semiconductor contact, *Phys. Rev. 71*, 717–727 (1947).
8. A. M. Cowley and S. M. Sze, Surface states and barrier height of metal–semiconductor systems, *J. Appl. Phys. 36*, 3212–3220 (1965).
9. S. Kurtin, T. C. McGill, and C. A. Mead, Fundamental transition in the electronic nature of solids, *Phys. Rev. Lett. 22*, 1433–1436 (1969).
10. T. C. McGill and C. A. Mead, Electrical interface barriers, *J. Vac. Sci. Technol. 11*, 122–127 (1974).
11. C. A. Mead, Metal–semiconductor surface barriers, *Solid-State Electron. 9*, 1023–1033 (1966).
12. M. Schlüter, Chemical trends of Schottky barriers: A reexamination of some basic ideas, *J. Vac. Sci. Tech. 15*, 1374–1376 (1978).
13. C. A. Mead and W. G. Spitzer, Fermi level position at metal–semiconductor interfaces, *Phys. Rev. 134*, A713–A716 (1964).
14. J. O. McCaldin, T. C. McGill, and C. A. Mead, Correlation for III–V and II–VI semiconductors of the Au Schottky barrier energy with anion electronegativity, *Phys. Rev. Lett. 36*, 56–58 (1976).
15. J. O. McCaldin, T. C. McGill, and C. A. Mead, Schottky barriers on compound semiconductors: The role of the anion, *J. Vac. Sci. Technol. 13*, 802–806 (1976).
16. W. E. Spicer, I. Lindau, P. Skeath, C. Y. Su, and P. Chye, Unified mechanism for Schottky barrier formation and III–V oxide interface states, *Phys. Rev. Lett. 44*, 420–423 (1980).
17. W. E. Spicer, I. Lindau, P. Skeath, and C. Y. Su, Unified defect model and beyond, *J. Vac. Sci. Tech. 17*, 1019–1027 (1980).
18. M. S. Daw and D. L. Smith, Energy levels of semiconductor surface vacancies, *J. Vac. Sci. Technol. 17*, 1028–1031 (1980).

19. R. H. Williams, V. Montgomery, and R. R. Varma, Chemical effects in Schottky barrier formation, *J. Phys. C 11*, L735–L738 (1978).
20. S. P. Kowalczyk, J. R. Waldrop, and R. W. Grant, Reactivity and interface chemistry during Schottky-barrier formation: Metals on thin native oxides of GaAs investigated by x-ray photoelectron spectroscopy, *Appl. Phys. Lett. 38*, 167–169 (1981).
21. R. W. Grant, J. R. Waldrop, S. P. Kowalczyk, and E. A. Kraut, Correlation of GaAs surface chemistry and interface Fermi-level position: A single defect model interpretation, *J. Vac. Sci. Technol. 19*, 477–480 (1981).
22. S. P. Kowalczyk, J. R. Waldrop, and R. W. Grant, Interfacial chemical reactivity of metal contacts with thin native oxides of GaAs, *J. Vac. Sci. Technol. 19*, 611–616 (1981).
23. J. C. Phillips, Chemical bonding at metal–semiconductor interfaces, *J. Vac. Sci. Technol. 11*, 947–950 (1974).
24. J. M. Andrews and J. C. Phillips, Chemical bonding and structure of metal–semiconductor interfaces, *Phys. Rev. Lett. 35*, 56–59 (1975).
25. L. J. Brillson, Transition in Schottky barrier formation with chemical reactivity, *Phys. Rev. Lett. 40*, 260–263 (1978).
26. L. J. Brillson, C. F. Brucker, A. D. Katnani, N. G. Stoffel, R. Daniels, and G. Margaritondo, Fermi level Pinning and Chemical structure of InP–Metal Interfaces, *Physics and Chemistry of Semiconductor Interfaces Conference*, Asilomar, Calif. (1982).
27. E. Hökelek and G. Y. Robinson, Schottky contacts on chemically etched p- and n-type indium phosphide, *Appl. Phys. Lett. 40*, 426–428 (1982).
28. J. M. Woodwall and J. L. Freeouf, GaAs metalization: Some problems and trends, *J. Vac. Sci. Technol. 19*, 794–798 (1981).
29. J. L. Freeouf and J. M. Woodall, Schottky barriers: An effective work function model, *Appl. Phys. Lett. 39*, 727–729 (1981).
30. H. A. Bethe, Theory of the Boundary Layer of Crystal Rectifiers, *MIT Radiation Laboratory Report 43-12* (1942).
31. C. R. Crowell, The Richardson constant for thermionic emission in Schottky barrier diodes, *Solid-state Electron. 8*, 395–399 (1965).
32. C. R. Crowell and S. M. Sze, Current transport in metal–semiconductor barriers, *Solid-State Electron. 9*, 1035–1048 (1966).
33. Y. A. Goldberg, E. A. Posse, and B. V. Tsarenkov, Mechanism of flow of direct current in GaAs surface barrier structures, *Sov. Phys.—Semicond. 9*, 337–340 (1975).
34. A. K. Srivastava, B. M. Arora, and S. Guha, Measurement of Richardson constant of GaAs Schottky barriers, *Solid-State Electron. 24*, 185–191 (1981).
35. J. M. Borrego, R. J. Gutmann, and S. Ashok, Richardson constant of Al– and Au–GaAs Schottky barrier diodes, *Appl. Phys. Lett. 30*, 169–171 (1977).
36. S. Ashok, J. M. Borrego, and R. J. Gutmann, Electrical characteristics of GaAs Schottky diodes, *Solid-State Electron. 22*, 621–631 (1979).
37. B. L. Smith, Au–(n-type) InP Schottky barriers and their use in determining majority carrier concentrations in n-type InP, *J. Phys. D 6*, 1358–1362 (1973).
38. G. G. Roberts and K. P. Pande, Electrical characteristics of Au/Ti–(n-type)InP Schottky diodes, *J. Phys. D 10*, 1323–1328 (1977).
39. G. S. Korotchenkov and I. P. Molodyan, Properties of surface-barrier M–n-InP structures, *Sov. Phys.—Semicond. 12*, 141–143 (1977).
40. G. Y. Robinson, A Study of Metal–Semiconductor Contacts on Indium Phosphide, Air Force Systems Command, Griffiths Air Force Base, New York, Interim Report RADC-TR-80-108 (1980).
41. F. A. Padovani and R. Stratton, Field and thermionic-field emission in Schottky barriers, *Solid-State Electron. 9*, 695–707 (1966).
42. C. R. Crowell and V. L. Rideout, Normalized thermionic-field (T-F) emission in metal–semiconductor (Schottky) barriers, *Solid-State Electron. 12*, 89–105 (1969).

43. A. M. Cowley, Depletion capacitance and diffusion potential of GaP Schottky diodes, *J. Appl. Phys.* 37, 3024–3032 (1966).
44. R. J. Stirn, Y. C. M. Yeh, E. Y. Wang, E. P. Ernst, and C. J. Wu, Recent improvements in AMOS solar cells, *Tech. Dig. Intern. Electron. Dev. Meeting*, IEEE, Paper 4.2 (1977).
45. A. M. Goodman, Metal–semiconductor barrier height measurement by the differential capacitance method—one carrier system, *J. Appl. Phys.* 34, 329–338 (1963).
46. R. H. Fowler, The analysis of photoelectric sensitivity curves for clean metals at various temperatures, *Phys. Rev.* 38, 45–56 (1931).
47. E. Hökelek and G. Y. Robinson, A comparison of Pd Schottky contacts on InP, GaAs, and Si, *Solid-State Electron.* 24, 99–103 (1981).
48. C. L. Anderson, C. R. Crowell, and T. W. Kao, Effects of Thermal excitation and quantum mechanical transmission on photothreshold determination of Schottky barrier height, *Solid-state Electron.* 18, 705–713 (1975).
49. R. Hackam and R. Harrop, Electrical properties of Ni low doped n-type GaAs Schottky diodes, *IEEE Trans. Electron Devices* ED-19, 1231–1238 (1972).
50. C. J. Madams, D. V. Morgan, and M. J. Howes, Outmigration of gallium from AuGaAs interfaces, *Electron. Lett.* 11, 574–574 (1975).
51. A. K. Sinha and J. M. Poate, Effect of alloying behavior on the electrical characteristics of n-GaAs Schottky diodes metalized with W, Au, and Pt, *Appl. Phys. Lett.* 23, 666–668 (1973).
52. G. Y. Robinson, Variation of Schottky-barrier energy with indiffusion in Au and Ni/Au–Ge films on GaAs, *J. Vac. Sci. Technol.* 13, 884–887 (1976).
53. W. Tantraporn, Determination of low barrier heights in metal–semiconductor films, *J. Appl. Phys.* 41, 4669–4671 (1970).
54. B. R. Pruniaux and A. C. Adams, Dependence of barrier height of metal semiconductor contact (Au–GaAs) on Thickness of semiconductor surface layer, *J. Appl. Phys.* 43, 1980–1982 (1972).
55. B. L. Smith and E. H. Rhoderick, Possible sources of error in deduction of semiconductor impurity concentrations from Schottky-barrier (C, V) characteristics, *J. Phys. D 2*, 465–467 (1969).
56. H. Grinolds and G. Y. Robinson, A study of Al/Pd$_2$Si contacts on Si, *J. Vac. Sci. Technol.* 14, 75–78 (1977).
57. H. R. Grinolds and G. Y. Robinson, Pd/Ge contacts to n-type GaAs, *Solid-State Electron.* 23, 973–985 (1980).
58. P. K. Vasudev, B. L. Mattes, E. Pietras, and R. H. Bube, Excess capacitance and non-ideal Schottky barriers on GaAs, *Solid-state Electron.* 19, 557–559 (1976).
59. B. Pellegrini and G. Salardi, Excess capacitance in metal–GaAs contacts as an effect of nonlinear dielectric susceptibility, *Solid-State Electron.* 21, 465–469 (1978).
60. J. A. Copeland, Diode edge effect on doping profile measurements, *IEEE Trans. Electron. Devices* ED-17, 404–407 (1970).
61. D. Kahng, Au–n-type GaAs Schottky barrier and its application, *Bell Syst. Tech. J.* 43, 215–224 (1964).
62. C. F. Genzabella and C. M. Howell, Gallium arsenide Schottky mixer diodes, *Symp. Gallium Arsenide Conf. Ser.* 3 (Un. Reading, England), Paper 18, 131–137 (1966).
63. B. L. Smith, Ph.D. Thesis, Manchester University (1969).
64. A. Y. Cho and P. D. Dernier, Single-crystal-aluminum Schottky barrier diodes prepared by molecular-beam-epitaxy (MBE) on GaAs, *J. Appl. Phys.* 49, 3328–3332 (1978).
65. B. L. Smith, GaAs Schottky diodes with linear log I/V behavior over eight decades of current, *Electron. Lett.* 4, 332–333 (1968).
66. G. B. Seirangyan and Y. A. Tkhorik, On the Schottky barrier height of metal–GaAs systems, *Phys. Status Solidi A* 13, K115–K118 (1972).

67. F. A. Padovani and G. C. Sumner, Experimental study of Au–GaAs Schottky diodes, *J. Appl. Phys. 36*, 3744–3747 (1965).
68. A. K. Sinha and J. M. Poate, Relative thermal stabilities of thin-film contacts to n-GaAs metalized with W, Au, and Pt, *Japan. J. Appl. Phys. Suppl. 2*, 841–844 (1974).
69. W. J. Devlin, RF sputtered Au–Mo contacts to n-GaAs, *Electron. Lett. 16*, 92–93 (1980).
70. P. Guetin and G. Schreder, Quantitative aspects of the tunneling resistance in n-GaAs Schottky barriers, *J. Appl. Phys. 42*, 5689–5698 (1971).
71. S. Guha, B. M. Arora, and V. P. Salvi, High temperature annealing behavior of Schottky barriers on GaAs with gold and gold–gallium contacts, *Solid-State Electron. 20*, 431–432 (1977).
72. J. Massies, P. Devoldere, and N. T. Linh, Silver contact on GaAs(001) and InP(001), *J. Vac. Sci. Technol. 15*, 1353–1357 (1978).
73. R. D. Baertsch and J. R. Richardson, A. Ag–GaAs Schottky-barrier ultraviolet detector, *J. Appl. Phys. 40*, 229–235 (1969).
74. K. J. Linden, Self-passivated GaAs/W mixer diode, *Solid-State Electron. 19*, 843–849 (1976).
75. A. K. Sinha and J. M. Poate, Effect of alloying behavior on the electrical characteristics of n-GaAs Schottky diodes with W, Au, and Pt, *Appl. Phys. Lett. 23*, 666–668 (1973).
76. C. Barret and A. Vapaille, Study of Pt–GaAs interface, *Solid-State Electron. 21*, 1209–1212 (1978).
77. E. Hökelek, A study of Schottky contacts on InP, Ph.D. Thesis, University of Minnesota, Minneapolis, Minn. (1982).
78. P. M. Batev, M. D. Ivanovitch, E. I. Kafedjiiska, and S. S. Simeonov, Schottky-barrier on W–GaAs contact, *Phys. Status Solidi A 45*, 671–675 (1978).
79. A. K. Sinha, Metalization scheme for n-GaAs Schottky diodes incorporating sintered contacts and a W diffusion barrier, *Appl. Phys. Lett. 26*, 171–173 (1975).
80. O. Yu. Borkovskaya, N. L. Dmitruk, R. V. Konakova, and M. Yu. Filatov, Influence of low temperature thermotreatment on the characteristics of Cr–GaAs and Au–Cr–GaAs Schottky diodes, *Electron. Lett. 14*, 700–701 (1978).
81. M. Hagio, H. Takagi, A. Nagashimo, and G. Kano, Barrier height change of Pt/Cr/n-GaAs Schottky contacts due to heat treatments, *Solid-State Electron. 22*, 347–348 (1979).
82. P. M. Batev, M. D. Ivanovitch, E. I. Kafedjiiska, and S. S. Simeonov, Schottky barrier at a Mo–GaAs contact, *Int. J. Electron. 48*, 511–517 (1980).
83. J. A. Calviello, J. L. Wallace, and P. R. Bie, High performance quasi-planar varactors for millimeter waves, *IEEE Trans. Election. Devices ED-21*, 624 (1974).
84. K. Kajiyama, S. Sakata, and O. Ochi, Barrier height of Hf/GaAs diode, *J. Appl. Phys. 46*, 3221–3222 (1975).
85. R. H. Williams, V. Montgomery, R. R. Varma, and A. McKinley, The influence of interfacial layers on nature of gold contacts to Si and InP, *J. Phys. D 10*, L253–L256 (1977).
86. J. S. Barrera and R. J. Archer, InP Schottky-gate field epect transistors, *IEEE Trans. Election. Devices ED-22*, 1023–1031 (1975).
87. O. Wada end A. Majerfeld, Low leakage nearly ideal Schottky barriers to n-InP, *Electron. Lett. 14*, 125–126 (1978).
88. D. V. Morgan and J. Frey, Increasing the effective barrier height of Schottky contacts to n-In$_x$Ga$_{1-x}$As, *Electron. Lett. 14*, 737–739 (1978).
89. A. Christou and W. T. Anderson, Jr., Material reactions and barrier height variations in sintered Al–InP Schottky diodes, *Solid-State Electron. 22*, 857–863 (1979).
90. R. J. Archer and J. Cohen, Control of Thin-Film Interface Barriers, Technical Report No. ALFAL-TR-70-256, Air Force Avionics Laboratory, Wright Patterson Air Force Base (1970).

91. H. B. Kim, A. F. Lovas, G. C. Sweeney, and T. M. S. Heng, Effects of heat treatment on metal–InP Schottky diodes characterized by secondary ion mass spectrometry, *International Symposium on Gallium Arsenide*, Edinburgh; *Inst. Phys. Conf. Ser. No. 336*, 145 (1977).
92. N. Szydlo and J. Oliver, Behavior of Au/InP Schottky diodes under heat treatment, *J. Appl. Phys. 50*, 1445–1449 (1979).
93. K. Hattori and Y. Izumi, The electrical characteristics of InP Schottky diodes, *J. Appl. Phys. 52*, 5699–5700 (1981).
94. J. S. Escher, L. W. James, R. Sankaran, G. A. Antypas, R. L. Moon, and R. L. Bell, Schottky-barrier height of Au/p-InGaAsP alloys lattice matched to InP, *J. Vac. Sci. Technol. 13*, 874–876 (1976).
95. J. S. Escher, P. E. Gregory, and T. J. Maloney, Hot electron attenuation length in Ag/InP Schottky diodes, *J. Vac. Sci. Technol. 16*, 1394–1397 (1979).
96. K. Kaminura, T. Suzuki, and A. Kunioka, Metal–insulator–semiconductor Schottky-barrier solar cells fabricated on InP, *Appl. Phys. Lett. 38*, 259–261 (1981).
97. H. G. White and R. A. Logan, GaP surface-barrier diodes, *J. Appl. Phys. 34*, 1990–1997 (1963).
98. B. L. Smith, Near ideal Au–GaP Schottky diodes, *J. Appl. Phys. 40*, 4675–4676 (1969).
99. B. L. Smith and M. Abbott, Minority carrier diffusion length in liquid epitaxial GaP, *Solid-State Electron. 15*, 361–369 (1972).
100. Yu. A. Goldberg, E. A. Posse, and B. V. Tsarenkov, Ideal GaP surface-barrier diodes, *Electon. Lett. 7*, 601–602 (1971).
101. K. Okamoto, C. E. C. Wood, and L. F. Eastman, Schottky barrier heights of molecular beam epitaxial metal–AlGaAs structures, *Appl. Phys. Lett. 38*, 636–638 (1981).
102. M. F. Millea, M. McColl, and A. H. Silver, Electrical characterization of metal/InAs contacts, *J. Electron. Mater. 5*, 321–340 (1976).
103. J. N. Walpole and K. W. Nill, Capacitance–voltage characteristics of metal barriers on pPbTe and pInAs, *J. Appl. Phys. 42*, 5609–5617 (1971).
104. M. L. Korwin-Pawlowski and E. L. Heasell, The properties of some metal–InSb surface barrier diodes, *Solid-State Electron. 18*, 849–852 (1975).
105. R. Chin, R. A. Milano, and H. D. Law, Schottky barrier height of Au on n-type $Ga_{1-x}Al_xSb$ ($0.0 \leq x \leq 0.65$), *Electron. Lett. 16*, 626–627 (1980).
106. A. A. Gutkin, M. V. Dmitriev, and D. N. Nasledov, Photosensitivity of Au–nGaP surface-barrier diodes in the 1.4–5.2 eV spectral range, *Sov. Phys.—Semicond. 6*, 429–433 (1972).
107. T. F. Lei, C. L. Lee, and C. Y. Chang, Metal/n-GaP Schottky barrier heights, *Solid-State Electron. 22*, 1035–1037 (1979).
108. C. R. Wronski, Effects of deep centers in n-type GaP Schottky barriers, *J. Appl. Phys. 41*, 3805–3812 (1970).
109. Y. Nannichi and G. L. Pearson, Properties of GaP Schottky diodes at elevated temperatures, *Solid-State Electron. 12*, 341–348 (1969).
110. L. P. Krukovskaya, L. S. Berman, A. ya.Vul', and A. Ya.Shik, Surface-barrier structure on gallium antimonide, *Sov. Phys.—Semicond. 11*, 1109–1110 (1977).
111. W. E. Spicer, P. W. Chye, P. R. Skeath, C. Y. Su, and I. Lindau, New and unified model for Schottky barrier III–V insulator states formation, *J. Vac. Sci. Technol. 16*, 1422–1433 (1979).
112. G. H. Parker and C. A. Mead, Energy–momentum relationship in InAs, *Phys. Rev. Lett. 21*, 605–606 (1968).
113. Y. A. Goldberg, T. Y. Rafiev, B. V. Tsarenkov, and Y. P. Yakovlev, Surface barrier structures of metal and n-type $Ga_{1-x}Al_xAs$ and their energy band diagram, *Sov. Phys.—Semicond. 6*, 398–401 (1972).

114. V. L. Rideout, Dependence of barrier height on energy gap in Au n-type GaAs$_{1-x}$P$_x$ Schotttky diodes, *Solid-State Electron.* 17, 1107–1108 (1974).
115. J. S. Best, The Schottky-barrier height of Au on n-Ga$_{1-x}$Al$_x$As as a function of AlAs content, *Appl. Phys. Lett.* 34, 522–524 (1979).
116. R. E. Allen and J. D. Dow, Unified theory of point-defect electronic states, core excitations, and intrinsic electronic states at semiconductor surfaces, *J. Vac. Sci. Technol.* 19, 383–387 (1981).
117. M. S. Daw, D. L. Smith, C. A. Swartz, and T. C. McGill, Surface vacancies in II–VI and III–V zinc blende semiconductors, *J. Vac. Sci. Technol.* 19, 508–512 (1981).
118. J. Barnard, H. Ohno, C. E. C. Wood, and L. F. Eastman, Double heterostructure Ga$_{0.47}$In$_{0.53}$As MESFETs with submicron gates, *IEEE Electron Devices Lett.* EDL-1, 174–176 (1980).
119. K. Kajiyama, Y. Mizushima, and S. Sakata, Schottky barrier height of n-In$_x$Ga$_{1-x}$As diodes, *Appl. Phys. Lett.* 23, 458–459 (1973).
120. T. F. Kuech and J. O. McCaldin, Composition dependence of Schottky barrier heights for Au on chemically etched In$_x$Ga$_{1-x}$P surfaces, *J. Vac. Sci. Technol.* 17, 891–893 (1980).
121. W. Keeler, W. J. Roth, and E. Fortin, photovoltaic effect and Schottky barriers in Au-In$_{1-x}$Ga$_x$Sb systems, *Can. J. Phys.* 58, 63–67 (1980).
122. T. P. Lee, C. A. Burrus, M. A. Pollock, and R. E. Nahory, High Speed Schottky-Barrier photodiode in LPE In$_x$Ga$_{1-x}$As for 1.0–1.1 μm Region, Developmental Research Conference. Paper IVa-3, Ottawa, Canada (June 24–26, 1975).
123. P. K. Bhattacharya and M. D. Yeaman, Enhanced barrier height of Au–In$_{1-x}$Ga$_x$As$_y$P$_{1-y}$ Schottky diodes, *Solid-State Electron.* 24, 297–300 (1981).
124. A. Y. C. Yu, H. J. Gopen, and R. K. Waits, Contacting Technology for GaAs, Final Technical Report, No. AFAL-TR-70-196, AFAL (AFSC), W-PAFB (1970).
125. L. L. Chang and G. L. Pearson, The solubilities and distribution coefficients of Zn in GaAs and GaP, *Phys. Chem. Solids* 25, 23–30 (1964).
126. Y. I. Nissim, J. F. Gibbons, and R. B. Gold, Non-alloyed ohmic contacts to n-GaAs by CW laser assisted diffusion from a SnO$_2$/SiO$_2$ source, *IEEE Electron Devices* ED-28, 607–609 (1981).
127. T. F. Deutsch, D. J. Ehrlich, R. M. Osgood, Jr., and Z. L. Liau, Ohmic contact formation on InP by pulsed laser photochemical doping, *Appl Phys. Lett.* 36, 847–849 (1980).
128. P. A. Pianetta, C. A. Stolte, and J. L. Hansen, Non-alloyed ohmic contacts to electron-beam-annealed Se-ion-implanted GaAs, *Appl. Phys. Lett.* 36, 597–599 (1980).
129. M. Helix, K. Vaidynanathan, B. Streetman, H. Dietrich, and P. Chatterjee, P. F. Plasma deposition of silicon nitride layers, *Thin solid Films* 55, 143–148 (1978).
130. T. Onuma, T. Hirao, and T. Sugawa, Study of encapsulants for annealing Si-implanted GaAs, *J. Electrochem. Soc.* 129, 837–840 (1982).
131. E. Yamaguchi, T. Nishioda, and Y. Ohmachi, Ohmic contacts to Si-implanted InP, *Solid-State Electron.* 24, 263–265 (1981).
132. P. A. Barnes, H. J. Leamy, J. M. Poate, and S. D. Ferris, Ohmic contacts produced by laser-annealing Te-implanted GaAs, *Appl. Phys. Lett.* 33, 965–967 (1978).
133. R. L. Mozzi, W. Fabian, and F. J. Pierkarski, Non-alloyed ohmic contacts to n-GaAs by pulse-electron-beam-annealed Se implants, *Appl. Phys. Lett.* 35, 337–339 (1979).
134. Z. L. Liau, N. L. DeMeo, J. P. Donnelly, D. E. Mull, R. Bradbury, and J. P. Lorenzo, Frabrication of Ohmic Contacts on p-type InP Using Ion Implantation and Laser Annealing, Paper presented at the Materials Research Society Metting, Cambridge, Mass. (November 1979).
135. H. Kressel and J. K. Butler, *Semiconductor Lasers and Heterojunction LEEDs*, Academic Press, New York (1977).

136. A. Y. Cho and J. R. Arthur, Molecular Beam Epitaxy, *Progr. Solid-State Chem.* 10, 157–191 (1975).
137. D. DeSimone, G. Wicks, and C. E. C. Wood, Doping Limits in MBE GaAs, Paper E.O, MBE Workshop, University of California, Santa Barbara (1981).
138. P. A. Barnes and A. Y. Cho, Non-alloyed ohmic contacts to n-GaAs by molecular beam epitaxy, *Appl. Phys. Lett.* 33, 651–653 (1978).
139. J. V. DiLorenzo, W. C. Niehaus, and A. Y. Cho, Non-alloyed and in situ ohmic contacts to highly doped n-type GaAs layers grown by molecular beam epitaxy for field effect transistors, *J. Appl. Phys.* 50, 951–954 (1979).
140. W. T. Tsang, In situ ohmic-contact formation to n- and p-GaAs by molecular beam epitaxy, *Appl. Phys. Lett.* 33, 1022–1025 (1978).
141. T. Sebestyen, M. Menyhard, and D. Szigethy, In situ measurements of arsenic losses during annealing of the usual evaporated contacts of GaAs Gunn diodes, *Electron. Lett.* 12, 96–97 (1976).
142. T. Sebestyen, H. Hartnagel, and L. H. Herron, New method for producing ideal metal–semiconductor ohmic contacts, *Electron. Lett.* 10, 372–373 (1975).
143. H. T. Mills and H. L. Hartnagel, Ideal ohmic contacts to InP, *Electron. Lett.* 11, 621–622 (1975).
144. S. Margalit, D. Fekete, D. M. Pepper, Chien-Ping Lee, and A. Yariv, Q-switched ruby laser alloying of ohmic contacts on gallium arsenide epilayers, *Appl. Phys. Lett.* 33, 346–347 (1978).
145. R. B. Gold, R. A. Powell, and J. F. Gibbons, Laser Alloying of Au–Ge Ohmic Contacts on GaAs, AIP Conference Proceedings No. 50; Laser–solid interactions and laser processing, *Am. Inst. Phys.* 1979 635–640.
146. W. T. Anderson, A. Christou, and J. F. Giuliani, Laser annealed Ta/Ge and Ni/Ge ohmic contacts to GaAs, *IEEE Electron. Devices Lett. EDL-2*, 115–118 (1981).
147. A. H. Oraby, K. Murakami, Y. Yuba, K. Gamo, and S. Namba, Laser annealing of ohmic contacts on GaAs, *Appl. Phys. Lett.* 38, 562–564 (1981).
148. C. P. Lee, J. L. Tandom, and P. J. Stocker, Alloying behavior of Au–Ge/Pt ohmic contacts to GaAs by pulsed electron beam and furnace heating, *Electron. Lett.* 16, 849–850 (1980).
149. R. D'Angelo and P. A. Verlangieri, Ohmic contacts on n-GaAs produced by spark alloying, *Electron. Lett.* 17, 290–291 (1981).
150. A. K. Sinha, T. E. Smith, and H. J. Levinstein, Sintered ohmic contacts to n- and p-type GaAs, *IEEE Trans. Electron. Devices ED-22*, 218–223 (1975).
151. J. Massies, J. Chaplait, M. Laviron, and N. T. Linh, Monocrystalline aluminum ohmic Contact to nGaAs by H_2S Adsorption, *Appl. Phys. Lett.* 38, 693–695 (1981).
152. R. Stall, C. E. C. Wood, and L. F. Eastman, Ultra low resistance ohmic contacts to n-GaAs, *Electron. Lett.* 15, 800–801 (1979).
153. R. A. Stall, C. E. C. Wood, K. Board, N. Dandekar, L. F. Eastman, and J. Devlin, A study of Ge/GaAs interface grown by molecular beam epitaxy, *J. Appl. Phys.* 52, 4062–4069 (1981).
154. J. M. Woodall, J. L. Freeouf, G. D. Pettit, T. Jackson, and P. Kirchner, Ohmic contacts to n-type GaAs using graded band gap layers of $Ga_xIn_{1-x}As$ grown by molecular beam epitaxy, *J. Vac. Sci. Technol.* 19, 626–627 (1981).
155. R. H. Cox and H. Strack, Ohmic contacts for GaAs devices, *Solid-State Electron.* 10, 1213–1218 (1967).
156. R. D. Brooks and H. G. Mathes, Spreading resistance between constant potential surfaces, *Bell Syst. Tech. J.* 50, 775–784 (1971).
157. L. E. Terry and R. W. Wilson, Metalization systems for Si integrated circuits, *Proc. IEEE* 57, 1580–1586 (1969).

158. E. Kuphal, Low resistance ohmic contacts to n- and p-InP, *Solid-State Electron.* 24, 69–78 (1981).
159. Y. K. Fang, C. Y. Chang, and Y. K. Su, Contact resistance in metal–semiconductor systems, *Solid–State Electron.* 22, 933–938 (1979).
160. W. Shockley, Research and Investigation of Inverse Epitaxial UHF Power Transistor, Final Technical Report, No. Al-TDR-64-207, AFAL (AFSC), W-PAFB (1964).
161. P. L. Hower, W. W. Hooper, B. R. Cairns, R. D. Fairmen, and D. A. Tremere, The GaAs field-effect transistor, in: *Semiconductors and Semimetals* (R. K. Willardson and A. C. Beer, eds.), Chapter 3, Vol. 7, Part A, pp. 147–200, Academic Press, New York (1973).
162. C-Y. Ting and C. Y. Chen, A study of the contacts of a diffused resistor, *Solid-State Electron.* 14, 433–438 (1971).
163. H. Murrmann and D. Widman, Current crowding on metal contacts to planar devices, *IEEE Trans. Electron. Devices ED-16*, 1022–1024 (1969).
164. H. H. Berger, Models for contacts to planar devices, *Solid-State Electron.* 15, 145–158 (1972).
165. H. H. Berger, Contact resistance and contact resistivity, *J. Electrochem. Soc. 119*, 507–514 (1972).
166. H. H. Berger, *Dig. Tech. Papers*, ISSCC, pp. 160–161 (1969).
167. G. K. Reeves, Specific contact resistance using a circular transmission line model, *Solid-State Electron.* 23, 487–490 (1980).
168. S. B. Schuldt, An exact derivation of contact resistance to planar devices, *Solid-State Electron.* 21, 715–719 (1978).
169. A. A. Immorlica, D. R. Decker, and W. A. Hill, Diagnostic pattern for GaAs PET material development and process monitoring, *IEEE Trans. Electron. Devices ED-27*, 2285 (1980).
170. G. K. Reeves and H. B. Harrison, Obtaining the specific contact resistance from transmission line model measurements, *IEEE Electron. Devices Lett. EDL-3*, 111–113 (1982).
171. B. Schwartz, ed., *Ohmic Contacts to Semiconductors*, Electrochemical Society, New York (1969).
172. V. L. Rideout, A review of the theory and technology for ohmic contacts to group III–V compound semiconductors, *Solid-State Electron.* 18, 541–550 (1975).
173. L. D. Libov, S. S. Meskin, D. N. Nasledov, V. E. Sedov, and B. V. Tsarenkov, Ohmic contacts of metals with GaAs (review), *Instrum. Expt. Tech. 4*, 746–753 (1965).
174. N. Braslau, J. B. Gunn, and J. L. Staples, Metal–semiconductor contacts for GaAs bulk-effect devices, *Solid-State Electron.* 10, 381–383 (1967).
175. A. M. Andrews and H. Holonyak, Jr., Properties of n-type Ge-doped epitaxial GaAs layers grown from Au-rich melts, *Solid-State Electron.* 15, 601–604 (1972).
176. G. Y. Robinson, Metallurgical and electrical properties of alloyed Ni/Au–Ge films on n-type GaAs, *Solid–State Electron.* 18, 331–342 (1975).
177. H. Paria and H. Hartnagel, Experimental evidence for GaAs surface quality affecting ohmic contact properties, *Appl. Phys.* 10, 97–99 (1976).
178. T. Hara and T. Inada, Trends in ion implantation in gallium arsenide, *Solid-State Technol.* 22, 69–74 (1979).
179. K. Ohata, T. Nozaki, and N. Kawamura, Improved noise performance of GaAs MESFETs with selectively implanted n^+ source regions, *IEEE Trans. Electron. Devices ED-24*, 1129–1131 (1978).
180. N. Braslau, Alloyed ohmic contacts to GaAs, *J. Vac. Sci. Technol.* 19, 803–807 (1981).
181. M. Heiblum, M. I. Nathan, and C. A. Chang, Characteristics of AuGeNi ohmic contacts to GaAs, *Solid-State Electron.* 25, 185 (1982).

182. Yu. Goldberg and B. V. Isarenkov, Dependence of resistivity of metal–gallium arsenide ohmic contacts on the carrier density, *Sov. Phys.—Semicond.* 3, 1447–1448 (1970).
183. W. D. Edwards, W. A. Hartman, and A. B. Torrens, specific contact resistance of ohmic contacts to gallium arsenide, *Solid-State Electron.* 15, 387–392 (1972).
184. R. S. Popovic, Metal-n-type semiconductor ohmic contact with a shallow N^+ surface layer, *Solid–State Electron.* 21, 1133–1138 (1978).
185. J. G. Werthen and D. R. Scifres, Ohmic contacts to n-GaAs using low temperature anneal, *J. Appl. Phys.* 52, 1127–1129 (1981).
186. H. J. Gopen and A. Y. C. Yu, Ohmic contacts to epitaxial p-GaAs, *Solid-State Electron.* 14, 515–517 (1971).
187. T. Sanada end O. Wada, Ohmic contacts to p-GaAs with Au/Zn/Au structure, *Japan. J. Appl. Phys.* 19, L491–L494 (1980).
188. H. Matino and M. Tokunaga, Contact resistance of several metals and alloys to GaAs, *J. Electrochem. Soc.* 116, 709–711 (1979).
189. H. Morkoç, T. J. Drummond, and C. M. Stanchak, Schottky barriers and ohmic contacts on n-type InP based compound semiconductors for microwave FETs, *IEEE Trans. Electron. Devices ED-28*, 1–7 (1981).
190. L. M. Schiavone and A. A. Pritchard, Ohmic contact to moderately resistive p-type InP, *J. Appl. Phys.* 46, 452–453 (1974).
191. F. A. Thiel, D. D. Bacon, E. Buehler, and K. J. Bachmann, Contacts to p-type InP, *J. Electrochem. Soc.* 124, 317–318 (1977).
192. L. P. Erickson, A. Waseem, and G. Y. Robinson, Charecterization of ohmic contacts to InP, *Thin Solid Films* 64, 421–426 (1979).
193. W. Tseng, A. Christou, H. Day, J. Davey, and B. Wilkins, Ohmic contacts to lightly doped n and p indium phosphide surfaces, *J. Vac. Sci. Technol.* 19, 623–625 (1981).
194. H. Temkin, R. J. McCoy, V. G. Keramidas, and W. A. Bonner, Ohmic contacts to p-type InP using Be–Au metalization, *Appl. Phys. Lett.* 36, 444–446 (1980).
195. A. J. Valois and G. Y. Robinson, Au/Be ohmic contacts to p-type indium phosphide, *Solid-State Electron.* 25, 973 (1982).
196. A. Piotrowska, P. Auvay, A. Guivarch, and G. Pelois, On the formation of binary compounds in Au/InP system, *J. Appl. Phys.* 52, 5112–5117 (1981).
197. K. K. Shih and J. M. Blum, Contact resistances of Au–Ge–Ni, Au–Zn and Al to III–V Compounds, *Solid-State Electron.* 15, 1177–1180 (1972).
198. M. Itoh, T. Itoh, Y. Yammamoto, and K. G. Stephens, Low resistance ohmic contacts containing Sb to GaP, *Solid-State Electron.* 23, 447–448 (1980).
199. J. Pfeifer, Ohmic contact to p-type GaP, *Solid-State Electron.* 19, 927–929 (1976).
200. T. F. Lei, C. L. Lee, and C. Y. Chang, Specific contact resistance of the Ni/Au–Ge/nGaP system, *Solid-State Electron.* 21, 385–391 (1978).
201. W. A. Brantley, B. Schwartz, V. G. Keramidas, G. W. Kamhlott, and A. K. Sinha, Gallium migration through contact metalizations on GaP, *J. Electrochem. Soc.* 122, 434–436 (1975).
202. W. A. Brantley, B. Schwartz, V. G. Keramidas, A. K. Sinha, and G. W. Kammlott, Modified contact metalizations for GaP to provide barrier action against gallium migration, *J. Electrochem. Soc.* 122, 1152–1154 (1975).
203. O. Ishihara, K. Nishitani, H. Sawano, and S. Mitsui, Ohmic contacts to P-type GaAs, *Japan. J. Appl. Phys.* 15, 1411–1412 (1976).
204. W. T. Anderson, Jr., A. Christou, and J. E. Davey, Development of ohmic contacts for GaAs devices using epitaxial Ge films, *IEEE J. Solid-State Circuits SC-13*, 430–435 (1978).
205. T. Kagawa and G. Motosugi, AlGaAsSb photodiodes lattice matched to GaSb, *Japan. J. Appl. Phys.* 18, 1001–1002 (1979).

206. G. Jung, Binary Ag–In ohmic contacts to GaAs and GaSb, *Electron Technol.* **8**, 63–84 (1975).
207. T. Kagawa and G. Motosugi, AlGaAsSb avalanche photodiodes for 1.0–1.3 μm wavelength region, *Japan. J. Appl. Phys.* **18**, 2317–2318 (1979).
208. V. Wrick and L. F. Eastman, private communication (1973).
209. H. T. Mills and H. L. Hartnagel, Ohmic contacts to InP, *Int. J. Electron.* **46**, 65–73 (1979).
210. G. Weimann and W. Schlapp, Ohmic contacts on indium phosphide, *Phys. Status Solidi A* **50**, K219–K223 (1978).
211. P. A. Barnes and R. S. Williams, Alloyed tin–gold ohmic contacts to n-type indium phosphide, *Solid-State Electron.* **24**, 907–913 (1981).
212. Y. Nakano, S. Takahashi, and Y. Toyoshima, Contact resistance dependence on InGaAsP layers lattice matched to InP, *Japan. J. Appl. Phys.* **19**, L495–L497 (1980).
213. H. H. Wieder, A. R. Clawson, D. I. Elder, and D. A. Collins, Inversion mode insulated gate $Ga_{0.47}In_{0.53}As$ field-effect transistors, *IEEE Electron. Devices Lett.* **EDL-2**, 73–74 (1981).
214. B.-L. Twu, A reproducible ohmic contact to n-type $GaAs_{0.6}P_{0.4}$, *Solid-State Electron.* **22**, 501–505 (1979).
215. C. A. Armiento, J. P. Donnelly, and S. H. Groves, p-n junction diodes in InP and $In_{1-x}Ga_xAs_yP_{1-y}$ frabricated by beryllium-ion implantation, *Appl. Phys. Lett.* **34**, 229–231 (1979).
216. O. Ishihara, K. Nishitani, H. Sawano, and S. Mitusi, Ohmic contacts to p-type GaAs, *Japan. J. Appl. Phys.* **15**, 1411–1412 (1976).
217. K. L. Klohn and L. Wandinger, *J. Electrochem. Soc.* **116**, 507 (1969).

3

The Deposited Insulator/III–V Semiconductor Interface

J. F. Wager and C. W. Wilmsen

1. Introduction

The development of a MISFET technology using the III–V compounds requires a gate insulator which is highly resistive, mechanically strong, and electronically stable and which produces a low interface state density. Grown oxides on III–V semiconductors have not yet demonstrated these attributes. For example, arsenic and phosphorus oxides rapidly absorb water from the atmosphere and arsenic oxides are thermodynamically unstable in the presence of GaAs, InAs, or InGaAs. The failure of the grown oxides to provide a suitable gate insulator dictates the use of deposited insulators. The use of a deposited insulator is, however, a compromise at best since the uniformity, thickness, and properties of a deposited insulator cannot be controlled as well as with a grown oxide. Also, the heteromorphic nature of the deposited insulator/III–V interface implies the possibility of a nonabrupt interfacial mismatch and its associated trap states. Thus, deposited insulators have inherent bulk and interface problems which require special attention.

Deposited insulators have four primary applications for III–V semiconductor technology: the gate insulator for MISFETs, a passivation layer, a cap for annealing, and an insulating layer for interconnects. This chapter will concentrate on the first two applications but all of the applications have similar requirements. As the technologies progress

---

J. F. Wager and C. W. Wilmsen ● Department of Electrical and Computing Engineering, Oregon State University, Corvallis, OR 97331, and Department of Electrical Engineering, Colorado State University, Fort Collins, CO 80523.

there will be an increasing overlap of requirement, for example, the need for a self-aligned gate demands that the gate insulator withstand the ion-implant anneal cycle.

At present only the indium-based materials have demonstrated surface accumulation of electrons and thus InP, InSb, and $In_{0.53}Ga_{0.47}As$ are now the main materials of interest for MISFET application. A significant problem limiting the usefulness of III–V MISFETs is trapping of carriers at the insulator/III–V semiconductor interface which causes the drain current to decrease after the MISFET is turned on. The rate of drain current decrease as well as the inversion mobility has been found to vary considerably with processing. In this chapter we provide general information on the chemistry of insulator/III–V interface formation in order that the reader can better understand how to improve the fabrication of III–V MISFET devices. The first section discusses the choice of insulator and deposition technique. This is followed by a discussion of interface formation. The last major section reviews the literature on MIS interfaces of InSb, GaAs, and InP.

2. General Overview of the Deposited Insulator/III–V Interface

Most deposited insulator/III–V interfaces are formed as a result of numerous processing steps. This section divides interface formation into a generic set of processing steps of surface preparation and pretreatment, and insulator deposition and postdeposition treatments, and points out how these can influence the final interface.

The formation of a deposited insulator/III–V interface inevitably begins with some form of wet chemical surface preparation step. The purpose of this process step is to treat the substrate so that the initial semiconductor surface is smooth, stoichiometric, undamaged, and free from unwanted oxidation and contamination. In preparing a substrate for insulator deposition, the surface is usually polished with bromine–methanol on a soft pad. If done properly this polish creates a smooth, highly reflective surface with very little oxide or surface contamination (see Chapter 7 for further details of surface oxidation). Since the polishing is normally carried out in air, an oxide begins to grow and contamination, such as hydrocarbons, collects on the surface. Processing in a dry box purged with N_2 might reduce the severity of undesirable oxidation and contamination. The polished samples are then often stored prior to insulator deposition. The oxide which grows during this storage time may then be removed by a short dip in an oxide etch, such as HF, HCl,

Br–MeOH, or hydrazine, with a subsequent water or methanol rinse. Exposure to air will again result in the growth of a thin oxide layer. Unless some intentional effort is made to modify the semiconductor surface prior to the deposition of the insulator, the electrical interface properties may very well be dominated by this naturally formed native oxide/contamination layer since this layer will otherwise be incorporated into the final interface. Although conventional wisdom dictates that surface oxidation and/or contamination is inherently bad, it is possible that certain types of oxides and/or impurities may actually yield superior electrical device characteristics. In fact, Meiners[1] has speculated that reduction of the native InP oxide using a hydrazine treatment in conjunction with carbon incorporation at the InP surface may stabilize the InP surface and be responsible for the dramatic reduction of surface state density after hydrazine treatment.

After wet chemical treatment, the water is placed in a deposition chamber which is then either evacuated or purged with an inert gas. In some cases further *in situ* processing is undertaken to remove surface contamination and oxides. This processing may be, for example, exposure to HCl vapor, H_2 gas, or a H_2 plasma. HCl should remove the oxide and some of the substrate which may roughen the surface. The H_2 gas or plasma may remove the oxide but it will also preferentially attack the substrate, for example, on InP, volatile PH_3 is formed leaving In metal on the surface.[2]

After *in situ* processing, the substrate is usually heated (it may also have been heated during the above processing), usually to a temperature between 200 and 350°C. If all of the oxide had been removed then the volatile column V elements, P, As, or Sb, may then be evaporated. However, most deposition systems are pumped by either a mechanical or diffusion pump in which case there is probably sufficient oxygen in the chamber to cause some oxidation which would seal the surface.

Table 1 is a list of the approximate temperatures at which the rates of oxidation or incongruent evaporation begin to become significant for InSb, GaAs, and InP. The pretreatment temperature should always be maintained below the incongruent evaporation temperature since the preferential loss of the column V element from the surface will result in surface defects.

Once the substrate is stabilized at the desired deposition temperature, the source gases are introduced and in some cases a plasma or photon excitation is initiated. The excited and unexcited gases impinge upon the surface where they may react with the substrate or its oxide. As an example, for SiO_2 chemical vapor deposition the impinging gases may include SiH_4, H_2, Si, O_2, and/or N_2O, all of which could alter the surface chemistry and texture.

Table 1. Approximate Temperatures Above Which Substantial Amounts of Oxidation or Incongruent Evaporation Will Occur for InSb, GaAs, and InP

Semiconductor	T (oxidation)	T (incongruent evaporation)
InSb	250°C[3]	300°C[4]
GaAs	350°C < T(ox) < 450°C[5]	657°C[6]
InP	350°C[7]	365°C[8]

After the insulator film has begun to form, there can be interdiffusion of the substrate and film which will further alter the interface. The processes of evaporation, oxidation, reaction, and interdiffusion are all strongly dependent upon temperature. Thus, if these processes are to be minimized, it is advantageous to deposit the insulator at as low a temperature as possible. Unfortunately, the insulating and mechanical properties of the deposited film degrade with decreasing temperature. Therefore, a compromise temperature is usually used.

A final process step that is sometimes performed after insulator deposition is that of postdeposition treatment. Postdeposition treatments usually consist of annealing at relatively high temperatures in a variety of gas ambients. Relatively little work has been reported on postdeposition treatments for insulator/III-V semiconductor interfaces. The usual purpose of this treatment is to reduce the interface state and fixed charge densities, densify the deposited insulator, remove moisture from the deposited insulator, or study the encapsulating properties of deposited insulators.

A brief review of the annealing of the SiO_2/Si interface will provide insight into the purpose of postdeposition treatments. Two types of postdeposition annealing treatments are typically used for thermally grown SiO_2/Si interfaces.[9] High-temperature (900-1100°C) annealing in an inert atmosphere (Ar or N_2, typically) decreases the fixed oxide charge and also may reduce the interface state density.[10] Lower-temperature (350-450°C) annealing in an ambient with hydrogen further decreases the interface state density. This low-temperature hydrogen anneal is particularly effective if it is accomplished after gate metalization.[9]

The exact physical mechanisms responsible for the reduction of the fixed oxide charge and the interface state density are still matters of some controversy. However, the general consensus at this time appears to be that high-temperature annealing allows the oxidation process near the interface to go further towards completion, thus reducing the fixed oxide charge. The hydrogen gas used in the low-temperature anneal is thought to somehow tie up "dangling bonds" at the interface and thus reduce the interface state density.

It is important to note that while postdeposition anneals are effective in reducing the fixed oxide charge and interface state density for the SiO_2/Si system, these quantities are already quite small even prior to the annealing treatments. For example, Johnson et al.[10] report that a combination of a high-temperature, inert ambient anneal and a low-temperature, forming gas anneal yielded an order-of-magnitude reduction in interface state density. If the SiO_2/Si can be considered as a guide, then it may be possible to reduce the interface state density of "as-deposited" insulator/III–V semiconductor interfaces by about one order of magnitude. However, it would not be reasonable to expect that postdeposition treatments will be a panacea for deposited insulator/III–V interfaces.

It has already been mentioned that postdeposition treatments are sometimes performed in order to densify deposited insulator films, particularly for CVD layers. Pliskin and Lehman[11] have examined SiO_2 films deposited on silicon by CVD, reactive sputtering, electron-beam evaporation, and thermal evaporation. They found that, in general, these "as-deposited" SiO_2 layers were porous, susceptible to moisture, strained, have faster etch rates, usually have different refractive indices, and exhibit differences in their infrared spectra compared to thermally grown SiO_2 layers. Densification of these films by heat treatments in excess of 800°C were found to produce films with excellent properties, comparable to thermal SiO_2. Annealing cycles may be necessary to obtain high-quality dielectrics for insulator/III–V semiconductor interfaces. However, the 800°C densification temperature is too high to be compatible with III–V semiconductor processing.

Encapsulation studies have some relevance with respect to the nature of postdeposition treatments. Various dielectrics have been investigated for use as encapsulants during high-temperature annealing after ion implantation. An annealing step is necessary to remove the lattice damage caused by ion implantation, as well as to activate the implanted impurities. The temperatures required for annealing are in excess of the incongruent evaporation temperatures of the semiconductor substrates. The purpose of the encapsulant is to prevent dissociation of the semiconductor surface while allowing the lattice damage to be annealed out and the implanted impurities to be activated.

The encapsulation studies of interest provide insight into the outdiffusion behavior of the semiconductor atoms—which atoms have high diffusion mobilities into a given insulator and at what temperatures noticeable outdiffusion will occur. These studies are invariably conducted at temperatures much higher than those used for postdeposition treatments, but the results are still relevant because the relative outdiffusion trends are expected to be similar at lower temperatures, only on a much smaller scale. Specific experimental results will be reviewed in another section of this chapter.

3. Choice of Insulator and Deposition Technique

In choosing an insulator for deposition onto a III–V compound semiconductor, the following bulk insulator properties are important to consider:

Mechanical, chemical, and physical stability
Electrical resistivity
Dielectric constant
Breakdown strength
Bulk traps

These requirements are now considered in more detail.

The deposited insulator should adhere well to the semiconductor substrate, be relatively scratch resistant, and be compatible with further integrated-circuit processing, such as masking, etching, diffusion, ion implantation, annealing, and metalization. The insulator should also be insensitive to environmental effects, such as moisture, and should be capable of withstanding appropriate thermal and electrical stressing. Although stability is an important requirement, it appears to be a requirement obtained fairly readily with deposited insulators.

The mechanisms responsible for and methods of minimizing dc leakage current in insulators, and thus obtaining high-resistivity dielectrics, have been considered previously by Harrop and Campbell.[12,13] Direct-current conduction can occur by electronic or ionic mechanisms. Electronic conduction may occur by injection of electrons or holes into the conduction or valence bands, respectively. For electrical conduction to occur, carriers must surmount an energy barrier. This barrier is directly related to the band gap of the insulator, E_g. Therefore, the best way to ensure that electronic dc leakage current is minimized is to choose an insulator with a wide band gap. Harrop and Campbell[12,13] have further observed that the insulator band gap increases as the mean atomic number per molecule for an insulator decreases. This implies that wide-band-gap insulators will be composed of relatively small atoms.

Electronic conduction may also occur via hopping transport from one trap site to another or via some other type of tunneling process involving insulator traps. Insulator traps may be caused by impurities or nonstoichiometry.[12,13] Also, some impurities may result in unwanted ionic conduction. Therefore, insulators should be stoichiometric and free of impurities in order to minimize the amount of both electronic and ionic dc leakage.

Insulator morphology is another important consideration for minimizing dc leakage. Insulators should be free of pinholes since they can effectively shunt dc current through the insulator. Polycrystalline dielectrics should also be avoided in favor of amorphous dielectrics since grain boundaries in polycrystalline dielectrics may also provide shunt paths for dc leakage current.

Low dc leakage is an extremely important requirement for insulator/III–V semiconductor applications and is one consideration which has often been overlooked. If the leakage current is too large, electron or hole accumulation may be impossible since these carriers will be drained from the semiconductor surface into the insulator. Also, electrical analysis of devices with leaky dielectrics is inaccurate and often misleading since surface potential and interface state analysis presumes the attainment of quasiequilibrium.[14] This condition is not satisfied when the leakage current is too large.[14] A reasonable criterion for leakage current would be the ability to perform quasistatic capacitance–voltage (C–V) analysis. This requires that the dielectric resistivity be greater than 10^{15} Ω-cm.

Figure 1 illustrates what is often encountered experimentally when characterizing insulator/III–V semiconductor interfaces using the C–V technique. This figure shows that the accumulation capacitance decreases at the higher measurement frequency. This frequency dependence of the oxide capacitance may be caused by the dispersive nature of the real portion of the insulator dielectric constant ε_{ox}, since the oxide capacitance C_{ox} is related to the insulator dielectric constant by

$$C_{ox} = \frac{\varepsilon_{ox} A}{d}$$

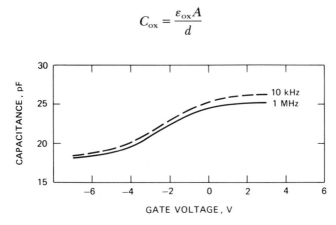

Figure 1. Illustration of frequency dispersion of the dielectric constant of a SiO_2/InP structure.

where A is the capacitor area and d is the insulator thickness. Alternatively, this type of dispersion could also be caused by a high density of interface states which cannot respond to the higher signal frequency.

To distinguish whether the frequency dispersion is caused by the dielectric properties of the deposited insulator or by interface states, the insulator may be deposited onto a metal substrate or a degenerate silicon substrate, which behaves similarly to a metal. Metal–insulator–metal (MIM) structures can then be fabricated and the frequency dispersion of the insulator may be measured directly. In the remainder of this section, we will ignore frequency dispersion caused by interface states and will concentrate on dispersion of the dielectric constant.

The frequency dispersion mechanism of interest for the relatively low-frequency range being considered here is usually termed interfacial polarization.[15,16] Interfacial polarization is associated with either mobile carriers or heterogeneous dielectrics. Electronic or ionic species may be relatively mobile within the dielectric and they may be transported within the insulator by the very low signal frequency. If a dielectric is heterogeneous with two phases of differing conductivity and dielectric constant (the InP anodic oxide composed of In_2O_3 and P_2O_5 is a good example of this), then interfacial polarization may occur due to accumulation of charge at the phase boundaries. The time associated with charge accumulation at the phase boundaries may be from less than a second to minutes so that this type of dispersion is observed only at very low frequencies.

Since interfacial polarization can be caused by heterogeneous insulators, these types of phase-separated dielectrics should be avoided. Many anodic, thermal, and plasma native oxides appear to fall into the category of heterogeneous dielectrics. Ionic and electronic transport can also give rise to interfacial polarization. Since ionic and electronic transport can often be associated with insulator impurities or nonstoichiometry, these should be avoided to minimize the dispersion of ε_{ox}.

A final consideration for minimizing the frequency dispersion of ε_{ox} involves the choice of the insulator band gap. A wider-band-gap insulator will tend to have less electronic and ionic conduction if it is free of impurities and stoichiometric since donor and acceptor traps associated with the transport will tend to be further removed from the conduction- and valence-band edges. Results of Harrop and Campbell[13] are reproduced in Fig. 2 and indicate a correlation for larger band gaps to have lower losses, and hence less frequency dispersion of the dielectric constant. This consideration is further evidence of the importance of choosing an insulator with a wide band gap.

Insulator breakdown is accompanied by appreciable flow of electron or ion current.[13] Therefore, the same considerations which will minimize

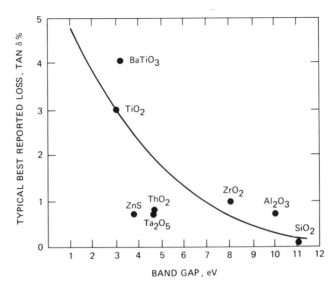

Figure 2. Correlation of the dielectric loss with band gap for various oxides (Ref. 13).

the dc leakage will also be appropriate for obtaining high-breakdown fields. This means that the insulator should have a wide band gap, be free of impurities, be stoichiometric, and also be amorphous rather than polycrystalline. Harrop and Campbell[12,13] have observed that the breakdown field is inversely proportional to the insulator dielectric constant. Further, they have observed that lighter insulators (i.e., the mean atomic number/molecule) have smaller dielectric constants. This suggests that the best dielectric breakdown properties will be obtained in insulators which are composed of relatively light atoms and which have small dielectric constants.

The final bulk insulator property of importance is that of bulk traps. Traps provide a mechanism for electrical conduction and can thus influence the electrical resistivity, breakdown strength, and dielectric dispersion properties of the insulator. Also, bulk insulator traps provide a potential source of electrical interface instabilities, a subject discussed later. Traps existing within the insulator may be associated with impurities, nonstoichiometry, defects, or strain within the insulator. Elimination of these problems by the deposition of contamination-free, stoichiometric insulators may improve the insulator properties markedly. However, it is also possible that some traps are inherent to the particular insulator and cannot be avoided. In fact, it can be shown that point defects, some of which may be active traps, are thermodynamically stable and some solids are actually most stable when they are nonstoichiometric.[17] Therefore, it is important to choose an insulator which does not have "intrinsic" traps located at regions within the band gap at which large amounts of charge exchange can occur.

In summary, the bulk insulator properties required for optimum III–V semiconductor MIS applications are the following:

Wide band gap
Low impurity concentration
Stoichiometric
Homogeneous
Amorphous
Free from traps within the band gap

Perhaps the most important bulk insulator property is band gap. Figure 3 from Harrop and Campbell[13] is a plot of band gap of various oxides as a function of mean atomic number per molecule. Clearly, the oxides with small atomic number produce the large-band-gap materials. Note that only a few oxides, BeO, SiO_2, Al_2O_3, MgO, and P_2O_5 (Table 4), have band gaps greater than 8 eV. Of these, BeO is highly toxic and P_2O_5 is soft and rapidly absorbs water. Therefore, even though there are many oxides, nitrides, fluorides, and inorganic compounds which could be used for insulators, very few will meet the necessary requirement of having a sufficiently wide band gap for III–V MIS applications at room temperature.

For III–V applications it is also very important to choose an appropriate insulator deposition technique. Some of the possible techniques

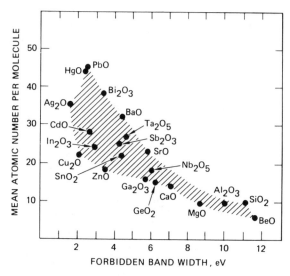

Figure 3. Mean atomic number per molecule vs. band gap for a variety of oxides (Ref. 13).

include:

> Chemical vapor deposition (CVD)
> Plasma-enhanced CVD (PECVD)
> Photon-assisted CVD (PACVD)
> Sputtering
> Thermal evaporation
> Reactive evaporation or sputtering
> Spin on and anneal

The factors for consideration in choosing the deposition technique are:

> What is the required substrate temperature?
> Does the deposition involve a surface or vapor-phase reaction?
> What is the depositing species? (i.e., In the deposition of SiO_2, is Si or SiO incident upon the surface?)
> Do the incident species react with the substrate or its oxides?
> Will the substrate atoms diffuse into the insulator?
> Will the insulator atoms diffuse into the substrate and possibly cause traps or doping states?

The most important insulator deposition parameter is the substrate temperature. High-quality insulators are usually deposited at relatively elevated temperatures. However, higher temperature also increases the rates of oxidation, surface reaction, diffusion, and substrate evaporation. One usually attempts to minimize these processes. Therefore, a compromise temperature is required. Other forms of energy, such as plasma or photon excitation, can be employed to reduce the need for high-temperature deposition. However, there are other effects associated with these excitations which also must be considered. In the following section, other factors influencing insulator/semiconductor interface formation are considered.

4. Interfacial Properties

4.1. Interfacial Reactions

Both thermodynamics and kinetics must be considered when investigating possible reactions between two elements or compounds. Equilibrium thermodynamic calculations provide a prediction of whether a reaction is feasible or not. However, a kinetic barrier may prevent the reaction from taking place. In these cases an excitation or catalyst is required for the reaction to proceed. Heat, light, and plasma ionization are common forms of excitation used to overcome chemical and diffusion barriers. A catalyst provides a reaction path around a chemical barrier, for example, water can be used in conjunction with elevated temperature to allow In_2O_3 to react with P_2O_5 to form $InPO_4$.

The sign of the Gibbs free energy of reaction, G_R, determines whether a chemical reaction is thermodynamically feasible for reactions occurring at constant T and P. If G_R is negative, the reaction may occur while a positive G_R indicates that the reaction should not occur. If several possible reactions are thermodynamically feasible, the reaction with the most negative G_R would be expected to dominate. For reactions occurring in the solid phase at relatively low temperatures, the T and P dependence of G_R may be ignored and G_R may be approximated by the enthalpy of reaction, H_R. H_R may be calculated using the equation

$$H_R = H_f \text{ (products)} - H_f \text{ (reactants)}$$

where H_f's are enthalpies of formation. This analysis neglects the kinetics of the reaction and assumes equilibrium is attained. It is, therefore, not infallible in predicting which reaction will dominate. However, equilibrium thermodynamic arguments have been very successful in accurately predicting the reaction products for various kinds of interfacial behavior.[18,19]

The utility of thermodynamic analysis will be demonstrated by examining which chemical reactions would be predicted to occur during the formation of various insulator/InP interfaces. The enthalpies of formation used in this analysis are listed in Table 2.

Table 2. Enthalpies of Formation of Various Compounds

Compound	H_f (kcal/mole)	References
AlN	−76.0	20
Al_2O_3	−390.0	20
AlP	−39.8	20
$AlPO_4$	−414.4	20
InN	−4.2	20
In_2O_3	−221.27	20
InP	−21.2	20
$InPO_4$	−300.0	21
P_3N_5	−71.4	20
P_2O_3	−196.0	20
P_2O_5	−356.6	20
Si_3N_4	−177.7	20
SiO	−105.0	20a
SiO_2	−217.0	20
SiP	−17.0	20b

First, consider the deposition via homogeneous nucleation in the gas phase of SiO_2 onto an InP substrate. Thermodynamic analysis predicts that SiO_2 in intimate contact with InP is stable and no reaction will occur. However, for many deposition techniques, the formation of SiO_2 proceeds via heterogeneous nucleation in which Si or SiO reacts with oxygen at the semiconductor surface. If Si lands upon the InP surface, thermodynamic analysis indicates that either no reaction will occur or perhaps the Si will replace the In and form SiP (the accuracy of the H_f's is not sufficient to predict this unequivocally). SiO will not react with the InP substrate, either. However, if oxygen is present to react with the Si or SiO and the InP substrate, the most probable reaction products would be SiO_2 and $InPO_4$. In other words, if oxygen is available, the substrate will tend to oxidize and stoichiometric SiO_2 will tend to form.

It is very unlikely that the semiconductor substrate will be clean and unoxidized upon the initiation of the insulator deposition. InP surfaces will tend to be oxidized to $InPO_4$ and the deposited insulator will actually interact with this compound. For SiO_2 deposition onto an oxidized $InPO_4$ surface, no reaction would be expected. However, if Si or SiO nucleate on the $InPO_4$ surface, SiO_2 will tend to form at the expense of the native oxide being reduced. Again, if enough oxygen is available, the interface will tend to be fully oxidized as SiO_2 and $InPO_4$.

Table 3 is a summary of the thermodynamic predictions for SiO_2, Si_3N_4, Al_2O_3, and AlN insulators deposited upon an InP substrate. If the insulator is deposited via homogeneous nucleation in the gas phase directly upon an oxide-free InP surface, no chemical reactions are predicted to occur. However, heterogeneous nucleation at the surface may cause interfacial compounds (i.e. SiP, AlP, P_2O_5, $InPO_4$, SiO_2, In_2O_3) to form, especially if oxygen is available for reaction. When a native oxide is present, various reactions may occur which reduce the native oxide. Notice that the Al compounds are predicted to react with the native oxide, regardless if they are homogeneously or heterogeneously nucleated, while reactions are predicted to occur for the Si compounds only when the deposition occurs via heterogeneous nucleation at the surface. Also, Al depositing species may result in the formation of an interfacial layer of $AlPO_4$. The band gap of this oxide is not known but may be estimated to be ~8–10 eV from Fig. 3. Perhaps it is the formation of this wide-band-gap insulator that is responsible for the early encouraging reports of Al_2O_3/InP interfaces.

If a nitrogen-containing species is used during pretreatment or insulator deposition, it is possible that a thin native nitride layer could be formed at the interface. Thermodynamic analysis has previously indicated (see Table 3) that InN or P_3N_5 formation is feasible for the homogeneous nucleation of AlN on a native $InPO_4$ surface since the

Table 3. A Summary of Thermodynamic Predictions for SiO_2, Si_3N_4, Al_2O_3 and AlN Deposition onto InP

Depositing species	Substrate	Reaction products	ΔH_R (kcal/g-at.)
CASE 1: Homogeneous nucleation/no native oxide			
SiO_2	InP	No Reaction	
Si_3N_4	InP	No Reaction	
Al_2O_3	InP	No Reaction	
AlN	InP	No Reaction	
CASE 2: Heterogeneous nucleation/no native oxide			
Si	InP	No reaction or SiP + In	+1
SiO	InP	No reaction	
Si or SiO + O_2(g)	InP	SiO_2 + $InPO_4$	
Al	InP	AlP + In	−6
Al + O_2(g)	InP	Al_2O_3 + $InPO_4$	−61
		or	
		$AlPO_4$ + In_2O_3	−59
		or	
		Al_2O_3 + P_2O_5 + In_2O_3	−54
CASE 3: Homogeneous nucleation/native oxide			
SiO_2	$InPO_4$	No reaction	
Si_3N_4	$InPO_4$	No reaction	
Al_2O_3	$InPO_4$	No reaction	
AlN	$InPO_4$	$AlPO_4$ + In_2O_3	−4
		or	
		$AlPO_4$ + InN	−5
		or	
		Al_2O_3 + In_2O_3 + P_3N_5	−4
CASE 4: Heterogeneous nucleation/native oxide			
Si	$InPO_4$	SiO_2 + reduce native oxide	
SiO	$InPO_4$	SiO_2 + reduce native oxide	
Si or SiO + O_2(g)	$InPO_4$	SiO_2 + $InPO_4$	
Al	$InPO_4$	Al_2O_3 + InP	−28
		or	
		$AlPO_4$ + In	−13
		or	
		Al_2O_3 + P_2O_5 + In	−11

formation of $AlPO_4$ or Al_2O_3 is energetically favorable. However, it appears that the formation of a native nitride layer at the interface is usually unlikely due to the unreactive nature of nitrogen compared to oxygen. For example, a clean InP surface can react with nitrogen via the chemical reaction,

$$3InP + 2N_2(g) = 3InN + P_3N_5$$

with an enthalpy of reaction of $H_R = -1.5$ kcal/g-at. In contrast, oxygen reacts with a clean InP substrate via the reaction,

$$InP + 2O_2(g) = InPO_4$$

with $H_R = -46.5$ kcal/g-at. The large difference in the H_R's indicates that there is a much larger thermodynamic driving force favoring oxidation over nitration. Indeed, Yamaguchi[22] has investigated the thermal nitration of InP and reports that it is extremely important to reduce the oxidant concentration in the nitration ambient in order to form a nitride rather than an oxide.

For an interfacial nitride to be beneficial to device operation, it should have a large band gap. The band gap of InN is 2.0 eV[23] and the band gap of GaN is 3.34 eV,[23] while it appears that the band gaps of other III–V nitrides have not been reported. The relatively small band gaps of InN and GaN suggest that charge exchange with the semiconductor would be very likely for either of these nitrides and electrical interface instabilities could result.

4.2. Interfacial Oxide

III–V native oxides are discussed in Chapter 7 within this volume and the reader is advised to consult that chapter for details of oxide growth and composition. For the present discussion of deposited insulator/III–V interfaces, it is desired to know under what conditions a native oxide will appear at the interface. The interfaces between SiO_2 and InP and InSb have been examined most extensively and these are discussed later in Section 5.

There are four opportunities available for the formation of native oxides:

1. After wet chemical surface pretreatments and exposure to an air environment.
2. During pretreatment in the deposition reactor. This is particularly important during high-temperature pretreatments in the presence of an oxidizing ambient. However, oxygen is highly reactive and additional oxidation may even occur when only a low partial pressure of an oxidizing species is present in the reactor.

3. During the initial portion of the deposition process, prior to when the insulator has sealed off the semiconductor surface. This is particularly true if the deposition gases include oxygen-containing species that have been excited.
4. During the later stages of the deposition or during high-temperature postdeposition anneals. This can occur via oxygen diffusion through the deposited insular or by chemical reduction of the deposited insulator if it is partially composed of oxygen. Again, high temperatures will tend to increase the possibility of this happening.

Most deposition systems contain significant partial pressures of O_2, H_2O, N_2O, and CO_2. This occurs because most systems are pumped to a background pressure of only 10^{-3}–10^{-6} Torr. Also, some form of oxidant is usually introduced to deposit the insulator, for example, SiO_2 deposition often is accomplished using $SiH_4 + O_2$ or N_2O. In addition, some form of excitation such as heat, light, or plasma is often employed in order to form a good insulating layer.

These two factors—the presence of an oxidant and an excitation—very often leads to the growth of an oxide on the substrate surface. This means that *in situ* removal of the air-grown native oxide does not guarantee that there is no interfacial oxide. This has clearly been shown for InP with pyrolytic and plasma deposition of SiO_2.[24,25] The thickness of the *in situ* grown oxide changes with the conditions within the deposition chamber. This is illustrated for InP in Fig. 4, which shows the profile of the interfacial oxide resulting from positioning the substrate at different distances from an oxygen plasma transfer tube. Photoenhancement also tends to increase the interfacial oxide thickness.

It is not readily apparent whether a thin native oxide is desirable or undesirable for insulator/III-V semiconductor MIS applications. The formation of thick, thermally grown native oxides during processing is clearly to be avoided since thermally grown native oxides have been shown to yield poor MIS devices. However, a thin (less than 20–30 Å) interfacial oxide could conceivably help to grade the interfacial mismatch between the insulator and the semiconductor, thereby minimizing the amount of interfacial strain and defect formation. The native oxide could also possibly act as an interdiffusion barrier or a surface encapsulant to seal the surface during low-temperature processing. On the other hand, the initial oxidation of III-V semiconductors after cleavage in UHV has been shown to cause Fermi-level pinning due to the creation of defects.[26,27] Therefore, it is not clear whether a thin native oxide will actually decrease or increase the interface state density. Also, a thin native oxide may be that source of interfacial oxide traps which could cause Fermi-level pinning and cause electrical interface instabilities.

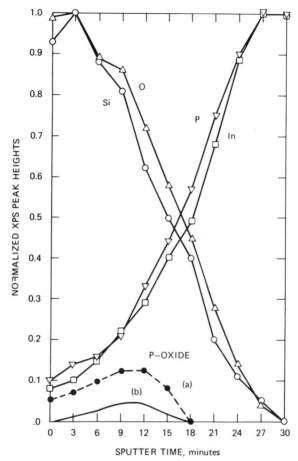

Figure 4. XPS depth profiles of PECVD SiO_2 deposited onto InP. The difference in the phosphorus oxide signal is due to exposure of the InP surface to a direct plasma [curve (a)] and a remote plasma [curve (b)](Ref. 25).

Morrison[28] has reported that mixed surface oxides (the native oxides of InSb and GaAs appear to be mixtures of In_2O_3, Sb_2O_3 and Ga_2O_3, As_2O_3, respectively) tend to be acceptorlike in nature.

Finally, native oxides may have small band gaps which, if the conduction or valence bands are located at the appropriate positions in the energy-band diagram, could also cause Fermi-level pinning and electrical interface instabilities. Table 4 is a list of the band gaps of various oxides which could possibly be present on the InSb, GaAs, or InP surfaces. The band gaps of $GaAsO_4$ and P_2O_5 are estimated by interpolation of the data of Harrop and Campbell[12,13] (Fig. 3) correlating mean atomic number/molecule vs. band gap. These estimates are admittedly crude.

Table 4. Band Gaps of Various Possible Native
Oxides of InSb, GaAs, and InP

Oxide	Band gap (eV)	References
As_2O_3	4.0	29
Ga_2O_3	4.4	29
$GaAsO_4$	~3–7	12, 13
In_2O_3	2.6	30
$InPO_4$	4.5	31, 32
P_2O_5	~8–10	12, 13
Sb_2O_3	4.2	29

This table indicates that, except for P_2O_5, all of the native oxides have relatively small band gaps. The small band gap would imply that charge exchange between the semiconductor and the native oxide conduction or valence bands would be possible.

4.3. Interdiffusion and Impurity Incorporation

Even if a native oxide or nitride does not exist at the insulator/III–V semiconductor interface, it is still unlikely that the interface will be totally abrupt. The electrochemical potential gradient existing at the interface tends to cause interdiffusion between the semiconductor and insulator. Brillson[18] has extensively studied Schottky-barrier formation of III–V and II–VI compounds and has observed that substantial interdiffusion occurs by both metal indiffusion and semiconductor outdiffusion. Conceptually, it is convenient to discuss indiffusion and outdiffusion separately, although they will often be intimately interrelated phenomena.

Indiffusion of the insulator atoms into the semiconductor substrate is associated with atomic or ionic transport via interstitial or vacancy defects. Insulator atoms penetrating into the semiconductor may act as dopants or traps at the semiconductor surface. For example, Si in bulk InP acts as a donor while O in bulk InP acts as a trap.[33] Of course, these atoms may not necessarily play the same role when located at an interface instead of the bulk. This may be particularly true when substantial interdiffusion occurs and the atom can form more complicated defect complexes, such as an impurity–vacancy pair. Indiffusion of the insulator atom could also result in a vacancy in the amorphous insulator network. It is not clear what effect an insulator vacancy would have on the electrical interface properties, but it could conceivably be a source of insulator traps.

Outdiffusion of semiconductor atoms into the insulator could result in the creation of vacancy defects at the semiconductor surface. Bulk cation vacancies (i.e., In, Ga) tend to be acceptors, while anion vacancies (i.e., As, P, Sb) tend to act as donors.[17] Of course, if substantial interdiffusion occurs, these vacancies could form other types of more complicated defect complexes. The outdiffusing semiconductor atoms will be incorporated into the insulator and are another possible source of insulator traps.

The diffusivities of various III–V elements through thermal SiO_2 films have been investigated at elevated temperatures (1100–1200°C).[34] The diffusivities of these elements in SiO_2 have a strong ambient and concentration dependence.[34] At these elevated temperatures, Ga diffuses very rapidly in SiO_2, while B, P, As, and Sb are slow diffusers. It is suspected that these trends will extrapolate to lower temperatures. There is evidence that In diffuses through deposited SiO_2 films at low temperatures.[35] The diffusivity of III–V elements in deposited SiO_2 films may be greater than in high-quality thermal SiO_2 layers.

This brief discussion of interdiffusion indicates that an abrupt interface appears to be most desirable since nonabrupt interfaces seem to be associated with semiconductor and insulator trap formation.

Another topic important to the interfacial layer is impurity incorporation. As mentioned previously, it is not clear that impurity incorporation at the interface need always be bad. However, for an impurity to enhance the electrical interface properties the impurity should either help grade the interfacial mismatch, neutralize insulator and/or semiconductor traps, or prevent interdiffusion. It is not clear which impurities, if any, will be successful in accomplishing these goals.

Table 5 is a list of how various bulk impurities act in InP. If these impurities act similarly at an InP interface, it would appear that Bi, Cr, Fe and O would be unwanted at an interface since they would lead to interfacial traps. Also, from previous considerations regarding insulator stoichiometry, it would appear that impurities would usually be undesirable within the insulator. Therefore, although there may be exceptions to the rule, it seems advisable to minimize the amount of interface contamination.

Table 5. The Dopant Type of Various Impurities in Bulk InP[33]

Type of dopant	Impurity
Acceptor	Be, C, Cd, Cu, Mg, Mn, Zn
Donor	Ge, S, Se, Si, Sn, Te
Trap	Bi, Cr, Fe, O

4.4. Surface Evaporation

Preferential evaporation of the column V element above the incongruent evaporation temperature should always be avoided. This can be understood in terms of the defect model[26,27] which asserts that defects at the semiconductor surface are responsible for interface states within the band gap and prevent the surface potential from being modulated by an applied gate voltage. Preferential evaporation of the column V element can create vacancy defects which are thought to be the source of Fermi-level pinning for a variety of III–V semiconductors. The evaporation of the column V element also leaves a residue of metallic In and/or Ga which can have a strongly degrading effect on the surface.

The oxides of the column V elements, As_2O_3, Sb_2O_3, and P_2O_5, are volatile, although $InPO_4$ is not.[36] The air-grown native oxide of InP is $InPO_4$ and thus thermal treatments in vacuum of the air-exposed InP surface do not greatly change the surface until the incongruent melting point (365°C)[8] is exceeded. The evaporation of the InP surface proceeds at a very slow rate ($\leq 10^{-5}$ monolayers/s below 365°C) until the temperature exceeds 500°C when this thin oxide ruptures and the underlying substrate rapidly evaporates.[37] The air-grown oxides of GaAs contain Ga_2O_3 and As_2O_3.[38] Heating in vacuum above ~475°C desorbs the As_2O_3 but leaves the Ga_2O_3.[38] Various chemical treatments can leave different surface oxides as described in the chapter on oxidation.

4.5. Energy of the Depositing Molecules

The energy associated with the insulator deposition process is another important consideration. Table 6 is a summary of estimates of the enthalpy of bulk defect formation for InSb, GaAs, and InP. Every defect will not necessarily lead to an interface trap within the semiconductor band gap. However, theoretical considerations have predicted that both vacancies[41] and antisite defects[42] can result in traps within the band gap.

Table 6. Van Vecten's[39,40] Estimates of the Enthalpy of Formation (eV) of Various Bulk Defects of InSb, GaAs, and InP

Semiconductor	$H(V_A^x)^a$	$H(V_B^x)^b$	$H(B_A A_B^x)^c$	$H_0(B_A)^d$	$H_0(A_B)^e$
InSb	2.12	2.12	0.54	0.27	0.27
GaAs	2.59	2.59	0.70	0.35	0.35
InP	3.04	2.17	1.30	0.42	0.89

$^a H(V_A^x)$ = enthalpy of formation of a neutral cation vacancy.
$^b H(V_B^x)$ = enthalpy of formation of a neutral anion vacancy.

To prevent the formation of semiconductor surface defects, and hence interface states, it is important to choose an insulator deposition technique which imparts only a small amount of energy into the surface. Table 7 lists typical energies associated with various types of deposition. Thermal evaporation and CVD appear to be appropriate choices while sputtering techniques would be expected to create defects which would prevent modulation of the surface potential.

Returning to Table 6, notice that these are bulk estimates. It is likely that the enthalpy of formation of surface or interface defects is even smaller.[43] Therefore, it would be expected that virtually any deposition process could cause defects to be formed. In fact, it apppears that even the enthalpy of adsorption of a metal on a semiconductor surface is sufficient for defect formation.[43] It appears, therefore, that while the deposition energy should be chosen to avoid the formation of large numbers of surface defects, even the most benign technique will result in some defects.

Since defects appear to be unavoidable (from Table 6 it would appear that antisite defects are particularly unavoidable), the best approach appears to be to choose a semiconductor which has defects located at energy positions within the band gap where they can do the least amount of harm. For example, it appears that the two main defects associated with GaAs interfaces[26,27] are located near midgap and prevent electron and hole accumulation. In contrast, the location and density of the predominant defects of InP[26,27] allow electrons to be readily accumulated at the surface of both n- and p-type InP.

One final consideration for choosing an insulator is that the lattice parameters associated with the insulator should be capable of matching to the semiconductor in an abrupt manner which would minimize the amount of interfacial strain and lessen the likelihood of defect creation.

Table 7. The Energy Associated with Various Deposition Techniques

Type of deposition	Energy (eV)
Thermal evaporation	0.05–0.2
CVD	0.1–1.0
Ion-beam sputtering (substrate)	~2–4 with tails to 20–30
Ion-beam sputtering (target)	50–2000
Diode sputtering	10–2000
Magnetron sputtering	10–2000

$^c H(B_A A_B^x)$ = enthalpy of formation of a neutral antistructure pair.
$^d H(B_A)$ = enthalpy of formation of an antisite defect, anion-on-cation site.
$^e H(A_B)$ = enthalpy of formation of an antisite defect, cation-on-anion site.

4.6. Interfacial Trapping and Instabilities

There are three typical tests which indicate the presence of electrical interface instabilities:

1. Hysteresis in the $C-V$ characteristics of MIS capacitors.
2. Hysteresis in the MISFET curve-tracer current–voltage characteristics.
3. Long-term drift in the drain voltage of a MISFET after the application of a gate voltage.

These electrical instabilities are usually associated with one of three sources:

1. Insulator polarization.
2. Ion drift.
3. Charge injection into the interface.

The first two sources of instability shift the $C-V$ curves in the same direction and in the opposite direction to that of the third source. While insulator polarization and ion drift are important problems for some insulator/III–V semiconductor systems, they are problems which have often been eliminated by careful device processing and will not be considered further in this section.

However, charge injection into the interface is probably the single most important problem plaguing most insulator/III–V semiconductor devices and is the one issue still needing to be resolved before these devices can become commercially viable.

This type of charge injection from the semiconductor into the interface was considered in detail by Heiman and Warfield,[44] who attribute the long-term drift observed as $C-V$ hysteresis of silicon MIS capacitors as being caused by tunneling of carriers into oxide traps. They also distinguish between "fast states" located directly at the oxide/semiconductor interface and "oxide traps" which are located within the oxide itself. Heiman and Warfield find that the time constant τ for a trap to capture an electron directly at the oxide/semiconductor interface is

$$\tau_t (x = 0) = \frac{1}{\sigma_n \bar{v} n_s}$$

while the time constant for an oxide trap at a distance x away from the interface to capture an electron is

$$\tau_t(x) = \tau_{t0} \, e^{2K_0 x}$$

where σ_n is the capture cross section for a trap directly at the oxide/semiconductor interface, $\bar{v}$ is the electron thermal velocity, n_s is the electron concentration at the oxide/semiconductor interface, and K_0 is the decay constant associated with tunneling. Notice that there are two physical sources which may be responsible for a very long interface trap time constant. First, if the electron surface concentration is small, as it is in depletion (especially for wide-band-gap semiconductors), then the trap time constants will become large. Second, if there are oxide traps which can trap electrons deep within the insulator, the time constant associated with this process will also be very large.

It is important to note that the $C-V$ hysteresis obtained with MIS capacitor structures may be fundamentally different than that observed with MISFET structures. In the $C-V$ measurement, a gate bias is applied which drives the semiconductor surface into depletions where depopulation of interface traps is an inherently slow process, even without invoking the exponential dependence of the cross section with distance into the oxide, due to the small n_s. For example, for an InP midgap trap located directly at the interface, $\tau \sim 80$ s (assuming $\sigma_n = 10^{-16}$ cm^2). In contrast, an InP MISFET operates in an electron accumulation mode with $n_s = 10^{17}$ cm^{-3} with an associated time constant of $\tau \sim 10$ ns. Therefore, within the framework of the Heiman–Warfield model, long-term MISFET drift must be associated with traps located away from the insulator/semiconductor interface.

The Heiman–Warfield model was extended by Koelmans and DeGraaff[45] to explain the slow decay of drain current for thin-film FETs. However, this model does not explain the temperature dependence of the drift observed in InP MISFETs. Fritzsche[46] suggested that this temperature dependence may be explained by assuming that only thermally excited electrons capable of surmounting an energy barrier can tunnel into oxide traps. Therefore, according to this model, some trapping states are located above the InP conduction band. These electrical instability models for InP MISFETs are discussed more fully in the experimental results section.

In summary, long-term instabilities may be due to one of the following:

1. Depletion of the semiconductor surface and the associated decrease in n_s.
2. The physical separation of the traps away from the insulator/semiconductor interface which decreases the capture cross section exponentially as a function of distance.
3. The activation of carriers over an energy barrier which implies that fewer carriers are able to interact with traps.

Now that the basic mechanisms responsible for this type of electrical instability have been outlined, it is appropriate to inquire into how the insulator chemistry may be associated with this instability. It will initially be assumed that no native oxide or interdiffusion layers exist at the interface, but that the insulator and semiconductor are abruptly matched.

As pointed out previously, insulator traps may be associated with impurities, nonstoichiometry, defects, or strain within the insulator. Elimination of these problems by the deposition of contamination-free, stoichiometric insulators may help to minimize electrical interface instabilities, but it is also possible that oxide traps are "intrinsic" to the insulator and cannot be avoided. If "intrinsic" insulator traps exist at energies corresponding to within the band gap of the semiconductor substrate, electrical instabilities may be impossible to prevent (again, assuming an abrupt insulator/semiconductor interface).

Figure 5 is a comparison of the energy-band diagram of InP with the energy-band diagrams and the reported trap locations of SiO_2, Al_2O_3, and Si_3N_4.[47-51] This diagram is meant to indicate only the approximate insulator trap locations with respect to the semiconductor band gap. It is not clear whether the traps are "intrinsic;" also the band-gap and trap

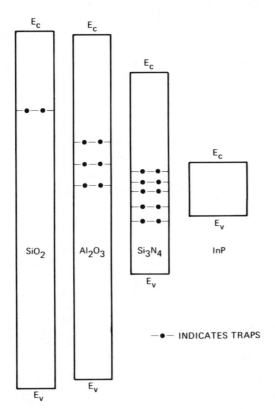

Figure 5. Approximate energy-band alignment positions and trap locations for SiO_2, Al_2O_3, and Si_3N_4 with respect to InP (data from Refs. 47–51).

locations may be dependent upon the deposition technique and the exact insulator composition. This is particularly true of Si_3N_4 which experimentally has been found to be silicon oxynitride in composition.[52]

Analysis of Fig. 5 suggests that SiO_2 is a good candidate for insulator/InP interfaces with small amounts of electrical interface instabilities since it has only one reported trap level and this level is located quite far from the semiconductor band gap. Al_2O_3 appears to have at least three bulk traps which are located within or near the band gap of InP. Therefore, unless some type of layer grades the interface, electrical instabilities would be expected to be associated with Al_2O_3 traps. Si_3N_4 appears to be an even worse candidate from this viewpoint since it has five reported trap levels, all within or near the InP band gap.

Now consider the possibility of having native oxide or interdiffusion present at the insulator/semiconductor interface. If the native oxide has a small band gap, it is more likely that the oxide conduction and/or valence band will be near enough in energy to that of the semiconductor that charge exchange may occur. This is one possible mechanism for electrical instabilities and can be avoided if the native oxide possesses a large band gap or a band gap which does not align with the semiconductor band gap. Another possibility is that the native oxide has some intrinsic (or extrinsic) traps which appropriately align with the semiconductor band gap to cause instabilities. It is very difficult to modify the native oxide chemistry, and hence its band structure, in any sort of controllable fashion. Hence, to a certain degree, these are intrinsic properties which depend exclusively on the semiconductor substrate chosen.

The effects of interdiffusion have been discussed previously with respect to trap formation. These traps could conceivably be located at appropriate locations in distance and energy to cause electrical interface instabilities.

5. Experimental Results

5.1. InSb

Eaton et al.[53] performed surface conductance measurements on p-type InSb at 77 K. The InSb surfaces were prepared by electrolytically etching in nitric acid and ethylene glycol. The surfaces were observed to be p-type after the surface treatment. They found that if the InSb sample was illuminated with white light at 77 K, the surface conductance measured in the dark after illumination was increased. This increase in the conductance persisted for at least three hours if the temperature was maintained at 77 K. However, if the illuminated sample was allowed to warm to room temperature and was then cooled again, the surface conductance would return to its previous value. They attributed this effect as being caused by electron excitation into the native oxide and subsequent trapping at localized states in the oxide. By charge neutrality, this negative trapped charge in the oxide induces the surface to be more p-type and hence the measured surface conductance is increased. Upon heating to room temperature, the electrons trapped in the native oxide are apparently thermally emptied from the traps.

Pagina[54,55] pursued this low-temperature, light-induced trapping of electrons further. He employed surface photoconductivity and field-effect measurements of p-type InSb surfaces etched with CP-4. His measurements were performed at temperatures of less than 100 K. He found that photons with energies greater than ~2 eV induced the same quasistable changes in the surface conductivity as were observed by Eaton et al.[53] He also found that photons of energy ~0.6 eV induced a change in conductivity which decayed with a time constant on the order of minutes. In agreement with Eaton et al.,[53] he interprets his results in terms of a native oxide which traps electrons and these electrons are "frozen" into the native oxide at temperatures less than 100 K. Figure 6 illustrates Pagina's model of a "real" InSb surface. Traps located near the native oxide conduction band are responsible for the quasistable

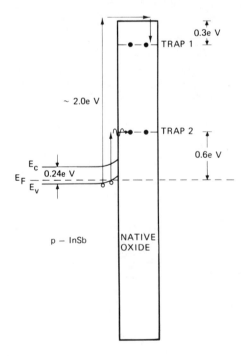

Figure 6. Pagina's model of a "real" InSb surface (Refs 54 and 55).

changes in the surface conductivity. Photons with ~2-eV energy induce electronic transitions from the InSb valence-band maximum to the native oxide conduction band and some of these electrons are trapped in the native oxide. The 0.6-eV photons excite electrons from the InSb valence-band maximum into the lower-energy native oxide traps. It is not totally clear why the higher-energy trap should induce a quasistable change in the surface conductivity, while the lower trap causes a reversible change. Pagina[55] suggests that perhaps the lower-energy traps are located closer to the InSb surface, while the higher-energy traps are predominantly filled at the outer native oxide surface.

Davis[56] measured the conductivity, photoconductivity, and field-effect mobility of etched (111) InSb surfaces at 77 K. He discovered that rinsing the InSb in a very dilute solution of Na_2S after etching reduced the surface state density by two orders of magnitude. He believes that the Na_2S rinse removes metal ions from the InSb surface and that these metal ions are responsible for the high surface state density. After the Na_2S treatment, he found two discrete surface states located near the valence- and conduction-band edges. The surface state density near the conduction band was estimated to be in the low $10^{11}/cm^2$ range, while the density was higher near the valence band. The surface state density was substantially lower within the midgap region.

Huff et al.[57,58] studied etched InSb surfaces using ac field-effect measurements at 113 K. Both p-type and n-type surfaces with 100, 110, 111A, and (111)B orientations were studied using a 2:2:3 mixture of $HF:HNO_3:H_2O$ as an etchant. Their results are in basic agreement with Davis[56] in that two discrete surface states were detected near to the conduction- and valence-band edges. These two states were always present, independent of surface orientation. All of these surfaces were found to be p-type and the amount of p-type surface character was directly related to the density of the surface state near the valence-band edge. This state was thereby concluded to be acceptorlike. They also observed that the density of surface states for surfaces cleaved in air was very similar to that observed with etched surfaces. They conclude from these observations that the surface state density is basically a property characteristic of the substrate material and relatively insensitive to surface treatment.

Sewell and Anderson[59] investigated the slow decay of the conductivity of SiO_x/InSb thin-film FETs. The SiO_x was deposited by evaporation of SiO. They found that the conductivity decay was nonexponential and was associated with tunneling of electrons into interfacial traps located within ~20 Å of the interface. By varying the insulator preparation parameters, they determined that the traps were associated with the stoichiometry of the insulator, rather than impurity related.

Aspnes and Studna[60] and Aspnes[61] applied spectroscopic ellipsometry to the study of chemically prepared InSb surfaces. He found that HCl, $NH_4:H_2O$, or $HCl:CH_3OH$ could effectively remove native InSb oxides, but that an overlayer still remained after polishing and stripping treatments. This overlayer apparently consists of a microscopically rough bulk material, possibly consisting of some excess Sb. The highest-quality surfaces were obtained after polishing in a dilute solution of bromine–methanol with no subsequent stripping rinses. He believes that methanol- or water-based stripping solutions react with the InSb surface.

Iwasaki et al.[62,63] have studied the room-temperature oxidation of InSb by XPS. They found that a moist environment enhances the rate of InSb oxidation and that the chemical shifts caused by wet oxidation are larger than that observed after dry oxidation. They report that the dry oxidation products are In_2O_3 and Sb_2O_3 over the temperature range of 20–250°C, but they do not identify the chemical composition observed after wet oxidation.

The native oxide present after bromine–methanol polishing, etching in a solution of lactic acid, HNO_3, and HF, and dipping in HF was characterized by Vasquez and Grunthaner.[64,65] This native oxide was found to be ~15–20 Å thick with a composition of In_2O_3 and Sb_2O_3 in a 3:1 ratio. The InSb substrate appeared to be stoichiometric, having an In to Sb ratio of 1:1.

Anderson et al.[66] applied a variety of surface science techniques to the characterization of SiO_xN_y layers deposited upon InSb. The SiO_xN_y was deposited by CVD at 250°C using SiH_4, O_2, and NH_3. The N concentration in the insulator was found to be about 5%. They detected In throughout the entire SiO_xN_y film, regardless if it was deposited upon InSb or Si substrates, although more In was found in the samples with the InSb as a substrate. When silicon nitride layers were coated on the back of the InSb substrates prior to the insulator deposition, the amount of In detected in the insulator film decreased significantly. Apparently 250°C was sufficient for the InSb back surface to thermally decompose, thus contaminating the deposited layers.

Langan and Viswanathan[67,68] conducted a detailed analysis of the SiO_2/InSb interface. SiO_2 was deposited onto InSb substrates by the pyrolytic decomposition of silane and oxygen. They found that when the SiO_2 was homogeneously nucleated in the gas phase, the electrical interface properties measured at 77 K were virtually ideal. However, when the SiO_2 was heterogeneously nucleated at the InSb surface, the electrical interface properties were substantially degraded, with more $C-V$ hysteresis and a larger interface state density. They suggested that the preservation of a thin native InSb oxide was crucial for obtaining these nearly ideal electrical properties. Heterogeneous nucleation was believed to reduce the native oxide and thereby degrade the electrical properties. Langan[67] suggests that the ultimate InSb interface would consist of a thin chemisorbed oxygen monolayer which grades the insulator/semiconductor transition while avoiding the back-bond breaking of the InSb substrate which is associated with oxidation.

Vasquez and Grunthaner[64,65] analyzed in detail the $SiO_2/InSb$ interfaces of Langan[67] which were deposited at 220°C via a heterogeneous surface reaction. They found that some additional thermal oxide (in addition to the oxide present after chemical etching) grows during the initial portion of the SiO_2 deposition yielding an outer In_2O_3 layer with some excess elemental Sb near the semiconductor substrate (see Fig. 7). They also found that the thin outer In_2O_3 layer was indeed reduced and explained this reduction via the chemical reaction

$$3SiH_4 + 2In_2O_3 \rightarrow SiO_2 + 6H_2 + 4In$$

They suggest that it is the reduction of this native oxide (leaving elemental In) and the thermal oxidation of the InSb (leaving elemental Sb) that may be responsible for the poorer electrical interface properties observed by Langan[67] with interfaces produced by heterogeneous nucleation.

In characterizing the bulk properties of CVD SiO_2 layers, Langan[67] found from infrared absorption measurements a small amount of Si_2O_3 bonding in the "as-deposited" films. Apparently, this Si_2O_3 causes frequency dispersion of the oxide capacitance. This dispersion is reduced substantially by annealing. Langan[67] also showed, by $C-V$ analysis,

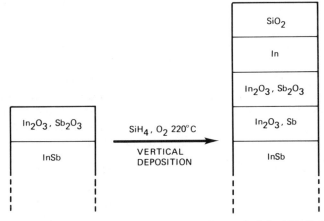

Figure 7. A model of the chemically etched InSb surface and of the $SiO_2/InSb$ interface after heterogeneous deposition of the SiO_2 (Refs. 64 and 65).

evidence of slow states in the "as-deposited" films. A vacuum bakeout for at least one hour was necessary to avert this problem. Water vapor was suspected to be the source of this problem since the CVD SiO_2 layer is known to be very porous and sensitive to moisture. Another solution to this problem was to sputter ~500 Å of SiO_2, which is not moisture sensitive, onto the CVD SiO_2.

The use of Al_2O_3 and SiO_2 as encapsulants for InSb was investigated by Schmid et al.[4] The SiO_2 was pyrolytically deposited at 320°C using silane and the Al_2O_3 was also pyrolytically deposited at 200°C using aluminum methylate. Backscattering measurements indicated that In or Sb outdiffuses into the Al_2O_3 at 500°C, while no difference could be detected between the annealed and unannealed SiO_2 sample.

5.2. GaAs

Flinn and co-workers[69-71] investigated the properties of (111) p- and n-type GaAs surfaces using surface photovoltage, field-effect, and photoconductance measurements. Both the 111A and 111B faces were examined and several chemical etches were employed, although no large differences in the surface properties could be detected. They were not able to observe a conductance minimum and conductance changes decayed to zero in less than a second after applying a dc voltage to the gate electrode. This implied that the total density of trapping states (both "fast" and "slow" states) was greater than 10^{12} cm^{-2}. The chemically etched GaAs surface was always found to be depleted with no applied bias. The barrier height for n-type GaAs surfaces was ~0.45 eV, while the barrier height for SiO_2/GaAs samples was 0.2–0.3 eV, independent of surface etchant and orientation. Etched surfaces had a higher density of "slow" states than the SiO_2/GaAs sample.

The etched surface of GaAs was investigated by Valahas et al.[72] using the pulsed-field-effect and photovoltage techniques. A conductance minimum could not be achieved with the field-effect measurements. Photovoltage measurements indicated that n-type GaAs surfaces were always depleted and saturation of the photovoltage could not be achieved. Similar results were obtained for different GaAs surface orientations, etchants, and gaseous ambient exposures. These measurements were interpreted in terms of a discrete trap state located at midgap with a large electron cross section.

Lagowski et al.[73] studied etched GaAs surfaces using surface photovoltage spectroscopy and the surface piezoelectric effect. A variety of etchants were used with n-type GaAs substrates. The GaAs surface was always found to be depleted with a barrier of about 0.55 eV. Surface state peaks were located ~0.72 and ~0.9–1.0 eV below the conduction band.

The surface potential position and surface state densities were relatively independent of surface etchant procedures and gaseous ambients.

Kreutz and Schroll[74] also investigated the etched GaAs surface. They used ac and pulsed-field-effect measurements and found that different etches yielded essentially the same results. The ac field-effect measurements were characterized by hysteresis, frequency dispersion, and the inability to obtain a conductance minimum. The pulsed-field-effect measurements also showed no conductance minimum. They interpret these measurements as indicating that the etched GaAs surface has a very high density of interface states which keep the surface potential pinned in depletion. Interface traps are responsible for the observed hysteresis and frequency dispersion.

Chang et al.[75] used AES and XPS to study the chemically etched, air-exposed GaAs surface. HCl- and NH_4OH-based solutions were both used for the chemical etching. The native oxide was identified to be Ga_2O_3 and As_2O_3 and was estimated to be ~10 Å thick. Carbon and oxygen were the dominant surface contaminants and S, Cl, Ca, and N were also sometimes detected on the GaAs surface in the 1% range after wet chemical etching. NH_4OH-etched surfaces appeared to have less contamination than HCl-etched surfaces.

Oda and Sugano[76] also studied chemically etched GaAs surfaces using AES. They also found carbon and oxygen to be the dominant surface contaminants. For HNO_3 and HNO_3:HF (3:1) etching solutions, there were large amounts of oxygen and small amounts of carbon on the surface while a HNO_3:HF (25:1) solution yielded more carbon but less oxygen.

Aspnes and Studna[77] used spectroscopy ellipsometry to characterize etched GaAs surfaces which were maintained in flowing N_2 after etching. They obtained their best results with HCl- or NH_4OH-based solutions to remove native oxides, a final Br-MeOH polish in a dilute (0.05%) solution after polishing in a stronger Br-MeOH solution, and stripping with Br-MeOH and H_2O.

Grunthaner et al.[78] used XPS to characterize the surface chemistry of GaAs oxidized in pure oxygen at room temperature. They found that this native oxide is composed of an oxide with an outer Ga to As surface concentration ratio of ~2:1, with an interfacial ratio of ~10:1. Elemental As is also detected at the native oxide/GaAs interface. They show evidence of two Ga oxide bonding states nearer to the semiconductor.

Iwasaki et al.[62,63] applied XPS to the study of the initial oxidation of GaAs. They found that water vapor enhances the rate of GaAs oxidation. However, the XPS chemical shifts indicate that oxidation of GaAs with wet or dry ambients yields the same native oxide chemical species, Ga_2O_3 and As_2O_3.

Various dielectrics have been investigated for compatibility with GaAs as an MIS insulating layer. These dielectrics include SiO_2,[79] Si_3N_4 (usually SiO_xN_y),[80-86] SiO_x,[87] Al_2O_3,[88] Ge_3N_4,[89,90] GaO_xN_y,[91] and Ta_2O_5 oxidized metal films.[92] These dielectrics have been deposited in a variety of ways including vacuum evaporation, CVD, PECVD, ion-beam sputtering, and rf ion plating. Although this represents an impressive variety of insulators and deposition techniques, the electrical properties obtained are usually remarkably similar. The interface state densities are typically quite high, the surface potential is limited to depletion, and electrical interface instabilities are invariably observed. Also, even though a substantial amount of work has been reported concerning different insulator deposition techniques and electrical characterization, very little work has been reported concerning the interfacial chemistry of these structures. The small amount that has been reported will now be reviewed.

Suzuki et al.[93] suggested that the native oxide of GaAs was responsible for the very high density of interface states which causes frequency dispersion of the apparent oxide capacitance. These conclusions were based upon their observation that GaAs surfaces which were rf sputter treated in N_2 prior to the deposition of GaO_xN_y dielectric layers had much less frequency dispersion of the apparent oxide capacitance than samples with GaO_xN_y in which the native oxide was not intentionally removed. However, although these results do imply improvement of the deposited insulator interface, they are not totally conclusive since the C-V analysis was performed at a maximum frequency of 500 kHz. It has been demonstrated that fast interface states on GaAs may respond to frequencies in excess of 1 MHz.[82]

Bayraktaroglu and Johnson,[85] and Clark and Anderson[86] also provide evidence that the native GaAs oxide degrades the electrical interface properties. Both groups employed PECVD silicon nitride as an insulator. Bayraktaroglu et al.[85] found that a H_2 or N_2 plasma treatment prior to the Si_3N_4 deposition seems to reduce the interface state density. Clark and Anderson[86] made the same observation and provide AES-depth-profiling evidence (Fig. 8) that the electrical interface properties do indeed correlate with a reduction of oxygen at the interface. Their best results were obtained using a H_2 plasma pretreatment and a 600°C postdeposition anneal. They suggest that the best dielectric/GaAs interfaces should be oxygen free and point out that this may be associated with the fact that oxygen is a deep trap in bulk GaAs.

Friedel and Gourrier[94] have investigated GaAs surfaces prepared by in situ H_2 and N_2 plasma treatment. They find that a short H_2 plasma treatment reduces C and N contamination left from wet chemical cleaning to levels below the Auger detection limit. Oxygen contamination is also reduced, although not completely eliminated. Surface photovoltage

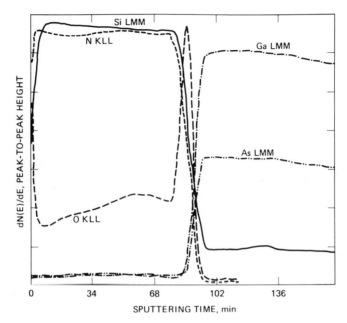

Figure 8. Auger depth profile of a Si_3N_4/GaAs sample in which the interfacial oxide was not completely removed during plasma processing (Ref. 86).

measurements indicate that the slow transients caused by long-time-constant states disappear after H_2 plasma processing. N_2 plasma treatment after H_2 plasma cleaning is characterized by two time regimes. For very short times (<10 s), a thin nitride layer appears to form. For exposures greater than 1 min, the nitride continues to grow and there is an associated depletion of As concentration at the surface.

Konig and Sasse[95] have employed XPS depth profiling to study the annealing behavior of CVD SiO_2/GaAs interfaces. The CVD SiO_2 was deposited at 427°C and the samples were annealed for 4 h at 667°C with flowing N_2. The unannealed samples exhibited very abrupt interfaces. The annealed samples, however, exhibited a much wider interfacial region consisting of Ga_2O_3, free As, and a small amount of As_2O_3. Also, some Ga was detected at the outer SiO_2 surface. The amount of Ga in the SiO_2 appears to vary with the GaAs substrate orientation. It was not determined whether the interfacial oxide formed due to the decomposition of the CVD SiO_2 or due to the unintentional incorporation of an oxygen partial pressure in the annealing ambient.

Grant et al.[96] studied a variety of deposited insulator/GaAs interfaces using $C-V$ and XPS analysis. They also conclude that oxygen-free insulators hold the most promise for GaAs MIS applications. They point out two major problems associated with oxides for GaAs MIS applications. First, photoemission measurements indicate that even a submonolayer coverage of oxygen onto a cleaved GaAs surface is sufficient to create enough defects to pin the Fermi level. Second, the GaAs native oxide is thermodynamically unstable and the following reaction tends to occur, yielding elemental As at the interface.

$$GaAs + As_2O_3 \rightarrow Ga_2O_3 + 4As$$

The existence of elemental As may be a major source of interface states. Therefore, Grant et al.[96] conclude that the most promising method of passivating GaAs surfaces would be to employ an *in situ* treatment to remove GaAs native oxides and to subsequently deposit a nonoxide insulator onto the freshly cleaned GaAs surface. They suggest that AlN is a good candidate since it has a wide band gap, has nearly the identical thermal expansion coefficient as GaAs, Al_{Ga} and N_{As} defects are electrically inactive, and AlN is also a III–V compound.

Several groups have studied the encapsulation properties of insulators on GaAs.[97-100] The annealing temperatures typically employed were 700–900°C and He backscattering, AES, and SIMS were used to study interdiffusion at the interface. These studies indicate that Ga outdiffusion and buildup at the interface are likely to occur if SiO_2 or SiO_xN_y are used as encapsulants. However, if AlN (actually AlO_xN_y) or Si_3N_4 (with only a small concentration of oxygen) are used as encapsulants, little or no interdiffusion is observed.

5.3. InP

Williams and McGovern[37] applied a variety of surface science techniques to the characterization of chemically etched n-type InP surfaces. Carbon and oxygen were observed to be the predominant surface contaminants. They were found to be strongly bonded to the surface and could not be removed by heat treatment until the InP surface actually decomposed of temperatures in excess of 500°C. Surface photovoltage measurements indicated that the chemically etched, n-type InP surface was slightly depleted with a band bending of about 0.1 V.

Clark et al.[101] used XPS to study various chemical etchants for 100 InP surface preparation. They found that mixtures of inorganic acids and H_2O_2 yielded surfaces which were phosphorus-rich, contaminated by substantial amounts of hydrocarbons, and predominantly oxide free, although the thin residual oxide is phosphorus oxide rich with almost no indium oxide. Br–MeOH, $Br_2HBr–H_2O_2$, or $NaOH–H_2O_2$ etchants were found to leave the surface much more oxidized than when inorganic acids were used.

Bertrand[102] also applied XPS to the study of chemically etched InP surfaces. She finds that HNO_3 leaves a surface much more oxidized than H_2SO_4, HCl, or Br etchants, which result in only a small amount of phosphorus oxide at the surface.

Clark et al.[101] and Bertrand[102] both observe that the chemical shifts associated with these native oxides are not consistent with In_2O_3 or P_2O_5, but are more characteristic of $InPO_4$, $InPO_4(xH_2O)$, $In(OH)_3$, or PO_4^{-3}.

Wager et al.[103] investigated the composition and thermal stability of bromine–methanol etched InP surfaces with XPS and UPS. The as-etched surfaces appeared to have a very thin surface oxide layer which was bonded predominantly as a phosphate. Heat treatments in a UHV environment contaminated with CO and H_2O vapor indicated that a significant amount of additional native oxide may grow in the temperature range 100–500°C, even when the partial pressure of the oxidizing species is very low. Room-temperature-grown native oxides were found to be thicker if exposed to a humid environment.

SiO_2,[1,46,104–111] Si_3N_4,[1,111] Al_2O_3[112–115] and Ge_3N_4[116] have been investigated as possible deposited insulators for InP MIS applications. The best electrical properties to date appear to have been achieved with SiO_2 or Al_2O_3. However, the electrical characteristics of InP MIS devices in general appear to be relatively similar and are invariably much superior to GaAs. InP devices typically have a relatively low density of interface states in the upper portion of the band gap which allows the surface potential to be readily modulated in this region, resulting in electron accumulation at the InP surface. Since the electrical properties of InP devices is reviewed in detail in another chapter of this volume, the work presented here will focus mainly on the interface chemistry.

Fritzsche[46,110] was able to substantially reduce the fast interface state density near the conduction band by adding HCl to the gas stream while depositing CVD SiO_2 onto InP. He believes that this reduction may either be due to a modification of the native oxide or that the HCl could be actually modifying the SiO_2, similar to HCl improvement of the thermally grown SiO_2 interface. Fritzsche also provides evidence of deep donor states located about 0.5 eV below the conduction-band edge which prevent the surface potential from being modulated below midgap. These states were not effected by either the addition of HCl or by annealing treatments. He points out that since these states are located near midgap, they will have long time constants but that n-channel MISFETs will not be influenced by these states as long as the device is not biased far enough into depletion to discharge these states. The one major problem associated with MISFETs reported by Fritzche was that of drain current drift as a function of time. This drift was not observed at 77 K. He attributed the temperature dependence of this drift to tunneling of thermally excited electrons in the distribution tail into oxide traps located at energies above the conduction-band edge. At low temperatures, the electron density in the high-energy distribution tail is very small and only very few electrons can tunnel into oxide traps. These oxide traps are believed to be located within the native InP oxide itself. Figure 9 illustrates Fritzsche's model of the SiO_2/InP interface.

Stannard[109] compared $C-V$ curves at 77 K of CVD SiO_2/InP samples deposited at 340 and 370°C. He found substantial amounts of $C-V$ hysteresis for samples deposited at 370°C, whereas samples deposited at 340°C showed virtually no $C-V$ hysteresis at 77 K. He interprets these results to indicate that decomposition of the InP surface above 350°C causes contamination of the insulator and interface trapping.

Wager and Wilmsen[5] have observed that the rate of thermal oxidation of InP increases dramatically above about 350°C. Also, Farrow[6] has found that heating in vacuum above 365°C results in a disproportionate loss of P_2 from the InP surface. Either of these mechanisms (or both) could be responsible for the degradation of the electrical properties of the CVD SiO_2 InP interface observed by Stannard[109] above 350°C.

The degradation of the electrical properties of CVD SiO_2/InP interfaces when substantial oxidation has occurred may be understood by considering a model of how InP oxidizes.[24] The thermal oxidation of InP at 400°C begins with the formation of a thin layer of $InPO_4$, about 30 Å thick, which covers the surface. Apparently, indium may readily diffuse through this thin layer while phosphorus cannot. Therefore, for oxide thicknesses greater than about 30 Å, the InP native oxide is a two-layer structure consisting of $InPO_4$ at the interface and primarily

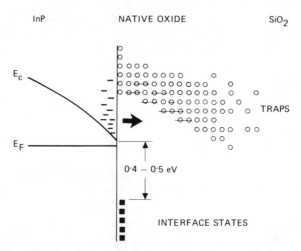

Figure 9. Fritzsche's model of the SiO_2/InP interface (Ref. 46).

In_2O_3 (with some $InPO_4$) at the outer surface of the oxide. Excess phosphorus which cannot easily diffuse through the $InPO_4$ layer builds up at the interface. If this model is correct and the rate-limiting step for the oxidation is actually phosphorus diffusion through the $InPO_4$ layer, then the thickness of the $InPO_4$ layer at the interface should be very sensitive to temperature.

If a thin thermal InP oxide (i.e., less than ~ 30 Å) is formed during the deposition of SiO_2, the composition of this oxide would be expected to be $InPO_4$. However, if the deposition temperature is high enough that a relatively thick native oxide (i.e., >30 Å) is formed at the interface, a complicated interface composed of In_2O_3, $InPO_4$, and excess phosphorus would be expected to form. The excess phosphorus could be a source of interface traps, while the In_2O_3 (a degenerate semiconductor with a band gap of 2.6 eV) could be associated with electrical interface instabilities.

Wilmsen et al.[35] applied XPS depth profiling to the study of the CVD SiO_2/InP interface. Deposition temperatures in excess of about 350°C resulted in substantial oxidation of the InP substrates and some In outdiffusion into the SiO_2. Samples deposited at temperatures below about 350°C had much more abrupt interfaces with much less interfacial oxide. However, there was still some evidence of In outdiffusion into the SiO_2 even at 340°C. Therefore, the degradation of the CVD SiO_2/InP electrical interface properties is apparently associated with substantial oxidation of the InP substrate, In outdiffusion into the SiO_2, and/or preferential loss of phosphorus at the InP surface.

Meiners[1] has used the PECVD technique to form SiO_2 and Si_3N_4 dielectric layers on InP substrates. The Si_3N_4 films consistently produced a surface potential modulation of 0.6 V, independent of the InP surface pretreatment. He attributes this to the fact that there is a substantial amount of H_2 present during the Si_3N_4 deposition and this activated H_2 may preferentially remove phosphorus from the InP surface. The SiO_2/InP electrical properties, however, were found to be dependent upon the InP surface pretreatment and process parameters. Bromine–methanol etched surfaces produced a surface potential modulation of about 0.4 V, while surfaces prepared with hydrazine yielded modulation up to 1.2 V. It is not clear, however, exactly what is responsible for the improvement in the electrical properties when the surface is treated with hydrazine. Meiners[1] notes that hydrazine is a strong reducing agent which probably reduces the native oxide present after wet chemical etching and air exposure. However, he suggests that perhaps it is the combination of the strong reducing agent acting in conjunction with a carbon-containing solvent which may stabilize the InP surface and prevent In outdiffusion into the SiO_2. Meiners[1] also found, similarly to Fritzsche,[110] that incorporation of HCl into the gas stream during the

SiO_2 deposition allowed the InP surface to be more heavily accumulated with electrons.

Wager and Wilmsen[24] used XPS to investigate the interface chemistry of PECVD SiO_2/InP samples fabricated by Meiners.[1] They found that a 10–25 Å thick $InPO_4$ layer was always present at the interface. This oxide was shown to be plasma grown and thicker than native oxides formed during CVD SiO_2 deposition at comparable deposition temperatures.

Meiners[117] subsequently investigated an indirect-PECVD technique for the deposition of SiO_2 in which the oxidizing mixture is excited in a separate chamber connected to the primary deposition chamber by a quartz transfer tube. Wager et al.[25] have studied the temperature dependence of the growth rate and composition of native InP oxides grown in this indirect-PECVD reactor. They found that oxides were thicker if exposed directly to excited oxygen from the transfer tube than when exposed to unexcited oxygen (Fig. 4). The thinner plasma oxides (as well as chemically etched air-grown oxides and thermal oxides) were found to be $InPO_4$ in composition. For oxide thicknesses greater than about 60 Å there was also some In_2O_3 present in the native oxide. Therefore, the plasma oxidation of InP appears to be similar to that of thermal oxidation, although the thickness of the intermediate $InPO_4$ layer appears to be greater for plasma-grown oxides. These results indicate that even though the indirect-PECVD technique should yield less interfacial oxide than with conventional PECVD, some additional native oxide will still probably grow during the initial portion of the SiO_2 deposition.

Okamura, Kobayashi, and co-workers[112–115] have used aluminum triisopropoxide to deposit Al_2O_3 as a gate insulator onto InP substrates. Their original MISFETs fabricated using Al_2O_3 exhibited drain current drift and hysteresis in the curve-tracer current–voltage characteristics. These instabilities were highly dependent upon temperature and dramatically disappeared as the temperature was lowered. They explained the temperature dependence of the instability in terms of a model similar to that of Fritzsche[46] in which electrons in the tail of the thermal distribution tunnel into oxide traps located above the conduction-band edge in energy. These traps were associated with the native oxide. To avoid this problem, they attempted to minimize the mount of native oxide by etching 1500 Å of the InP surface *in situ* prior to the Al_2O_3 deposition and then depositing the Al_2O_3 in a H_2 carrier gas to minimize the amount of oxide formed during the deposition. Devices made using this technique displayed virtually no drain current drift or curve-tracer hysteresis. This result suggests that the native InP oxide is a source of electrical interface instabilities and should be avoided.

Okamura and Kobayashi[114] subsequently reported intentionally growing a thermal oxide (~100 Å at 350–450°C) prior to depositing Al_2O_3 as part of a MISFET fabrication process. They observed an improvement in the drain current drift characteristics compared to Al_2O_3 deposited without an *in situ* HCl etch. However, the effective mobility of these devices was found to be much less than when no thermal oxide was intentionally grown. The improvement in the drain current drift is difficult to explain for 100 Å of thermal oxide since this oxide would be expected to have excess phosphorus at the interface and In_2O_3 within the bulk oxide.

Kobayashi *et al.*[115] also studied the deposition temperature dependence of Al_2O_3 on the electrical properties of InP MISFETs. HCl etching and the H_2 carrier gas were also used for device fabrication in this work. They found that the drain current increased, rather than decreased, as a function of time for deposition temperatures less than 330°C. They explain that this may be due to polarization within the Al_2O_3 when it is deposited at these lower temperatures. Carbon contamination within the dielectric was found to correlate with this anomalous increase in the current drift and they speculate that organic molecules are incorporated into the Al_2O_3 due to the incomplete decomposition of the source gas. Deposition temperatures above about 350°C were found to give effective mobilities much smaller than that found for lower deposition temperatures. This could be due to the preferential loss of phosphorus from the InP surface or to substantial oxidation of the InP substrate.

Lile and Taylor[118] have studied the current decay of SiO_2/InP MISFETs. Figure 10 is a comparison of their results with those of Fritzsche[46] and Okamura and Kobayashi.[112] Notice that the results of Lile and Taylor[118] are equivalent to the improved devices obtained by Okamura and Kobayashi.[112] Since no intentional efforts were made to avoid the formation of native oxide (indeed, a plasma-grown oxide is probably present at the interface),[24,25] they question the conclusion of Okamura and Kobayashi[112] that complete removal of the native oxide is beneficial. They suggest that the native oxide may play a central role in the electrical stability of InP MISFETs, and that an appropriately grown native oxide may yield an optimum interface.

Wager *et al.*[31,32] have employed a variety of surface analytical techniques to characterize native oxides grown in an indirect-PECVD reactor. They estimate the band gap of $InPO_4$ to be approximately 4.5 eV. They also find evidence of In_2O_3 at the outer surface of the native oxide. Figure 11 is an approximate energy-band diagram of the native oxide/InP interface. It was suggested that tunneling of electrons from the InP conduction band through the $InPO_4$ and into the In_2O_3 conduction band could account for electrical interface instabilities observed with InP MIS

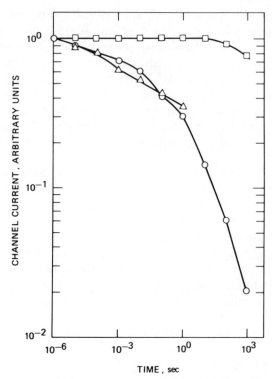

Figure 10. Normalized channel current *vs.* time after the application of a voltage step to a variety of InP enhancement-mode FETs. (○) from Ref. 112 with no HCl etch; (△) from Ref. 46; (□) from Ref. 118. The □ curve is also representative of the upper range of reported values for the improved devices in Ref. 112.

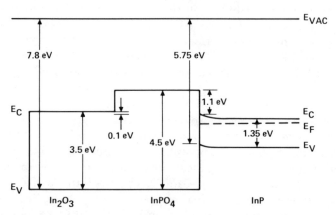

Figure 11. Approximate energy-band diagram for the native oxide/InP interface for a sample treated in a plasma reactor (Ref. 32).

devices. This suggests, in agreement with Lile and Taylor,[118] that appropriate modification of the native oxide may lead to optimum device behavior.

Waldrop et al.[119] used XPS to study the surface reaction of Al with the native oxide of InP left after wet chemical treatment. They find that the native oxide is reduced according to the following chemical reaction:

$$2Al + 2InPO_4 \rightarrow Al_2O_3 + P_2O_5 + 2In$$

Thus, if Al_2O_3 deposition occurs by heterogeneous nucleation of Al at the native oxide surface, the native oxide will be reduced.

Dautremont-Smith and Feldman[120] used Rutherford backscattering detected at grazing angle under channeling conditions to investigate InP surface damage induced by rf sputtering and PECVD of SiO_2. They found that rf sputtering always resulted in surface damage, over the entire range of power densities used. In contrast, no damage was observable due to PECVD.

6. Concluding Remarks

In this chapter we have attempted to point out various considerations which appear to be important for III–V MIS applications. Also, we have reviewed pertinent work regarding InSb, GaAs, and InP MIS applications. This work indicates that, to a large extent, the properties of the dielectric/III–V semiconductor interface are determined by the choice of the semiconductor itself. InSb appears to be an appropriate choice for low-temperature, infrared applications and InP (but not GaAs) appears to be a good candidate for room-temperature, high-speed MIS applications.

The second important consideration is the choice of an appropriate insulator. The two most promising insulators appear to be SiO_2 and Al_2O_3, mainly because of their wide band gaps. Some materials which have been investigated as possible insulators are not appropriate simply because their band gaps are not wide enough. The band gap appears to be the simplest and most important criterion for choosing an insulator. However, other factors are important, such as the trap structure of the insulator, the stoichiometry and impurity content of the insulator, and the deposition energy associated with the insulator formation.

The crucial problem remaining to be resolved before room-temperature deposited insulator/III–V semiconductor devices are technologically viable is that of electrical interface instabilities. It is not yet clear whether the native oxide, interdiffusion, impurities, or something else is responsible for this problem. It would appear that a combination of chemical and electrical characterization of deposited insulators, "real" III–V semiconductor surfaces, and their associated interfaces may provide insight into the nature and prevention of these instabilities.

This discussion has been limited to binary III–V semiconductors. However, it appears that ternary and quaternary III–V compounds may also be suitable for deposited insulator/III–V semiconductor MIS applications. Recent results by Gardner et al.[121] using CVD $SiO_2/In_{0.53}Ga_{0.47}As$ structures are very encouraging. MIS capacitors displayed evidence of accumulation and inversionlike behavior with very little C–V hysteresis. The interface state density was estimated by G–V analysis to have a minimum in the 10^{10} cm^{-2} eV^{-1} range. Depletion mode MISFETs had peak effective mobilities of 5200 cm^2 V^{-1} s^{-1}.

Acknowledgments

The authors wish to acknowledge the support of ONR and ARO in the writing of this chapter.

References

1. L. G. Meiners, Electrical properties of SiO_2 and Si_3N_4 delectric layers on InP, *J. Vac. Sci. Technol.* **19**, 373–379 (1981).
2. R. P. H. Chang, C. C. Chang, and S. Darack, Hydrogen plasma etching of semiconductors and their oxides, *J. Vac. Sci. Technol.* **20**, 45–50 (1982).
3. M. L. Korwin-Pawlowski and E. L. Hesell, Thermal oxide layers on indium antimonide, *Phys. Status Solidi (A)* **27**, 339–346 (1975).
4. K. Schmid, H. Ryssel, H. Muller, K. H. Wiedeburg and H. Betz, Properties of Al_2O_3 surface layers on InSb, investigated by backscattering techniques, *Thin Solid Films* **16**, S11–S12 (1973).
5. S. P. Murarka, Thermal oxidation of GaAs, *Appl. Phys. Lett.* **26**, 812–814 (1975).
6. R. F. C. Farrow, The evaporation of GaAs under equilibrium and nonequilibrium conditions using a modulated beam technique, *J. Phys. D.* **7**, 1693–1701 (1974).
7. J. F. Wager and C. W. Wilmsen, Thermal oxidation of InP, *J. Appl. Phys.* **51**, 180–181 (1980).
8. C. T. Foxon, J. A. Harvey, and B. A. Joyce, The evaporation of InP under Knudsen (equilibrium) and Langmuir (free) evaporation conditions, *J. Phys. Chem. Solids* **34**, 2436–2448 (1973).
9. Y. C. Cheng, Electronic states at the silicon–silicon-dioxide interface, *Progr. Surf. Sci.* **8**, 181–218 (1977).
10. N. M. Johnson, D. K. Biegelsen, and M. D. Moyer, Characteristic defects at the Si–SiO_2 interface, in: *The Physics of MOS Insulators* (G. Lucovsky, S. T. Pantelides, and F. L. Galeener, eds.) pp. 311–315, Pergamon Press, New York, (1980).
11. W. A. Pliskin and H. S. Lehman, Structural evaluation of silicon oxide films, *J. Electrochem. Soc.* **112**, 1013–1019 (1965).
12. P. J. Harrop and D. S. Campbell, Selection of thin film capacitor dielectrics, *Thin Solid Films* **2**, 273–292 (1968).

13. P. J. Harrop and D. S. Campbell, in: *Handbook of Thin Film Technology* (L. I. Maissel and R. Glang, eds.), McGraw-Hill, New York (1970).
14. F. H. Hielscher and H. M. Preier, Non-equilibrium C–V and I–V characteristics of metal–insulator–semiconductor capacitors, *Solid–State Electron.* 12, 527–538 (1969).
15. C. P. Smyth, *Dielectric Behavior and Structure*, McGraw-Hill, New York (1955).
16. N. P. Bogoroditskii and V. V. Pasynkov, *Properties of Electronic Materials*, Boston Tech. Pub. Cambridge, Mass. (1972).
17. R. A. Swalin, *Thermodynamics of Solids*, Wiley, New York (1962).
18. L. J. Brillson, Advances in understanding metal–semiconductor interfaces by surface science techniques, *J. Phys. Chem. Solids* 44, 703–733 (1983).
19. S. P. Kowalczyk, J. R. Waldrop, and R. W. Grant, Interfacial chemical reactivity of metal contacts with thin native oxides of GaAs, *J. Vac. Sci. Technol.* 19, 611–616 (1981).
20a. G. Samsonov (ed.), *The Oxide Handbook*, Plenum Press, New York (1973).
20b. O. Kuaschewski, E. Evans, and C. B. Alcock, *Metallurgical Thermochemistry*, 4th ed. Pergamon Press, Oxford (1967).
21. G. P. Schwartz, W. A. Sunder, and J. E. Griffiths, The In–P–O phase diagram: Construction and applications, *J. Electrochem. Soc.* 129, 1361–1367 (1982).
22. M. Yamaguchi, Thermal nitridation of InP, *Japan. J. Appl. Phys.* 19, L401–L404 (1980).
23. R. E. Bolz and G. L. Tuve (eds.), *CRC Handbook of Tables for Applied Engineering Science*, 2nd ed., CRC Press, Cleveland, Ohio (1976).
24. J. F. Wager and C. W. Wilmsen, Plasma-enhanced chemical vapor deposited SiO_2/InP interface, *J. Appl. Phys.* 53, 5789–5797 (1982).
25. J. F. Wager, W. H. Makky, C. W. Wilmsen, and L. G. Meiners, Oxidation of InP in a plasma-enhanced chemical vapor deposited realtor, *Thin Solid Films* 95, 343–350 (1982).
26. W. E. Spicer, P. W. Chye, P. R. Skeath, C. Y. Su, and I. Lindau, Nature of interface states at III–V insulator interfaces, *Inst. Phys. Conf. Ser. No. 50*, 216–233 (1980).
27. W. E. Spicer, I. Lindau, P. Skeath, and C. Y. Su, Unified defect model and beyond, *J. Vac. Sci. Technol.* 17, 1019–1027 (1980).
28. S. R. Morrison, *The Chemical Physics of Surfaces*, Plenum Press, New York (1977).
29. R. H. Bube, *Photoconductivity of Solids*, Wiley, New York (1960).
30. R. L. Weiher and R. P. Ley, Optical properties of indium oxide, *J. Appl. Phys.* 37, 299–302 (1966).
31. J. F. Wager, C. W. Wilmsen, and L. L. Kazmerski, Estimation of the bandgap of $InPO_4$, *Appl. Phys. Lett.* 42, 589–591 (1983).
32. J. F. Wager, K. M. Geib, C. W. Wilmsen, and L. L. Kazmerski, Native oxide formation and electrical instabilities at the insulator/InP interface, *J. Vac. Sci. Technol. B* 1, 778–781 (1983).
33. K. J. Bachmann, Properties, preparation, and device applications of indium phosphide, *Ann. Rev. Mater. Sci.* 11, 441–484 (1981).
34. M. Ghezzo and D. M. Brown, Diffusivity summary of B, Ga, P, As, and Sb in SiO_2, *J. Electrochem. Soc.* 120, 146–148 (1973).
35. C. W. Wilmsen, J. F. Wager, and J. Stannard, Chemical vapour deposited SiO_2–InP interface, *Inst. Phys. Conf. Ser. No. 50*, 251–257 (1980).
36. C. W. Wilmsen, Chemical composition and formation of thermal and anodic/III–V compound semiconductor interfaces, *J. Vac. Sci. Technol.* 19, 279 (1981).
37. R. H. Williams and I. T. McGovern, Surface characterization of indium phosphide, *Surf. Sci.* 51, 14–28 (1975).
38. R. W. Grant, J. R. Waldrop, S. P. Kowalczyk, and E. A. Kraut, 'Correlation of GaAs surface chemistry and interface Fermi-level position: A single defect model interpretation' *J. Vac. Sci. Technol.* 19, 477–480 (1981).

39. J. A. Van Vecten, Simple theoretical estimates of the Schottky constants and virtual-enthalpies of single vacancy formation in zinc-blende and wurtzite type semiconductors, *J. Electrochem. Soc. 122*, 419–422 (1975).
40. J. A. Van Vecten, Simple theoretical estimates of the enthalpy of antistructure PVR formation and virtual-enthalpies of isolated antisite defects in zinc-blende and wurtzite type semiconductors, *J. Electrochem. Soc. 122*, 423–429 (1975).
41. M. S. Daw and D. L. Smith, Energy levels of semiconductor surface vacancies, *J. Vac. Sci. Technol. 17*, 1028–1031 (1980).
42. J. D. Dow and R. E. Allen, Surface defects and Fermi-level pinning in InP, *J. Vac. Sci. Technol. 20*, 659–661 (1982).
43. P. W. Chye, I. Lindau, P. Pianetta, C. M. Garner, C. Y. Su, and W. E. Spicer, Photoemission study of Au Schottky-barrier formation on GaSb, GaAs, and InP using synchrotron radiation, *Phys. Rev. B 18*, 5545–5558 (1978).
44. F. P. Heiman and G. Warfield, The effects of oxide traps on the MOS capacitance, *IEEE Trans. Electron. Devices ED-12*, 167–178 (1965).
45. H. Koelmans and H. C. DeGraaff, Drift phenomena in CdSe thin film FET's, *Solid-State Electron. 10*, 997–1005 (1967).
46. D. Fritzsche, Interface studies on InP MIS inversion FET's with SiO_2 gate insulation, *Inst. Phys. Conf. Ser. No. 50*, 258–265 (1980).
47. R. Williams, Photoemission of electrons from silicon into silicon dioxide, *Phys. Rev. 140*, A569–575 (1965).
48. V. J. Kapoor and R. A. Turi, Charge storage and distribution in the nitride layer of the metal–nitride–oxide–semiconductor structures, *J. Appl. Phys. 52*, 311–319 (1981).
49. V. J. Kapoor and S. B. Bibyk, Energy distribution of electron trapping defects in thick-oxide MNOS structures, in: *The Physics of MOS Insulators*, G. Lucovsky, S. T. Pantelides, and F. Galeener (eds.), pp. 117–121, Pergamon Press, New York (1980).
50. E. Harari and B. S. H. Royce, Trap structure of pyrolytic Al_2O_3 in MOS capacitors, *Appl. Phys. Lett. 22*, 106–107 (1973).
51. G. W. Gobeli and F. G. Allen, Photoelectric threshold and work function, in: *Semiconductors and Semimetals, Vol. 2*, R. K. Willardson and A. C. Beer (eds.), pp. 263–280, Academic Press, New York (1966).
52. D. M. Brown, P. V. Bray, F. K. Heumann, H. R. Philipp, and E. A. Taft, Properties of $Si_xO_yN_z$ films on Si, *J. Electrochem. Soc. 115*, 311–317 (1968).
53. G. K. Eaton, R. E. J. King, F. D. Morten, A. T. Partridge, and J. G. Smith, Surface conductance on P-type InSb at 77 K, *J. Phys. Chem. Solids 23*, 1473–1477 (1962).
54. H. Pagina, Surface photoconduction in' p-InSb single crystals, *Phys. Status Solidi 28*, K89–92 (1968).
55. H. Pagina, Field effect and surface photoconductivity studies on InSb single crystals, *Phys. Status Solidi 34*, 121–134 (1969).
56. J. L. Davis, Surface states on the $(\overline{111})$ surface of indium antimonide, *Surf. Sci. 2*, 33–39 (1964).
57. H. Huff, S. Kawaji, and H. C. Gatos, Field effect measurements on the A and B {111} surfaces of indium antimonide, *Surf. Sci. 5*, 399–409 (1966).
58. H. Huff, S. Kawaji, and H. C. Gatos, Electronic configuration of indium antimonide surfaces, *Surf. Sci. 10*, 232–238 (1968).
59. H. Sewell and J. C. Anderson, Slowstates in $InSb/SiO_x$ thin film transistors, *Solid-State Electron. 18*, 641–649 (1975).
60. D. E. Aspnes and A. A. Studna, Preparation of high quality surfaces on semiconductors by selective chemical etching, *J. Vac. Sci. Technol. 20*, 488–489 (1982).
61. D. E. Aspnes, Dielectric function and surface microroughness measurements of InSb by spectroscopic ellipsometry, *J. Vac. Sci. Technol. 17*, 1057–1060 (1980).

62. H. Iwasaki, Y. Mizokawa, R. Nishitani, and S. Nakamura, Effects of water vapor and oxygen excitation on oxidation of GaAs, GaP, and InSb surfaces studied by X-ray photoemission spectroscopy, *Japan. J. Appl. Phys.* 18, 1525 (1979).
63. H. Iwasaki, Y. Mizokawa, R. Nishitani, and S. Nakamura, X-ray photoemission study of the oxidation process at cleaved (100) surfaces of GaAs, GaP, and InSb, *Japan. J. Appl. Phys.* 17, 1925–1933 (1978).
64. R. P. Vasquez and F. J. Grunthaner, XPS study of interface formation of CVD SiO_2 on InSb, *J. Vac. Sci. Technol.* 19, 431–436 (1981).
65. R. P. Vasquez and F. J. Grunthaner, Chemical composition of the SiO_2/InSb interface as determined by X-ray photoelectron spectroscopy, *J. Appl. Phys.* 52, 3509–3514 (1981).
66. G. W. Anderson, W. A. Schmidt, and J. Comas, Composition, Chemical bonding and contamination of low temperature SiO_xN_y insulating films, *J. Electrochem. Soc.* 125, 424–430 (1978).
67. J. D. Langan, Ph.D. Thesis, Study and Characterization of Semiconductor Surfaces and Interfaces, University of California at Santa Barbara (1979).
68. J. D. Langan and C. R. Viswanathan, Characterization of improved InSb interfaces, *J. Vac. Sci. Technol.* 16, 1474–1477 (1979).
69. I. Flinn and d. C. Emmony, Interface characteristics of Ge_3N_4–(n-type)GaAs MIS devices, *Phys. Lett.* 6, 1107–1109 (1963).
70. I. Flinn and M. Briggs, Surface measurements on gallium arsenide, *Surf. Sci.* 2, 136–145 (1964).
71. I. Flinn, Surface properties of n-type gallium arsenide, *Surf. Sci.* 10, 32–57 (1968).
72. T. M. Valahas, J. S. Sochanski, and H. C. Gatos, Electrical characteristics of gallium arsenide "REAL" Surfaces, *Surf. Sci.* 26, 41–53 (1971).
73. J. Lagowski, I. Baltov, and H. C. Gatos, Surface photovoltage spectroscopy and surface piezoelectric effect in GaAs, *Surf. Sci.* 40, 216–226 (1973).
74. E. W. Kreutz and P. Schroll, Field effect on real GaAs surfaces, *Phys. Status Solidi A* 53, 499–508 (1979).
75. C. C. Chang, P. H. Citrin, and B. Schwartz, Chemical preparation of GaAs surfaces and their characterization by Auger electron and X-ray photoemission spectroscopies, *J. Vac. Sci. Technol.* 14, 943–952 (1977).
76. T. Oda and T. Sugano, Studies on chemically etched silicon, gallium arsenide, and gallium phosphide surfaces by Auger electron spectroscopy, *Japan. J. Appl. Phys.* 15, 1317–1327 (1976).
77. D. E. Aspnes and A. A. Studna, Chemical etching and cleaning procedures for Si, Ge, and some III–V compound semiconductors, *Appl. Phys. Lett.* 15, 316–318 (1981).
78. P. J. Grunthaner, R. P. Vasquez, and F. J. Grunthaner, Chemical depth profiles of the GaAs/native oxide interface, *J. Vac. Sci. Technol.* 17, 1045–1051 (1980).
79. H. Becke, R. Hall, and J. White, Gallium arsenide MOS transistors, *Solid-State Electron.* 8, 813–823 (1965).
80. J. E. Foster and J. M. Swartz, Electrical characteristics of the silicon nitrios–gallium arsenide interface, *J. Electrochem. Soc.* 117, 1410–1417 (1970).
81. J. A. Cooper, E. R. Ward, and R. J. Schwartz, Surface states and insulator raps at the Si_3N_4-GaAs interface, *Solid-State Electron.* 15, 1219–1227 (1972).
82. L. G. Meiners, Electrical properties of the gallium arsenide–insulator interface, *J. Vac. Sci. Technol.*. 15, 1402–1407 (1978).
83. C. R. Zeisse, L. J. Messick, and D. L. Lile, Electrical properties of anodic and pyrolytic dielectrics on gallium arsenide, *J. Vac. Sci. Technol.* 14, 957–960 (1977).
84. B. Bayraktaroglu, W. M. Theis, and F. L. Schuermeyer, GaAs surface passivation using Si_3N_4: interface characteristics, *Inst. Phys. Conf. Ser. No. 50*, 280–286 (1980).

85. B. Bayraktaroglu and R. L. Johnson, Silicon-nitride–gallium-arsenide MIS structures produced by plasma enhanced deposition, *J. Appl. Phys. 52*, 3515–3519 (1981).
86. M. D. Clark and C. L. Anderson, Improvements in GaAs/plasma-deposited silicon nitride interface quality by predeposition GaAs surface treatment and post deposition annealing, *J. Vac. Sci. Technol. 21*, 453–456 (1982).
87. Y. Sato, The properties of the interface between gallium arsenide and silicon oxides, *Japan. J. Appl. Phys. 7*, 595–599 (1968).
88. K. Kamimura and Y. Sakai, The properties of GaAs–Al_2O_3 and InP–Al_1O_3 interfaces and the fabrication of MIS field-effect transistors, *Thin Solid Films 56*, 215–223 (1979).
89. G. D. Bagratishvili, R. B. Dzhanelidze, N. I. Kurdiani, Y. I. Pashintsev, O. V. Saksaganski, and V. A. Skorikov, GaAs/Ge_3N_4/Al structures and MIS field-effect transistors based on them, *Thin Solid Films 56*, 209–213 (1979).
90. K. P. Pande, M. L. Chen, M. Yousuf, and B. Lalevic, Interface characteristics of Ge_3N_4–(n-type)GaAs MIS devices, *Solid-State Electron. 24*, 1107–1109 (1981).
91. J. Nishizawa and I. Shiota, GaO_xN_y-based multiple insulating layers on GaAs surfaces, *Inst. Phys. Conf. Ser. No. 50*, 287–292 (1980).
92. R. K. Smeltzer and C. C. Chen, Oxidized metal film dielectrics for III–V devices, *Thin Solid Films 56*, 75–80 (1979).
93. N. Suzuki, T. Hariu, and Y. Shibata, Effect of native oxide on the interface property of GaAs MIS structures, *Appl. Phys. Lett. 33*, 761–762 (1978).
94. P. Friedel and S. Gourrier, Interactions between H_2 and N_2 plasmas and a GaAs (100) surface: Chemical and electronic properties, *Appl. Phys. Lett. 42*, 509–511 (1983).
95. U. Konig and E. Sasse, XPS study of annealed SiO_2/GaAs interfaces, *J. Electrochem. Soc. 130*, 950–952 (1983).
96. R. W. Grant, K. R. Elliott, S. P. Kowalczyk, D. L. Miller, J. R. Waldrop, and J. R. Oliver, GaAs Surface Passivation for Device Applications, AFWAL-TR-82-1081, Final Report (July 1982).
97. J. Gyulai, J. W. Mayer, I. V. Mitchell, and V. Rodriquez, Outdiffusion through silicon oxide and silicon nitride layers on gallium arsenide, *Appl. Phys. Lett. 17*, 332–334 (1970).
98. J. S. Harris, F. H. Eisen, B. Welch, J. D. Haskell, R. D. Pashley, and J. W. Mayer, Influence of implantation temperature and surface protection on tellurium implantation in GaAs, *Appl. Phys. Lett. 21*, 601–603 (1972).
99. R. D. Pashley and B. M. Welch, Tellurium-implanted Λ^+ layers in GaAs, *Solid-State Electron. 18*, 977–981 (1975).
100. K. V. Vaidyanathan, M. J. Helix, D. J. Wolford, B. G. Streetman, R. J. Blattner, and C. A. Evans, Jr., Study of encapsulants for annealing GaAs, *J. Electrochem. Soc. 124*, 1781–1784 (1977).
101. D. T. Clark, T. Fok, G. G. Roberts, and R. W. Sykes, An investigation by electron spectroscopy for chemical analysis of chemical treatments of the (100) surface of n-type InP epitaxial layers for Langmuir film deposition, *Thin Solid Films 70*, 261–283 (1980).
102. P. A. Bertrand, XPS study of chemically etched GaAS and InP, *J. Vac. Sci. Technol. 18*, 28–33 (1981).
103. J. F. Wager, D. L. Ellsworth, S. M. Goodnick, and C. W. Wilmsen, Composition and thermal stability of thin native oxides on InP, *J. Vac. Sci. Technol. 19*, 513–518 (1981).
104. L. Messick, InP/SiO_2 MIS structure, *J. Appl. Phys. 47*, 4949–4951 (1976).
105. L. Messick, D. L. Lile, A. R. Clawson, A microwave InP/SiO_2 MISFET, *Appl. Phys. Lett. 32*, 494–495 (1978).
106. D. L. Lile, D. A. Collins, L. G. Meiners, and L. Messick, n-Channel inversion-mode InP MISFET, *Electron. Lett. 14*, 657–659 (1978).
107. L. G. Meiners, D. L. Lile, and D. A. Collins, Inversion layers on InP, *J. Vac. Sci. Technol. 16*, 1458–1461 (1979).

108. L. G. Meiners, Capacitance voltage and surface photovoltage, *Thin Solid Films 56*, 201–207 (1979).
109. J. Stannard, Transient capacitance in GaAs and InP MOS capacitors, *J. Vac. Sci. Technol. 16*, 1508–1512 (1979).
110. D. Fritzsche, InP SiO_2 MIS structures with reduced interface state density near conduction band, *Electron. Lett. 14*, 51–52 (1978).
111. J. Woodward, D. C. Cameron, L. D. Irving, and G. R. Jones, The deposition of insulators onto InP using plasma-enhanced chemical vapour deposition, *Thin Solid Films 85*, 61–69 (1981).
112. M. Okamura and T. Kobayashi, Slow current-drift mechanism in n-channel inversion type InP-MISFET. *Japan. J. Appl. Phys. 19*, 2143–2150 (1980).
113. M. Okamura and T. Kobayashi, Improved interface in inversion type InP-MISFET by vapor etching technique, *Japan. J. Appl. Phys. 19*, 2151–2156 (1980).
114. M. Okamura and T. Kobayashi, Current drifting behaviour in InP MISFET with thermally oxidized InP/InP interface, *Electron. Lett. 17*, 941–942 (1981).
115. T. Kobayashi, M. Okamura, E. Yamaguchi, Y. Shinoda, and Y. Hirota, Effect of pyrolytic Al_2O_3 deposition temperature on inversion-mode InP metal–insulator–semiconductor field-effect transistor, *J. Appl. Phys. 52*, 6434–6436 (1981).
116. K. P. Pande and S. Pourdavoud, Ge_3N_4/InP MIS structures, *IEEE Electron. Devices Lett. EDL-2*, 182–184 (1981).
117. L. G. Meiners, Indirect plasma deposition of silicon dioxide, *J. Vac. Sci. Technol. 21*, 655–658 (1982).
118. D. L. Lile and M. J. Taylor, The effect of interfacial traps on the stability of insulated gate devices on InP, *J. Appl. Phys. 54*, 260–267 (1983).
119. J. R. Waldrop, S. P. Kowalczyk, and R. W. Grant, Correlation of Fermi-level energy and chemistry at InP (100) interfaces, *Appl. Phys. Lett. 42*, 454–456 (1983).
120. W. C. Dautremont-Smith and L. C. Feldman, Surface structural damage produced in InP (100) by RF plasma or sputter deposition, *Thin Solid Films 105*, 187–196 (1983).
121. P. D. Gardner, S. Y. Narayan, S. Colvin, and Y. Yun, $G_{8.47}In_{0.53}As$ metal insulator field-effect transistors (MISFETs) for microwave frequency applications, *RCA Rev. 42*, 542–556 (1981).

4

Electrical Properties of Insulator–Semiconductor Interfaces on III–V Compounds

L. G. Meiners

1. Introduction

An enormous amount of theoretical and experimental work has been published on metal–insulator–semiconductor (MIS) structures which employ thermally oxidized silicon. The technology of these devices has reached a level of perfection which permits the formation of insulator–semiconductor interfaces in which essentially no extraneous charge-trapping mechanisms are present. The understanding of such structures has reached a level which permits the design and large-scale technological applications of MIS transistors and MIS integrated circuits. Although several other semiconductors appear to be superior to silicon for certain MIS applications, such as microwave logic and signal processing, the understanding of the surface properties of the alternative semiconductors is in a primitive state. The dielectrics that have been tried on such semiconductors always have exhibited larger amounts of charge trapping at the interface and greater frequency dispersion of the electrical properties of the insulator than those attainable on thermally oxidized silicon. In fact, the surface properties are, in many cases, so poor that comparison between experiment and theoretical models developed for silicon can be very difficult and often confusing. An obvious need exists to examine, in detail, the results of measurements and analysis applied to semiconductors other than silicon, to compare the characteristic aspects of the surfaces

---

L. G. Meiners ● Electrical Engineering and Computer Sciences Department, University of California, San Diego, La Jolla, CA 92093.

produced with our present technology with those of ideal surfaces and those obtained on silicon MIS structures.

Consider the case of a parallel-plate capacitor composed of a metal–insulator–semiconductor (MIS) sandwich. Ideally, when an electric potential is applied between the metal and the semiconductor, the component of the electric displacement vector perpendicular to the surface is continuous across the semiconductor–dielectric boundary, and the electric field within the semiconductor is uniquely related to the field in the insulator. As will be shown, the surface electric field and surface charge can then be uniquely related to such experimentally measurable quantities as the surface capacitance and surface conductivity by the solution of Poisson's equation.

However, the condition of continuity of electric displacement is found to apply only for very specially prepared semiconductor–insulator interfaces. It is much more common for the electric displacement to be discontinuous because of the presence of charges trapped at the interface between the semiconductor and insulator. Furthermore, the magnitude and the polarity of this charge will be a function of the surface electric field and no longer will each elementary charge applied to the metal gate attract a mobile charge of opposite sign in the semiconductor bulk. Interpretation of surface capacitance and surface conductivity measurements thus become considerably more complicated if interfacial charging mechanisms are present and have to be taken into account.

2. Theoretical Background

The purpose of this section is to describe the physical model necessary for understanding the remainder of the chapter. This model is based on a mathematical solution of Poisson's equation for the space-charge region present at the semiconductor surface. Once one admits the possibility of a space-charge region in a semiconductor, it follows that an electric field exists within and an electric potential exists across the space-charge layer. If the value of this potential can be determined experimentally, then for nondegenerate semiconductors the use of the Boltzmann distribution function allows calculation of the carrier densities in the space-charge region. An analytical expression can be obtained for the surface electric field in a semiconductor as a function of the total potential drop across the surface region. From this expression equations may be derived for the parametric variations of the differential surface capacitance and the total surface charge. These calculated quantities can be compared with experimental measurements. The surface conductivity and surface photovoltage may also be calculated although these may have to be obtained by numerical techniques. Such calculations will also be discussed.

Space-charge regions can extend to significant depths (10–10,000 Å depending on the impurity doping density) in semiconductors. This is due to the fact that the mobile charge density is lower than that in metals by a factor of 10^5–10^{10} for doping densities of 10^{14}–10^{19} cm^{-3}. The ability of the mobile charge to screen the interior of the semiconductor from external electric fields increases with increasing charge density.

2.1. Differential Surface Capacitance

Consider a situation in which a steady electric field is applied perpendicularly to a semiconductor surface by a metal plate which does not quite touch the semiconductor. At equilibrium, no current will flow between the metal plate and the semiconductor. For the present time it is assumed that charge may exist only in the bulk and not at the semiconductor surface. From Gauss's law we know that the charge will redistribute itself until at equilibrium, all of the electric field lines will be terminated on charges in the vicinity of the surface. If this were not the case, then the nonzero field within the bulk of the crystal would cause a steady-state

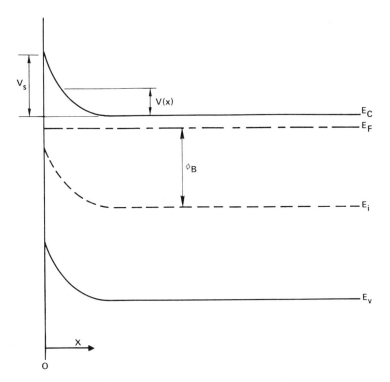

Figure 1. Energy bands of an n-type semiconductor in the surface region in depletion.

flow of current; this would redistribute the charge until the current would be reduced to zero.

We wish now to solve Poisson's equation for the potential distribution within the semiconductor for a fixed external electric field. Consider the band picture of an n-type semiconductor surface as shown in Fig. 1. The potential energy plot is for a surface in which electrons are depleted from the surface. In this diagram E_c is the conduction-band energy minimum in the bulk, E_v the bulk valence-band energy maximum, E_F the Fermi energy, E_i the Fermi energy for the intrinsic material, and $V(x)$ the electrostatic potential within the semiconductor. The difference between E_i and E_F deep in the crystal we define as the bulk potential, V_B.

When the static electric field is directed away from the semiconductor (negative charge on the metal plate), a positive space charge will be produced in the semiconductor. This can be either the fixed charge of the ionized donor impurities, or that of the mobile valence-band holes. This is the situation in Fig. 1 in which the energy bands bend up at the surface. The electrostatic potential at the surface (V_s) is, by convention, negative in this case. If the electric field is directed toward the surface (positive charge on the metal plate) then the space charge will consist of conduction-band electrons. For this situation the bands bend downward as shown in Fig. 2 and the surface potential V_s is by convention positive.

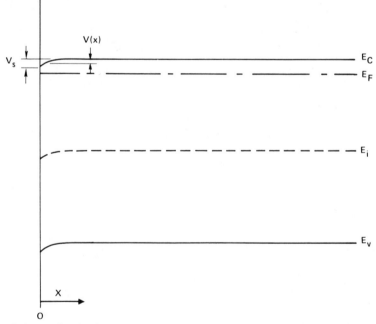

Figure 2. Energy bands of an n-type semiconductor in the surface region in accumulation.

It is usually adequate to consider analytically the case of nondegenerate carrier distributions (particles spaced far enough apart so as to be noninteracting) for the electrons and holes. This allows the use of Maxwell–Boltzmann statistics in calculating the mobile carrier densities. This formalism is described by the equation

$$\frac{D_1}{D_2} = \frac{e^{-E_1/kT}}{e^{-E_2/kT}} \tag{1}$$

where $D_{1,2}$ and $E_{1,2}$ represent the density and energy, respectively, of identical particles at two separate points in the crystal. In the present case the carrier densities are given by

$$n(x) = n_B \exp[qV(x)/(kT)] \tag{2a}$$

and

$$p(x) = p_B \exp(-qV(x)/kT) \tag{2b}$$

where n_B and p_B are the bulk electron and hole densities, respectively, and $V(x)$ is defined to be zero deep in the bulk material. As a further constraint, the electron and hole densities are related to each other by the mass action law, namely

$$n_B p_B = n_i^2 \tag{3}$$

where n_i is the intrinsic carrier concentration usually produced by thermally generated electronic transitions, and the assumption is made that all impurities are and remain ionized. This will be the case of shallow impurity levels at room temperature. The volume charge density within the semiconductor is then given by

$$\rho = q(-n + p + N_{Di} - N_{Ai}) \tag{4}$$

where N_{Di} and N_{Ai} are the ionized donor and acceptor densities, respectively. The one-dimensional Poisson equation

$$\frac{d^2 V}{dx^2} = -\rho/\varepsilon_s \tag{5}$$

must be solved for the above charge distribution with the boundary conditions that

$$V|_{x=0} = V_s \quad \text{and} \quad V|_{x=\infty} = 0 \tag{6}$$

Since $d^2 V/dx^2 = 0$ far into the bulk, from Eqs. (4) and (5) one has the result that

$$p_B - n_B + N_{Di} - N_{Ai} = 0 \tag{7}$$

or

$$N_{Di} - N_{Ai} = n_B - p_B$$

The charge density can then be written as

$$\rho = q(-n_B e^{eV/kT} + p_B e^{-qV/kT} + n_B - p_B) \qquad (8)$$

or

$$\rho = q[n_B(1 - e^{qV/kT}) - p_B(1 - e^{-qV/kT})]$$

Then

$$\frac{d^2 V}{dx^2} = \frac{q}{\varepsilon_s}[p_B(1 - e^{-qV/kT}) - n_B(1 - e^{qV/kT})] \qquad (9)$$

Use of the integrating factor dV/dx within Eq. (9) leads to

$$\int_x^\infty \frac{d^2 V}{dx^2}\frac{dV}{dx}dx = \frac{q}{\varepsilon_s}\int_x^\infty [p_B(1 - e^{-qV/kT}) - n_B(1 - e^{qV/kT})]\frac{dV}{dx}dx \qquad (10)$$

which reduces to

$$\frac{1}{2}\left(\frac{dV}{dx}\right)^2\bigg|_x^\infty = \frac{q}{\varepsilon_s}\int_V^0 [p_B(1 - e^{-qV/kT}) - n_B(1 - e^{qV/kT})]dV \qquad (11)$$

Integration of this expression gives the electric field

$$E = -\frac{dV}{dx} = \pm\left(\frac{2kT}{\varepsilon_s}\right)^{1/2}[p_B(e^{-qV/kT} + qV/kT - 1) + n_B(e^{qV/kT} - qV/kT - 1)]^{1/2} \qquad (12)$$

where the positive sign is taken for $V_s < 0$ ($\vec{E}$ points away from surface), and the negative sign is taken for $V_s > 0$. The total charge per unit area in the surface region is given by Gauss's law

$$Q_s = \varepsilon_s E_s \qquad (13)$$

where the electric field at the surface is

$$E_s = -\frac{dV}{dx}\bigg|_{x=0} \qquad (14)$$

or

$$Q_s = \pm(2\varepsilon_s kT)^{1/2}[p_B(e^{-qV_s/kT} + qV_s/kT - 1) + n_B(e^{qV_s/kT} - qV_s/kT - 1)]^{1/2} \qquad (15)$$

where the positive sign applies when $V_s < 0$ and vice versa. One can define a differential capacitance per unit area of the semiconductor surface C_D and a bit of algebra gives

$$C_D \equiv \frac{\partial Q_s}{\partial V_s} = \left(\frac{q^2 \varepsilon_s}{2kT}\right)^{1/2}$$
$$\times \frac{|p_B(1 - e^{-qV_s/kT}) + n_B(e^{qV_s/kT} - 1)|}{[p_B(e^{-qV_s/kT} + qV_s/kT - 1) + n_B(e^{qV_s/kT} - qV_s/kT - 1)]^{1/2}} \quad (16)$$

All of the above results apply to either n- or p-type material.

The reader can see at this point that the differential capacitance for a particular semiconductor is a function only of its impurity doping, temperature, and surface potential. Thus an experimental measurement of the parametric variations of the differential capacitance of the space-charge layer of such an ideal semiconductor allows a determination of the corresponding variations of the surface potential. One can then calculate the total surface charge, carrier densities, and electric field, and avoid the far more difficult experimental problem of measuring these quantities directly.

A plot of Q_s as a function of surface potential is shown for 10^{15} n-GaAs in Fig. 3. This also illustrates the terminology in common use. Positive

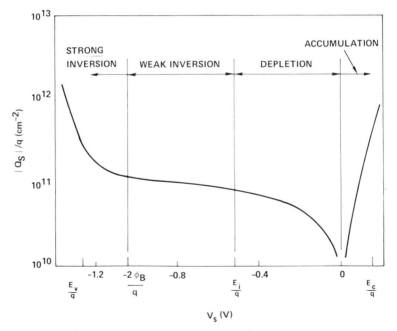

Figure 3. Surface charge density Q_s vs. V_s for 10^{15} cm^{-3} n-type GaAs at 300 K.

values of V_s correspond to an excess of surface electrons called accumulation. Negative values of V_s, up to the value of the Fermi level, in intrinsic material, correspond to the situation in which electrons are driven away from the surface; consequently, the electric field terminates on the ionized donors. This is termed the depletion regime. Further charges in the gate voltage which make the hole density at the surface equal to the electron density in the bulk lead to weak surface inversion. More negative values of V_s produce a larger density of surface holes. Such a surface layer of carriers whose sign is opposite to that of the charge carriers in the bulk is called an inversion layer. A plot of the differential capacitance calculated from Eq. (16) for this same material is shown in Fig. 4. As expected this curve rises steeply when the surface density of carriers becomes larger than the bulk value.

A question not yet touched upon here is the condition when Eq. (16) corresponds to a physically measurable capacitance. In a very general sense the space-charge capacitance is the equivalent circuit element which

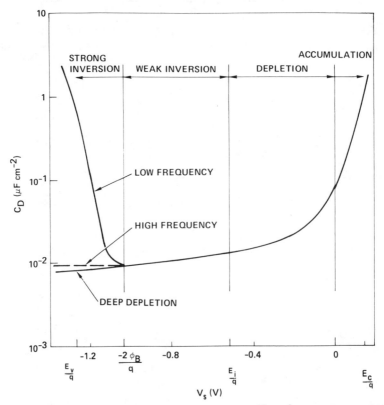

Figure 4. Differential surface capacitance C_s vs. V_s for 10^{15} cm^{-3} n-type GaAs at 300 K.

represents the flow of charge carriers in the surface region in response to an external electric field. In order to appear capacitive in an ac applied field, this current must decay to zero in a time interval much smaller than the inverse of the ac frequency, otherwise the ac current will have a real (loss) and imaginary (capacitive) component. Currents which are in phase with the applied ac field may be described as a conductive component. Equation (16) holds only if the real or conductive component of the current can be ignored. The frequency at which this occurs depends on the material type and the surface electric field.

Equation (16) for the differential capacitance gives essentially a low-frequency curve because equilibrium statistics were used for the carrier population. In the accumulation regime for the material described in Fig. 4, a space-charge perturbation should come to equilibrium in a time τ_0 of the order of $\tau_0 = \varepsilon_s \rho$ which for 10^{15} cm^{-3} n-GaAs is approximately 10^{-12} s. This capacitance should then remain constant for all practical measurement frequencies. A depleted n-type surface can be approximated by assuming that the electrons are totally absent from a slab extending to a depth w from the surface. The space charge then consists only of the ionized donor impurities which have a volume density N_D. The thickness of this layer may be obtained from the requirement that $Q_{sc} = N_D w$ using Eq. (15) to obtain the total space charge Q_{sc}. Fluctuations in charge that result from small changes in surface potential then occur in a narrow region at depth w from the surface. Since this charge is in good communication with the semiconductor bulk, equilibrium should be restored in a time interval of the same order as the bulk relaxation time.

In inversion the situation is quite different. Hofstein and Warfield[1] and Heiman[2] have treated this problem and find that normally the largest source of minority carriers for the inversion layer is generation of carriers within the depletion region. The generation rate g of electron–hole pairs in the depletion layer is given by

$$g = \frac{n_i}{2\tau} \qquad (17)$$

where τ is the minority-carrier lifetime. From a very general point of view, one would expect that the inversion layer would come to equilibrium when the total elapsed time multiplied by the carrier generation rate was equal to the bulk impurity doping. Thus equilibrium is expected when

$$t = \frac{N_D}{g} = \tau \frac{N_D}{n_i} \qquad (18)$$

For GaAs at 300°C, $n_i = 1.94 \times 10^6$ cm^{-3} and $\tau \simeq 10^{-9}$ s. Thus for $N_D = 10^{16}$ cm^{-3}, $t \simeq 10$ s.

This time is quite long and thus the assumption of equilibrium statistics for the minority carriers is in general not valid. Attention must therefore be given to the frequency response of the space-charge layer. Among the first things one might wish to calculate is the high-frequency capacitance, that is, the capacitance in the limit of zero dispersion at high frequencies. Rigorously, the approach that would be taken is to repeat the earlier described calculations assuming that the minority carriers are not able to respond to the frequency of the measurement voltage.[3] However, a good approximation can be obtained[4] by taking the high-frequency capacitance to be constant in the inversion region at the value obtained when the surface potential is equal to twice the bulk potential $(-2\phi_B/q)$. This treatment is satisfactory for the purposes of the present discussion.

A somewhat different type of surface capacitance measurement can be made by driving the surface quickly into depletion with an applied voltage pulse. In this case the minority-carrier terms can be neglected in Eqs. (12), (15), and (16). Since the terms due to majority-carrier accumulation can be neglected also, the result for n-type material is then

$$E_s = \mp \left(\frac{2qn_B}{\varepsilon_s}\right)^{1/2} \left(-V_s - \frac{kT}{q}\right)^{1/2} \tag{19}$$

$$C_D = \left(\frac{q\varepsilon_s n_B}{2}\right)^{1/2} \left(-V_s - \frac{kT}{q}\right)^{1/2} \tag{20}$$

$$Q_s = \pm (2\varepsilon_s n_B q)^{1/2} (-V_s - kT/q)^{1/2} \tag{21}$$

The above equations comprise what is called the depletion approximation; they apply in either one of two cases: the first is the above-mentioned one produced with pulsed gate bias, and the second is the one which would be created if the minority carriers were transported away from the surface to the metal gate electrode, if the metal electrode were in direct contact with the semiconductor. The resulting structure is called a Schottky[5] barrier or metal–semiconductor junction. The approximation for the minimum in the high-frequency capacitance can be obtained from Eq. (21) upon the substitution of $\phi_B = kT \ln(n_B/n_i)$ for V_s. Recall that the basis for this approximation is that the space-charge capacitance measured at high frequency (HF) saturates at a minimum value given when $V_s = 2\phi_B/q$. This leads to

$$C_D\Big|_{\min}^{HF} \simeq \left(\frac{4\varepsilon_s kT \ln(n_B/n_i)}{q^2 n_B}\right)^{1/2} \tag{22}$$

The discussion so far has centered around only the differential capacitance of the semiconductor space-charge layer. In an actual measurement the capacitance of the dielectric region between the metal electrode and the semiconductor must also be considered. In practice the dielectric employed is not usually air but rather a thin insulating layer formed on the semiconductor surface. For the present, this dielectric will be considered to be homogeneous and to be independent of the frequency of the applied voltage so that the capacitance of the resulting structure will be

$$C_m = \frac{C_i C_D}{C_i + C_D} \tag{23}$$

where C_m is the measured capacitance and C_i is the capacitance of the dielectric layer given by

$$C_i = \frac{\varepsilon_i}{d} \tag{24}$$

where ε_i and d are the dielectric constant and thickness of the insulator, respectively. The electric field in the insulator is related to the electric field in the semiconductor by Gauss's law

$$\varepsilon_i E_i = \varepsilon_s E_s \tag{25}$$

and the voltage drop across the insulator is

$$V_i = E_i d = \frac{\varepsilon_s}{\varepsilon_i} E_s d \tag{26}$$

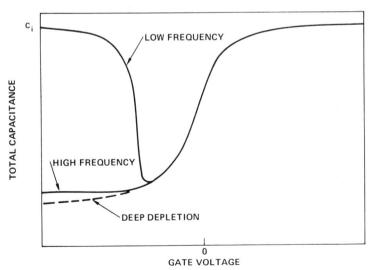

Figure 5. Differential capacitance of an MIS diode constructed with an n-type semiconductor.

The total voltage drop across the MIS structure is then

$$V_g = V_i + V_s \qquad (27)$$

The gate bias-dependent capacitance of an MIS structure constructed with the semiconductor of Fig. 3 would then appear as shown in Fig. 5.

2.2. Surface States

The previous treatment of the fields and charges within the semiconductor ignored the effects of charge which may be trapped at the interface between the insulator and the semiconductor. In practice these cannot be ignored and may, in some instances, dominate the measurements made on MIS devices. The surface of a semiconductor can be assumed to contain energy states not present in the bulk material. These may be related to the discontinuity of the crystal lattice and the perturbation of the band structure of electrons in the bulk relative to the surface or to defects and impurities localized upon it. The corresponding energy levels can be charged or discharged as a function of the measurement frequency and bias voltage. An equivalent circuit for an MIS device, which includes the effect of surface charging, is shown in Fig. 6 where C_{ss} represents a frequency- and surface-potential-dependent capacitor related to the surface states. From a qualitative point of view, one would expect the time constants for surface charging to be a function of the surface potential. The trapping rate of the surface states should be proportional to the product of the surface carrier concentration and the number of surface traps.

The total amount of charge trapped at the semiconductor–insulator interface can be determined experimentally by comparing the measured C–V characteristics to those predicted by the equations previously developed. The technique first used for doing this, proposed by Terman,[6] consists of comparing the measured high-frequency capacitance curve to that predicted by theory. In order for this technique to be

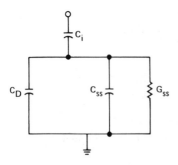

Figure 6. Small signal equivalent circuit model for the effect of surface states on the admittance of an MIS diode.

valid, the experimenter must first determine that the measurement frequency is high enough so that the surface states no longer respond and, therefore, that the corresponding surface state capacitance is zero.

In order to apply Terman's method the insulator capacitance and the semiconductor doping must be determined accurately. An accurate determination of C_i requires biasing the surface by the applied potential such that $C_{sc} + C_{ss} \gg C_i$. The measured capacitance is then equal to C_i, and the differential capacitance of the space-charge layer can be determined as a function of the gate voltage from Eq. (23). This procedure allows calculation of the corresponding surface potential V_s from Eq. (16) and semiconductor space charge Q_{sc} from Eq. (15) if the impurity doping and temperature are known. Since $V_g = V_i + V_s$, the voltage across the insulator can be determined as well as the total charge Q_T:

$$Q_T = V_i C_i = Q_{sc} + Q_{ss} \tag{28}$$

Thus

$$Q_{ss} = (V_g - V_s)C_i - Q_{sc} \tag{29}$$

The derivative of the surface state charge with respect to surface potential is commonly called the surface state density N_{ss}, that is,

$$N_{ss} = \frac{1}{q} \frac{\partial Q_{ss}}{\partial V_s} \tag{30}$$

This method yields the number of states per eV per cm^2 of all surface states, except those whose time constant is larger than the frequency of the gate bias sweep.

A different technique for measuring the same quantity, first proposed by Berglund,[7] consists of measuring the differential capacitance at a very low frequency. The measured capacitance is

$$C_m = \frac{dQ_T}{dV_g} \tag{31}$$

Recall that

$$dQ_T = C_i \, dV_i \tag{32}$$

so that

$$C_m = C_i \frac{dV_i}{dV_g} \tag{33}$$

Using

$$dV_g = dV_i + dV_s \tag{34}$$

then

$$C_m = C_i\left(1 - \frac{dV_s}{dV_g}\right) \qquad (35)$$

or

$$\frac{dV_s}{dV_g} = 1 - C_m/C_i \qquad (36)$$

This equation can be integrated to give

$$V_s(V_g) = \int_{-\infty}^{V_g} (1 - C_m/C_i)\, dV_g + \Delta \qquad (37)$$

where Δ is an additive constant. The change in surface potential between any two values of gate voltage can be determined by integrating the area between 1 and C_m/C_i on a normalized C–V curve. The constant Δ must be determined by a separate measurement which gives the surface potential unambiguously for at least one value of gate voltage or alternatively the carrier density must be determined accurately. Once the surface potential vs. gate voltage relationshp is known, then the surface state density can be determined from Eq. (30) in the same way as that used in the Terman method described earlier.

2.3. Surface Conductance

The change in conductance which occurs at a semiconductor surface in response to a change in surface potential may be calculated using a minor extension of the formalism developed for the space-charge capacitance. The quantity of interest is that defined by the equation

$$\Delta\sigma = q(\mu_e \Delta n + \mu_n \Delta p) \qquad (38)$$

where $\Delta\sigma$ is the change in conductance in mhos, μ_e and μ_n are the electron and hole mobilities, respectively, and Δn and Δp are the deviations in the total numbers of electrons and holes from their bulk values. These are defined by

$$\Delta n = \int_0^\infty (n - n_B)\, dx \qquad (39)$$

and

$$\Delta p = \int_0^\infty (p - p_B)\, dx \qquad (40)$$

Electrical Properties of Insulator–Semiconductor Interfaces

The conductance change, $\Delta\sigma$, is that which would be measured by attaching leads to a square sample and the units are commonly expressed as mhos/square or mhos/□. With the substitution of Eqs. (2) and (12) into Eq. (39), Eq. (39) becomes

$$\Delta n = \left(\frac{\varepsilon_s}{2kT}\right)^{1/2}$$
$$\times \int_{V_s}^{0} \frac{n_B(1 - e^{qV/kT})dV}{[p_B(e^{-qV/kT} + qV/kT - 1) + n_B(e^{qV/kT} - qV/kT - 1)]^{1/2}} \quad (41)$$

Substitution of Eqs. (2) and (12) into Eq. (40) yields

$$\Delta p = \left[\frac{\varepsilon_s}{2kT}\right]^{1/2}$$
$$\times \int_{V_s}^{0} \frac{p_B(e^{-qV/kT} - 1)dV}{[p_B(e^{-qV/kT} + qV/kT - 1) + n_B(e^{qV/kT} - qV/kT - 1)]^{1/2}} \quad (42)$$

In all probability neither of these integrals has an analytical solution; however, they may be evaluated numerically quite simply. The result of such a computation for 10^{15} cm^{-3} n-GaAs at 300 K, assuming an electron

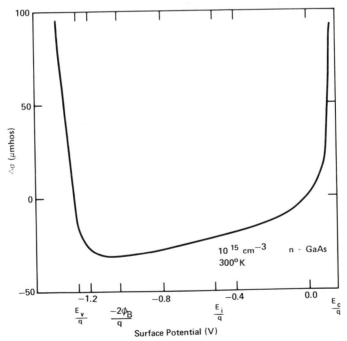

Figure 7. Surface conductance vs. V_s for 10^{15} cm^{-3} n-type GaAs at 300 K.

mobility, $\mu_e = 6000 \text{ cm}^2 \text{ V}^{-1} \text{ s}^{-1}$, and a hole mobility, $\mu_h = 400 \text{ cm}^2 \text{ V}^{-1} \text{ s}^{-1}$, is shown in Fig. 7. The surface conductance passes through zero for two values of surface potential. As may be expected it has a zero crossing when $V_s = 0$, and also crosses zero in the inversion region. A useful feature of this effect is that the conductance minimum occurs for a unique value of surface potential. If a minimum in conductance can be obtained experimentally, then this gives a reference point for determining the value of the surface potential for one value of the applied bias. Comparison of the calculated and experimental conductance curves then allows the determination of the surface potential values that are achievable.

3. Gallium Arsenide

The most intensively studied III–V semiconductor is GaAs. The successful development of the Gunn diode, the junction LED and laser, and the Schottky-barrier FET have all prompted an intensive effort on the growth and characterization of bulk and epitaxial GaAs.[8,9] The ready availability of high-quality GaAs substrate material and the potential advantages to be gained if the high carrier mobility of bulk electrons in this material could be also obtained in the surface layers have, in combination, led to the investigation of GaAs MIS structures in a large number of laboratories.

The literature on this topic includes work on the free surface, on deposited dielectrics, and on anodically or thermally grown native oxides. Unfortunately, a simple chronological description of the work on GaAs–dielectric interfaces would do more to confuse the reader than enhance his or her understanding of this subject. The reason for this is that the surface state density in GaAs is much higher than in silicon and these states can respond to frequencies greater than those observed experimentally on silicon MOS devices. In consequence, the $C-V$ data obtained on GaAs MOS structures are unlike those ever observed on silicon devices. However, by assuming that the data obtained on GaAs can be interpreted in a manner similar to that of Si, a number of unwarranted and erroneous conclusions about the surface properties of GaAs have been reached by some investigators. The most commonly made mistake can be understood by considering the data in Fig. 8 obtained on an n-type GaAs crystal in which the insulating layer was formed by anodic oxidation.[10] If this were a silicon device one might assume that the surface reached inversion with negative gate bias since the capacitance curve has become flat as predicted by the theory and as illustrated in Fig. 4.

From the minimum value of capacitance in inversion, the doping density in the semiconductor could be calculated and given this parameter

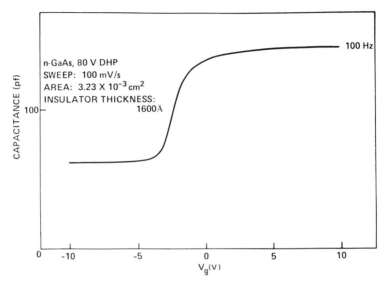

Figure 8. Room-temperature $C-V$ characteristic of anodized dielectric on n-type GaAs taken at 100 Hz.

as well as the oxide capacitance an analysis of the data could be performed by Terman's [6] method. The surface state density dependence on surface potential calculated in this manner would suggest that the Fermi level at the surface may be swept through the entire band gap from the valence band to the conduction band. This is guaranteed from the built-in assumptions because: (1) inversion was assumed in the choice of the doping density parameter and (2) a high-frequency capacitance curve has been assumed in performing the Terman [6] analysis (i.e., no surface states respond at the small signal measurement frequency). If under these conditions, the value of the device capacitance saturates with positive bias at the value of the oxide capacitance then the surface must be in accumulation and the Fermi level at the surface would be near the conduction-band edge. However, if motion of the surface state charge contributes to the measured capacitance with positive gate bias then it is no longer possible to calculate either the position of the surface potential with any confidence or the distribution of surface states in the band gap.

Capacitance–voltage data obtained over an extended frequency range for the GaAs device in Fig. 8 are shown in Fig. 9. Several features are noted:

1. The dip in the quasistatic $C-V$ curve is very shallow and integration of the area between this curve and a horizontal line through C_{ox} to yield the V_s vs. V_g relationship as indicated by Berlung [7]

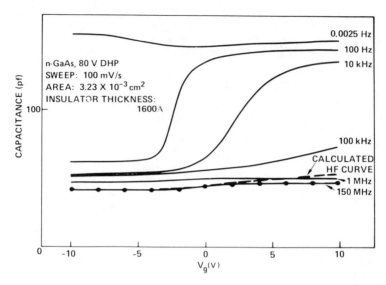

Figure 9. C–V characteristics of the device of Fig. 8 taken over an extended frequency range.

    gives a total surface potential change of 0.4 V; this is not consistent with the attainment of both accumulation and inversion of the surface since E_g, GaAs = 1.43 eV and $V_s|_{acc} - V_s|_{inv} \simeq 1.2$ V.
2. The high-frequency curves become nearly flat as the frequency is raised; if the Terman analysis of the 1-MHz data is performed using the value of carrier concentration obtained from a Hall measurement, the total surface potential change is again $\simeq 0.4$ V, in agreement with the quasistatic data.

The quasistatic and high-frequency measurement thus seem to simultaneously predict that the total surface potential change is severely limited yet do not allow calculation of the Fermi-level pinning position in the band gap unless the doping is known accurately. Determination of the doping density from deep-depletion measurements seems to be the most straightforward means of doing this for several reasons. If the doping density is calculated from a measured carrier density one has to rely on separate measurements of the Hall factor, effective mass, degeneracy of states in the donor or acceptor levels, and location of these levels with respect to the band edges. In addition, the substrate material is often inhomogeneous. A Hall measurement averages out these inhomogeneities, whereas a deep-depletion measurement probes the material only under the capacitor gate and seemingly gives a more meaningful value for use in the surface parameter calculations.

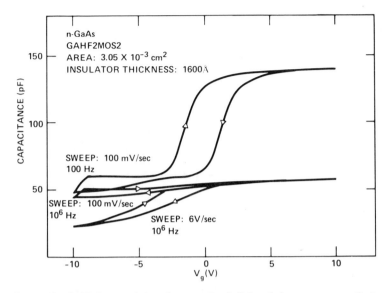

Figure 10. $C-V$ characteristics of an anodized dielectric layer on n-type GaAs.

If the doping density for the device of Fig. 10 is calculated from the equation:

$$N = 2\left(q\varepsilon_s \frac{d(1/C_m^2)}{dV}\right)^{-1} \qquad (43)$$

and the surface potential is calculated from Eq. (16), the result is that the Fermi level at the surface is restricted to a region from $\simeq 0.6$ eV below the CBM to $\simeq 1.0$ eV below the conduction band minimum (CBM). It is therefore clear that there is no accumulation of electrons for this sample. Less certain is the issue of whether or not inversion is possible. Taken at face value, this measurement indicates that an inversion layer does not form since the Fermi level does not approach closely enough to the valence band maximum (VBM). One can argue, however, that the cumulative error of the measurements might be enough to give a 0.1–0.2 eV error in the surface potential in which case the surface may actually invert. We will return to this question later in the discussion.

Capacitance–voltage data from 0.0025 Hz (quasistatic) to 150 MHz are shown in Fig. 11 for a p-type (100) GaAs sample which had an anodically formed dielectric annealed at 400°C for 12 min in forming gas.[11] The room-temperature 1-MHz $C-V$ curve rises with negative gate bias indicating that the surface is accumulating. However, when the measurement frequency is increased further the $C-V$ characteristic becomes nearly flat in the same manner as the data on the n-type sample,

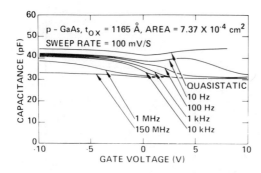

Figure 11. C–V characteristics of an anodized dielectric layer on p-type GaAs.

indicating that accumulation does not actually occur. Again the dip in the quasistatic C–V curve is very shallow, indicating a total change in surface potential of ~ 0.4 V. Calculation of the surface potential modulation in the same manner as discussed for the n-type specimen indicates that the Fermi level at the surface can be swept over a range from 0.46 eV above the VBM to 0.92 eV above the VBM for electric fields of the order $\pm 10^6$ V/cm. This result implies that neither accumulation nor inversion is occurring in the p-type sample.

It is well known that for silicon MOS device processing, conditions which, for example, produce an n-type surface, will produce this result regardless of the carrier type in the bulk of the semiconductor. In this example, the surface on n-type material would be accumulated and the surface on p-type material would be inverted. A similar trend is indicated for GaAs. This can perhaps be understood by considering that for a typical bulk doping density of an MIS device of 10^{16} cm^{-3} the number of impurity atoms at the surface is 4.6×10^{10} cm^{-2}. If this is compared to the total number of atoms at the surface of $\simeq 10^{14}$ cm^{-2} or even a typical interface state density on imperfect surfaces of 10^{12} cm^{-2}, it can be seen that bulk doping plays a secondary role in determining the quantity and the polarity of charge carriers in the surface region. With this in mind, it seems unreasonable to suppose that the n-type GaAs sample can be inverted while at the same time no accumulation is seen on the p-GaAs sample. The n- and p-type samples were prepared using the same processing steps and based on our experience with silicon MOS devices it seems highly likely that the same type of surface would be produced for each sample. The conclusion is therefore that for surface electric fields of the order of $\pm 10^6$ V/cm the position of the Fermi level at the surface is restricted to the lower two-thirds of the band gap and that neither accumulation nor inversion layers can be induced with static gate bias.

This argument is reinforced by considering the data of Kohn and Hartnagel[12] on n- and p-type MOS capacitors with anodically formed oxides which show the effect of lowered temperature on the 1-kHz C-V measurements. The data on the n-type sample shown in Fig. 12 look like a well-behaved curve on silicon at room temperature and above. However, as the temperature is lowered the curves become almost flat except for a large deep-depletion loop which occurs when the gate bias potential is swept from positive to negative voltages. The deep depletion loop occurs because electron emission from the surface traps is reduced as the temperature is lowered and the bias sweep rate used in this measurement is too fast to permit the surface traps and the semiconductor bulk to remain in equilibrium. The C-V data on the p-type sample shown in Fig. 13 are exactly analogous to those on the n-type sample. Again the curve goes flat at 77 K except for a deep-depletion loop caused by the fast ramp rate. One significant difference is that for this p-type surface 7 MHz is not a high-frequency measurement at room temperature.

It has been consistently observed in our work that the surface state charge on p-type GaAs samples responds at higher frequencies than the surface state charge in n-type samples prepared in a similar fashion. This

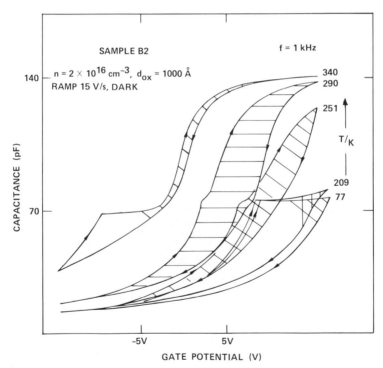

Figure 12. Effect of lowered temperature on the C-V characteristic of an n-type GaAs MIS diode (after Kohn and Hartnagel[12]).

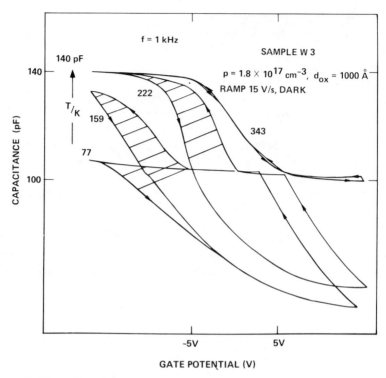

Figure 13. Effect of lowered temperature on the $C-V$ characteristic of a p-type GaAs MIS diode (after Kohn and Hartnagel[12]).

is consistent with the observation that the quiescent surface potential for n-type MOS devices is 0.8 V, whereas the quiescent band bending for p-type devices is $\simeq 0.6$ V. The amplitudes of the small signal currents which flow during the differential capacitance measurements are controlled by the emission and absorption of majority carriers from surface traps. The emission rate is controlled by the height of the local potential well in which the trapped charge resides; it is only indirectly related to the semiconductor barrier height. The absorption rate, however, is directly proportional to the current flow from the semiconductor bulk to the surface; it is exponentially related to the barrier height. In GaAs the lower barrier height of the p-type material permits a larger current to flow between the semiconductor bulk and the surface traps and results in a surface state capacitance (C_{ss}) which responds at higher frequencies in this material. Qualitatively, the effect of increasing temperature on the frequency response of the interfacial charge of GaAs MIS devices is illustrated in Fig. 14. The peak in the spectral density curve ($\partial N_{ss}/\partial \omega$ vs. ω) shifts to higher frequencies as the temperature is raised; however, the

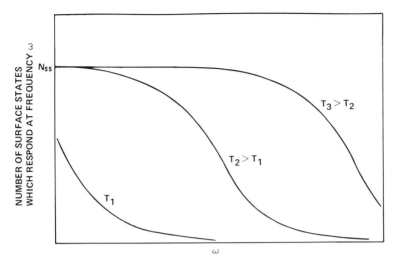

Figure 14. Effect of increased temperature on the frequency response of the interfacial charge of GaAs MIS diodes.

total surface state density ($N_{ss} = \int_0^\infty (\partial N_{ss}/\partial \omega)d\omega$) seems not to be a strong function of temperature.

There is some evidence that dielectric deposition procedures which increase the surface disorder on the semiconductor or cause the semiconducting material at the surface to be nonstoichiometric also cause the surface states to respond at higher frequencies. Figures 9, 15, and 16 depict C-V measurements on MIS capacitors formed using n-type GaAs substrate material on which the dielectric had been formed by either anodizing the surface at 8 μmA/cm^2 in 0.03 M aqueous ammonium dihydrogen phosphate and annealing at 400°C for 20 min in 10H$_2$–90He; by the pyrolytic deposition of silicon nitride at 600°C in a SiH$_4$–NH$_3$ process; or by low-energy reactive sputtering of silicon nitride using a Si target and an argon/nitrogen atmosphere. These processes would be expected to produce varying amounts of damage to the substrate during the insulator formation. This would be determined in part by the maximum temperature of the substrate during the process and by the maximum kinetic energy of the particles to which the surface has been exposed. The least damaged surface would then be the one formed by aqueous anodization; the next might be the one which had a pyrolytically deposited layer at 600°C; and the most heavily damaged interface would be the one exposed to the sputtering process. At very high frequency (150 MHz) the C-V characteristics of these devices are nearly identical indicating that the total surface potential modulation and total surface density for these devices are nearly identical. At intermediate frequencies

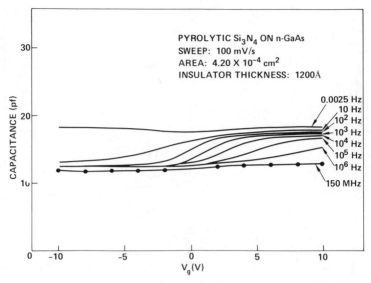

Figure 15. $C-V$ measurements of n-type GaAs with a dielectric consisting of pyrolytic silicon nitride deposited at 600°C.

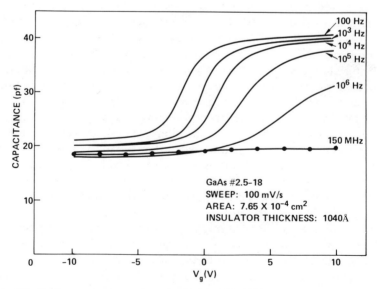

Figure 16. $C-V$ measurements of n-type GaAs with a dielectric consisting of reactively sputtered silicon nitride at low energy.

the surface states of the more heavily damaged surfaces give a larger response, especially for positive gate bias on the n-type substrates.

In view of the results presented above, rather specific conditions must be observed before it can be argued convincingly that either inversion or accumulation layers can be induced on the GaAs surface. If the arguments are based on $C-V$ measurements on MIS diodes, then the $C-V$ curve must rise up to the value of the insulator capacitance when the gate electrode is biased in the direction to induce accumulation. This measurement must be performed at a frequency which is sufficiently high so that the capacitance is no longer frequency dependent in order to ensure that the surface states do not respond. At room temperature this frequency may be quite high (100 MHz or more) if the surface is heavily damaged. If $C-V$ measurements are performed at low temperatures (77 K), then a 1-MHz measurement will probably suffice.

As previously discussed the uncertainties involved in accurately determining the substrate doping density make it risky to claim that inversion layers have been observed based on $C-V$ measurements alone. For this purpose surface transport measurements on either MISFET or four-terminal gate van der Pauw structures are needed. In addition, in order to measure the entire surface state population using this technique it is necessary to measure the surface conductivity as a function of the dc gate bias rather than using pulsed or ac techniques.

3.1. Chemically Clean Surface

The term chemically clean surface is used to represent those surfaces which have been freshly etched or cleaved and are covered with a thin native oxide layer which immediately forms on GaAs upon its exposure to air. Surface conductivity measurements are usually chosen to characterize such surfaces because dielectric spacers must be used to induce the perpendicular electric field. The contacting surfaces of the semiconductor and dielectric spacers can never be perfectly flat, hence the air gap between the spacers and the semiconductor is a function of the force holding the two materials together. This force increases as the magnitude of the electric field is increased making the net capacitance of air gap plus dielectric spacers a U-shaped function of the gate voltage. The variability of the insulator capacitance makes it difficult if not impossible to analyze the $C-V$ curve of the total device but does not significantly affect the surface conductance measurement.

Gerlich[13] performed surface conductivity measurements on n-GaAs (0.16 Ω-cm) at room temperature using a metal gate electrode and an unnamed dielectric spacer placed on one side of the sample to induce surface potential changes in the semiconductor. He observed a monotonically increasing surface conductivity as the voltage on the gate electrode of a sample etched in a $H_2SO_4/H_2O/H_2O_2$ (3:1:1) solution was increased from negative to positive.

The reader will recall from the discussion of surface conductivity that it is possible to determine experimentally the relationship between gate voltage and surface potential only to within an arbitrary constant unless a minimum is observed in the surface conductivity vs. gate voltage curve. Thus Gerlich could not determine if the surface was accumulated or depleted. In a similar device, which used mylar as a dielectric spacer and $HF/HNO_3/H_2O$ (1:3:2) for the etchant, Pilkuhn[14] observed a minimum in conductance as the gate voltage was swept at a rate of 0.8 Hz. The total conductance change observed was 7×10^{-8} mho/□ when ideally the total change should have been 1×10^{-4} mho/□. This result indicates that the surface potential variations obtained were very small and the surface was nearly pinned. Flinn and Briggs[15] performed surface photovoltage and conductance measurements on $\simeq 3 \times 10^{15}$ cm^{-3} n- and p-type single-crystal GaAs in a configuration employing transparent metal gates and polythene dielectric spacers on both faces of a 0.02-cm-thick semiconducting filament. They observed that any dc voltage step applied to the field electrode produced a conductance change which decayed to zero in less than a second, and from this concluded that for sufficiently long times all of the induced charge was trapped in surface states with a density greater than 10^{12} cm^{-2}. They were not able to obtain saturated photovoltage measurements on either n- and p-type material and concluded that for n-type material the zero-bias surface potential, V_{s0}, is $\simeq -0.5$ V and on p-type material $V_{s0} \simeq 0.15$ V. Although somewhat inconclusive the above experiments do lead to three important inferences. The Fermi levels at the surfaces of chemically clean n- and p-type GaAs devices are nearly pinned by surface states of density greater than 10^{12} cm^{-2} eV^{-1}. The pinning position is at least 0.5 V below the CBM for n-type material and at least 0.15 eV above the VBM for p-type material. Relatively slow surface states might be responsible for the pinning since a conductance minimum was observed when the gate bias was swept at 0.8 Hz and because tenths of seconds were required for the surface to reach equilibrium after application of a dc gate bias. It will become apparent as the discussion continues that these same trends are observed when dielectric layers are formed on GaAs.

3.2. Native Oxides

The native oxide prepared by thermal oxidation of GaAs is too conductive to allow meaningful $C-V$ measurements. Butcher and

Sealy[16] oxidized (110) n-GaAs in dry flowing O_2 at 510°C and obtained films with electrical resistivities of 10^9 and 10^{11} Ω-cm. Rutherford backscattering measurements on the grown oxides indicated that the films consisted mostly of gallium and oxygen in a 2:3 ratio. Films thinner than 500 Å exhibited Schottky-barrierlike I-V characteristics.

Highly insulating ($\rho = 10^{15}$–10^{16} Ω-cm) native oxide layers have been formed on GaAs by anodic oxidation in solution. Layers have been formed by anodization in a solution of: AGW (1 part of 3% aqueous tartaric or citric acid mixed with two to four parts of propylene glycol,[10–12,17–23] hydrogen peroxide in water,[18,19] $KMnO_4$ in dimethylformamide,[18] potassium dichromate in ethylene glycol,[24] and ammonium pentaborate in ethylene glycol.[22,25] Gallium arsenide native oxide layers have also been grown anodically in both dc and rf-generated oxygen plasmas.[26–33] The resistivity of the layers grown in this manner is considerably less— typically 10^{11}–10^{12} Ω-cm. The electrical properties of insulating layers on GaAs prepared either by plasma anodization or by anodization in solution are all very similar. The layers as grown exhibit large unstable hysteresis loops in their C-V characteristics and no apparent well-defined value of oxide capacitance.[21] When the layers are annealed at temperatures between 350 and 450°C the C-V curves lose most of their hysteretic behavior and become more stable. For this reason most of the reported data concern annealed specimens. The data in Figs. 9 and 11 are characteristic of those seen on the anodized GaAs specimens. A high-frequency curve on n-type material is typically obtained at 1 MHz, whereas such behavior on p-type substrates is not obtained even at 15.8 MHz.[19]

3.3. Deposited Insulators

Workers at RCA[24–37] reported the first C-V data on GaAs surfaces coated with a deposited dielectric material. By pyrolytically depositing silicon dioxide on GaAs at 730°C they were able to produce an MIS device whose capacitance could be changed by an applied gate voltage. At 1 MHz these C-V data are similar to those of Fig. 16. Given the high deposition temperature and the low measurement frequency (1 MHz) it is extremely likely that the surface states contribute to the measured device capacitance. This would render invalid a Terman analysis of the data. The most conclusive indication of a high interface state density for this surface is the fact that MOSFETs prepared using this dielectric exhibited a transconductance which at dc dropped to ~15% of its value at 120 Hz.[37] Foster and Swartz[38] constructed Si_3N_4-GaAs MIS devices by pyrolytically depositing Si_3N_4 using a silane–ammonia process with deposition temperatures in the range 650–750°C. They report C-V measurements on $n = 4 \times 10^{16}$ cm^{-3} and $p = 1.4 \times 10^{17}$ cm^{-3} specimens in the frequency range from 5 kHz to 1.5 MHz. The measured capacitance values continued to drop as the frequency was increased to 1.5 MHz indicating that the surface states were still responding. Klose et al.[39] deposited layers on n-GaAs substrates by pyrolyzing $SiCl_4$ and N_2H_4 vapors using either N_2 or Ar as the transport gas. They show $C-V$ curves taken at 3, 5, and 10 MHz which coincide for all values of gate voltage and which appear to show evidence of accumulation with positive gate bias. The fact that the 3-, 5-, and 10-MHz $C-V$ curves coincide indicates that a high-frequency capacitance has been achieved and that a Terman analysis of the data is valid. Unfortunately, further follow-up work has not been reported from this group. Ito and Sakai[41] report gate-modulable conducting surface layers in MIS transistors constructed on (111)-oriented p-type GaAs substrates and employing SiO_2, Al_2O_3, and double-layer SiO_2–Al_2O_3 gate dielectrics. The transistors they report are normally off and the source–drain current is increased when positive voltage is applied to the gate. This fact combined with a maximum field-effect mobility of $2240\ cm^2\ V^{-1}\ s^{-1}$ indicates the conduction is due to electrons. The fabrication process they employed required annealing at 700°C for 5 min and insulator growth at 680°C. It is known that the surface of GaAs material can convert to give an n-type layer during prolonged high-temperature processing.[41] Thin n-type layers have been purposely grown on semi-insulating GaAs substrates to produce a type of normally off MOS transistor.[42] In such a device the normally occurring 0.8-eV surface barrier can completely deplete carriers from the n-type layer if it is sufficiently thin. Positive voltages on the gate decrease the width of the depletion layer and allow source–drain current to flow. Although such a device can be normally off, it is not a true inversion-mode transistor and the maximum current is smaller than in an inversion device. Ito and Sakai indicate that their $C-V$ measurements on these interfaces are similar to those of other workers in that they see "abnormal frequency dispersion." This rather conflicting evidence makes it difficult to conclude whether or not they have actually observed inversion on the p-type GaAs surface.

Aluminum oxide layers have been grown on GaAs substrates by using molecular beams of Al and O_2[43–48] and by the pyrolytic decomposition of aluminum isopropoxide in argon at 300–500°C. The electrical properties of these interfaces were evaluated by $C-V$ measurements, the results of which look qualitatively like the data of Figs. 9 and 11. Neither inversion or accumulation seems to be occurring.

CVD SiO_2 layers[49] have been grown on $n = 6 \times 10^{16}\ cm^{-3}$ (100)-oriented GaAs wafers and the electrical properties have been evaluated by means of $C-V$ and $G-V$ measurements. The behavior of the $C-V$ data is similar to that in Fig. 9. The authors have used a 0.2-Hz small signal capacitance measurement in calculating the surface potential *vs.* gate voltage relationship in the manner proposed by Berlung.[7] At the same time the gate voltage was swept at a rate of 3.75×10^{-5} Hz. The Berglund analysis is only valid if the number of charges which respond to the capacitance measuring frequency is the same as the number which responds to change of gate bias by the slowly varying gate ramp voltage. If this is not the case, larger variations in surface potential will be predicted than are actually occurring. This has led the present investigators to calculate a peak in surface state density near midgap and a monotonically decreasing density as the valence-band edge is approached. This calculated peak is probably an artifact due the surface traps which are responding between 0.2 and 3.75×10^{-5} Hz.

Suzuki *et al.*[50] deposited gallium oxynitride layers on $n = 1.1 \times 10^{16}$ cm^{-3} (111)B-oriented GaAs surfaces which had been either chemically etched or sputter etched in N_2 immediately before deposition. Based on *C–V* measurements they see an increase in the response of the surface states at high frequencies when the surface was sputter etched beforchand.

Bayraktaroglu *et al.*[51] and Bagratishvili *et al.*[52] have deposited germanium nitride layers on GaAs substrates by evaporating either Ge_3N_4 powder[51] or germanium[52] in a hydrazine ambient. Pande *et al.*[53] have deposited Ge_3N_4 layers on $n = 5 \times 10^{16}$ cm^{-3} (100)-oriented GaAs substrates by evaporating germanium in a 20 mTorr nitrogen ambient in which a slow discharge was maintained. Pande *et al.* see accumulationlike *C–V* behavior at 10^5 Hz and Bayraktaroglu *et al.* see accumulationlike *C–V* behavior at frequencies as high as 10 MHz. It is difficult to conclude whether or not accumulation was actually being observed. Although, the authors do not discuss the matter, Ge_3N_4 is normally a more conducting material than either SiO_2 or Si_3N_4. This increased conduction can give rise to reduced hysteresis and increased response of the surface states at high frequencies; also the position of Fermi level in the dielectric is in question.

If asked to give a few short words of advice to anyone attempting measurement of the surface properties of GsAs MOS devices, I would include the following. If at all possible, start with a high-quality dielectric (one with a bulk resistivity $> 10^{16}$ cm at low fields). Rather anomalous charge drifting effects can occur even in the high-frequency measurements in the presence of significant charge leakage through the dielectric. Also charge leakage prevents obtaining the quasistatic data which are probably the most unambiguous measurement for obtaining the total change in surface potential with applied gate bias. High-frequency measurements unless taken at very high frequencies (100 MHz) can be misleading since the surface states can respond at such high frequencies on GaAs. An alternative is to cool the devices to 77 K or less to reduce the carrier generation rate in the depletion layer and perform the C–V measurements at 1 MHz. Calculations of the surface potential changes from both the quasistatic measurements and the high-frequency measurements should give the same result. Given the number of false starts that have been reported in the literature it is probable that no one working in the field will believe that accumulation or inversion have been achieved until both n- and p-channel normally off MISFETs operating at dc have been demonstrated.

GaAs surfaces which have been exposed to air or metalized are depleted with a somewhat larger barrier height on n-type material than on p-type material. This trend is continued as dielectric layers are either grown or deposited on the GaAs surface. The surface state density on MIS devices is typically a U-shaped curve with a minimum value of about 2×10^{12} cm^{-2} eV^{-1} which rapidly rises to the 10^{14} range at 0.2–0.3 eV on either side of the minimum.

4. Indium Antimonide

Early work on the surface properties of InSb crystals was performed using either mica or mylar insulating sheets to support a field electrode for making surface conductivity measurements. Eaton et al.[54] performed such measurements at 77 K on Czochralski-pulled Ge-doped (p-type, $\mu_h = 7000$ cm^2 V^{-1} s^{-1}) crystals with resistivities ranging from 0.3 to 14 Ω-cm. Prior to measurement, the surface was etched in a mixture of nitric acid and ethylene glycol. With a 100-μm-thick mica insulator and ±300 V on the gate electrode they achieved values of the normal electric field of ±3 × 10^4 V cm. They observed changes in the surface conductivity ranging from 3×10^{-10} to 5×10^{-9} mho/($\square$-V). If these are compared to the calculated value of $\Delta\sigma_s/\Delta V = 4 \times 10^{-8}$ mho/($\square$-V) which is obtained if surface charging is neglected, then the ratio of the change in trapped charge to the change in mobile charge is approximately 15. This ratio will be referred to subsequently as the interfacial quality factor. They observed a peak in the $\Delta R/R_0$ vs. V curve for one sample indicating that the p-type surface could be inverted by applying a negative gate voltage. Davis[55] performed ac and dc surface conductance measurements at 79 K on single-crystal n- and p-type samples which had been etched in CP-4 (75 ml HF, 150 ml HNO$_3$, 75 ml acetic acid, 2 ml Br) and rinsed in 10^{-6} M Na$_2$S (aq). This trace doping reduced N_{ss} by a factor of 100. The conductance response of the n-type samples saturated for frequencies greater than 10^4–10^5 Hz. From this he was able to equate the field-effect mobility to the mobility of the surface carriers (since the surface states no longer responded) and calculate the excess carrier concentrations n_s and p_s from the dc conductivity measurement. He concluded that a fast-state population of $N_{s2} = 2 \times 10^{15}\,\text{cm}^{-2}$ exists near the conduction-band edge and another (N_{s1}) with density in excess of $7 \times 10^{15}\,\text{cm}^{-2}$ exists near the valence-band edge. A quality factor of 24 is calculated in the band gap from his measurements. (See the section on surface states for a discussion of the quality factor.) Interestingly, one would predict

$$\frac{N_{s1}}{N_{s2}} \simeq \left(\frac{m_h}{m_e}\right)^{3/2} = (15)^{3/2}$$

if the interfacial charging were linearly related to the surface electric field. Davis reports $N_{s1}/N_{s2} = 17$ in good agreement with this picture.

Following the work of Dewald[56] on the chemistry of InSb anodization, Mueller and Jacobson[57] studied the photocontrolled surface conductance of $p = 2 \times 10^{15}\,\text{cm}^{-3}$ single-crystal InSb which had been etched in CP-4 and anodized in 0.1 N KOH until the oxide layer was 0.1 μm thick. At 77 K the samples had an n-type inversion layer which could be changed to a p^+ accumulation layer if the surface was exposed to photons in the 0.8–3.5 eV energy range. They interpreted this effect as being due to photoemission of electrons from the semiconductor into the oxide. The negative-charge layer in the oxide would induce a positive surface charge in the InSb.

Chang and Howard[58] reported the first $C-V$ measurements on the oxide formed by anodization of InSb. Czochralski crystals of (111) orientation with $p = 10^{15}\,\text{cm}^{-3}$ and $n = 8 \times 10^{14}\,\text{cm}^{-3}$ at 77 K were polished, etched in HF, and anodized in 0.1 N aqueous KOH for 15 min at constant voltage. Capacitance–voltage measurements were performed at 77 K on the MOS diodes which had been formed by evaporating gold dots on top of the anodized layers. They observed saturation of the $C-V$ curves with both positive and negative gate bias on each type of sample. They concluded from this that both accumulation and inversion were achievable. This is supported by the fact that 1-MHz and 1-KHz capacitance measurements on the p-type samples showed very little dispersion in the accumulation–depletion regime. Both n- and p-type samples had a positive charge in the oxide at zero gate bias, $Q_{ss0}/q = 0.7\text{--}3.7 \times 10^{11}\,\text{cm}^{-2}$. The anodized layers had a rather low breakdown strength of 10^6 V/cm.

Huff et al.[59] measured the large signal ac field-effect at 225 Hz of $n = 6 \times 10^{13}\,\text{cm}^{-3}$ and $p = 5.5 \times 10^{14}\,\text{cm}^{-3}$ crystals which had been etched in HF (40):HNO$_3$:H$_2$O (2:2:3), rinsed in deionized water and subsequently treated in either aqueous sodium sulfite or copper nitrite, or in chlorine gas. Aluminum field plates on each side of the sample were isolated using 0.005-in. mylar spacers. At 113 K they reported field-effect mobilities of 11,000 cm^2 V^{-1} s^{-1} for electrons in an inversion layer and 700 cm^2 V^{-1} s^{-1} for holes in an accumulation layer. They found, in agreement with Davis,[55] that the surface state density at the conduction-band edge is less than that at the valence-band edge. The ratio N_{ssV}/N_{ssC} varied from 1.6 to 5 depending on the sample preparation and orientation. Chang and Howard[60] performed a series of experiments in which the faces of (110), (111), and ($\overline{111}$) n- and p-type crystals were anodized in 0.1 N aqueous KOH. C-V measurements on MOS diodes employing gold gate electrodes indicated a zero-bias oxide charge $Q_{ss0}/q = 4 \times 10^{11}$–$9 \times 10^{11}$ cm^{-2}—positive for all orientations. Lile and Anderson[61] investigated the properties of evaporated polycrystalline InSb layers employing an anodized Al$_2$O$_3$ insulator and aluminum field plate. A Terman[6] analysis of C-V measurements at 77 K on $n \simeq 10^{16}$ cm^{-3} layers yielded a midgap density of surface states $N_{ss(min)} = 4.5 \times 10^{12}$ cm^{-2} eV^{-1} and a *negative* zero-bias oxide charge $Q_{ss0}/q = 1.5 \times 10^{12}$ cm^{-2}.

Komatsubara *et al.*[62] reported surface conductance and C-V measurements on 3-terminal MIS structures formed by evaporating SiO and aluminum sequentially onto $p \simeq 10^{13}$ cm^{-3} single-crystal InSb. They calculate the surface carrier concentration from the C-V data and use this result to obtain the field-effect mobility of electrons in the surface inversion layer from the surface conductivity data. Unfortunately, a well-defined value of oxide capacitance is not apparent from their C-V measurement and quantitative analysis of the data can only be considered speculative. Komatsubara *et al.*[63] repeated these measurements using the native oxide prepared by anodization in 0.1 N aqueous KOH and an aluminum field plate. The C-V measurements indicate a positive oxide charge Q_{ss0} and a field-effect mobility of electrons in the inversion layer $\mu_{FE} \simeq 5000$ cm^2 V^{-1} s^{-1} at 4.2 K. Hung and Yon[64] performed C-V measurements at 77 K on (111)-oriented $n = 7.5 - 9.5 \times 10^{14}$ cm^{-3} InSb samples which had been anodized in either 0.1 N aqueous KOH or a solution of KNO$_2$ in tetrahydrofurfuryl alcohol. Evaporated gold dots completed the MIS structure. Capacitance measurements at 1 MHz indicated that the dielectric constant of oxide formed in 0.1 N KOH was not constant as a function of thickness. Measurements on layers a few hundred angstroms thick indicated $\varepsilon_{ox} \simeq 16$, while $\varepsilon_{ox} \simeq 10$ for layers 4500 Å thick. Layers formed in KNO$_2$ solutions had a dielectric constant $\varepsilon_{ox} \simeq 14$. The C-V measurements indicated a *negative* oxide charge at zero gate bias $Q_{ss0}/q = -10^{11}$ to -10^{12} cm^{-2} which became increasingly negative if positive voltage was applied to the gate electrode. Electron microprobe analysis of the anodized layers indicated that oxides formed in KOH solutions were depleted of antimony, whereas the anodic oxide grown in the KNO$_2$/tetrahydrofurfuryl alcohol solution had a stoichiometric indium/antimony ratio.

Korwin-Pawlowski and Heasell[65] performed 100-kHz C–V measurements at 77 K on anodized MOS capacitors formed on $p \simeq 10^{14}$ cm^{-3} (77 K) (211)-oriented substrates. The semiconductor surface was prepared by etching in CP-4A, followed by rinsing in either H-100 (70 g KOH + 4 g tartaric acid + 8 g ethylenediamine tetraacetic acid + 78 ml H$_2$O mixed before use with 30% H$_2$O$_2$ in a volume ratio of 5:2) or 10^{-5} M Na$_2$S in deionized water. Following a deionized water rinse, the samples were anodized at constant current (1 mA/cm^2) in either 0.1 N KOH or 2-KNO$_2$ in tetrahydrofurfuryl alcohol. Gold gate electrodes were used. Previous work by Henneke[66] had shown that CP-4 etched surfaces were indium rich and the H-100 etch appeared to restore stoichiometry. The surfaces treated with H-100 gave a negative oxide charge $Q_{ss0} = -4.75 \times 10^{12}$ cm^{-2} either with or without the Na$_2$S flush. If the surface was etched in CP-4 and treated only with the aqueous Na$_2$S solution before anodization, a *positive* interfacial charge $Q_{ss0} \simeq 1.5 \times 10^{11}$ cm^{-2} was observed. They also fabricated MOS diodes employing silicon monoxide vacuum-evaporated layers and observed that the interfacial charge was *negative* ($Q_{ss0} = -6.8 \times 10^{11}$ cm^{-2} regardless of whether the surface was treated in CP-4, H-100, and Na$_2$S or CP-4 and Na$_2$S).

Kim[67] reported the first C–V and G–V measurements on InSb MIS structures employing chemical-vapor-deposited (CVD) layers. Details of the method of depositing the silicon oxynitride (SiON) were not disclosed except to the extent that the devices consisted of NiCr metal gates on 1500-Å SiON layers deposited on chemically etched n- and p-type surfaces. From ac conductance measurements, he calculated, for n-type surfaces, $N_{ss} = 3.3 \times 10^{12}$ cm^{-2} and, for p-type surfaces, $N_{ss} = 1.3 \times 10^{12}$ cm^{-2}. It is also implied that the hysteresis in the C–V measurements is less than that observed on anodically prepared oxides. Q_{ss0} was *positive* for both n- and p-type surfaces.

Etchel and Fischer[68] performed capacitance measurements at 77 K in the frequency range from 10 to 10^5 Hz on (211)-oriented Czochralski-grown $n = 15 \times 10^{14}$ cm^{-3} (77 K) substrates which had been anodized in 0.1 N aqueous KOH. Substrates were first electropolished in 4% HNO$_3$ in glycol, rinsed in deionized water, and then anodized at constant current in the range 0.1–1.0 mA/cm^2. The frequency dispersion of the capacitance curves indicated that the samples could be both accumulated and inverted. Approximately 3-V hysteresis was observed in the C–V surves when the gate voltage was swept between ±10 V. The charge stored in the insulator varied linearly with the maximum voltage applied to the device. This charge became more negative with each bias cycle. Q_{ss0} was *positive*. They observed a voltage shift in the C–V plots of 0.48 V when the gate electrode was switched from Al to Au. Au inverted the n-type material, whereas Al accumulated it.

Fufiyasu et al.[69] investigated the properties of borosilicate glass-anodized InSb oxide double-layer structures. Single-crystal (111)-oriented $p = 10^{14}$–10^{15} cm^{-3} at 77 K slices were lapped, polished, and etched in CP-4, anodized in 0.1 N KOH for a few seconds, and etched in dilute HF. Samples were then etched for 1 min in 0.1 N aqueous KOH to a thickness of about 100 Å. CVD borosilicate glass was then deposited in an ambient of SiH$_4$, B$_2$H$_6$, O$_2$, N$_2$, and argon at a temperature of 300°C. Layers 300–1000 Å thick had a resistivity of 10^{13} Ω-cm and a breakdown strength of 10^7 V/cm. Gate electrodes were a mixture of Al and In. For a gate voltage swing of ±8 V, the hysteresis in the C–V curves was reduced from 3 to 5 V for the anodized layers to 0.2–0.5 V for the double layers. The surface state density of these devices was calculated from the Terman[6] method to be in the mid-(10^{12}–10^{13}) eV^{-1} cm^{-2} range. Nakagawa and Fujisada[70] performed 1-MHz C–V measurements at 77 K on $n = 1.5 \times 10^{14}$ cm^{-3} and $p = 4.4 \times 10^{14}$ cm (111) B-oriented substrates anodized to a thickness of 3000–4000 Å. The n-type sample was anodized in 0.1 N aqueous KOH and the p-type sample was anodized in AGW (2 parts of ethylene glycol to 1 part of dilute citric acid with the pH 5, adjusted by ammonia). Q_{ss0} was *positive* for both samples and the surface state density as calculated from the Terman[6] method was in the 10^{12}–10^{13} eV^{-1} cm^{-2} range.

Heime and Pagnia[71] performed C–V and G–V measurements at 5 K in the frequency range from 5 to 50 kHz on anodized n-type wafers with doping density $N_D \simeq 2 \times 10^{15}$ cm^{-3}. Surfaces were prepared by lapping, etching in CP-4A (HNO$_3$:HF:CH$_3$COOH = 5:3:3), rinsing in deionized water, and anodizing in 0.1 N KOH at a constant current of 0.1 mA/cm^2 to a thickness of 1000 Å. Evaporated aluminum field plates completed the device. They were able to fit their C–V, G–V data to a model which includes an interlayer capacitance 2–10 times greater than C_{ox} with a specific resistivity $P = 10^6$ Ω-cm. They were able to simultaneously fit the C–V and G–V curves in the accumulation regime; however, when the effect of this interlayer was used to remove the background from the conductance curves they could not obtain a simultaneous fit in the depletion regime.

Langan and Viswanathan[72] report quasistatic and 1-MHz C–V measurements at 77 K on pyrolytic LTCVD (low-temperature CVD) SiO$_2$ layers deposited on (111)B-oriented $N_D = 5.3 \times 10^{14}$ cm^{-3} InSb. Silicon dioxide was deposited in a SiH$_4$–O$_2$ ambient at 220°C followed by the sputter deposition of Al gate electrodes. A *positive* interfacial charge was characteristic of these devices, as well as an apparent surface state density in the 10^{10} eV^{-1} cm^{-2} range. The LTCVD layers possessed a granularity with a mean particle size of 1500 Å. They emphasized the importance of not exposing the InSb surface to the reducing action of unreacted silane in obtaining good interfacial properties. XPS data indicated that a portion of the thin native oxide layer present on the InSb before SiO_2 reaction was reduced to elemental indium if the SiH_4–O_2 reaction was allowed to occur at the InSb surface. This was accompanied by degradation of the electrical properties of the MOS devices.

In summary, the preponderance of published data on the electrical properties of InSb surfaces treats the anodized oxide. This material is characterized by a dielectric constant between 10 and 16, a low resistivity ($\rho = 10^{13}$ Ω-cm), and large hysteresis and dc drift behavior. It should be noted that no one has reported quasistatic C–V measurements on these layers simply because the dc conductance is too large. Only two groups[64,65] report a negative value of zero-bias interfacial charge Q_{ss0} with this dielectric. More typically Q_{ss0} is positive so that n-type samples are accumulated and p-type samples are inverted. However, the fact that charges of both polarities are observed at the interface indicates the sensitivity of this processing technique to impurities in either the preanodization etches and rinse baths or in the anodization solution itself. Even though the surface state densities observed are rather high (10^{12}–10^{13} cm^{-2} eV^{-1}), the narrow energy band gap of InSb (0.25 eV) almost always allow both accumulation and inversion to be achievable. At $N_{ss} = 10^{13}$ cm^{-3} eV^{-1} approximately $E_g N_{ss} = 2.5 \times 10^{12}$ cm^{-2} charges must be swept out in going from the conduction band to the valence band. With a 1000-Å gate oxide and a dielectric constant of 10, a total gate voltage swing of $\simeq 4.5$ V is required. This corresponds to an electric field in the dielectric of 4.5×10^5 V/cm, well below its breakdown strength. Seemingly, the most promising dielectric material for use with InSb is CVD silicon oxide. Kim[67] and Langan and Viswanathan[72] report improved C–V data using this material with the surface state density less than 10^{11} cm^{-2} eV^{-1} for some surfaces. Positive Q_{ss0} is reported for this dielectric also. Field-effect mobilities of carriers in depletion layers of 10,600 cm^2 V^{-1} s^{-1} and 680 cm^2 V^{-1} s^{-1} have been reported (at 113 K)[59] for electrons and holes, respectively. A mobility of 5000 cm^2 V^{-1} s^{-1} (at 77 K) has been reported[63] for electrons in an inversion layer.

5. Indium Phosphide

At the moment the most attractive candidate III–V material for the development of an MOS technology is InP. This is based on its acceptably large band gap for room-temperature applications ($E_g \simeq 1.3$ eV), its high saturated drift velocity for electrons, and foremost of all the fact that it has by nature properties amenable to the modulation of the electron density at the surface. The surfaces of n-type material tend to be near flat band, whereas p-type surfaces tend to be inverted. In addition, the amount of interfacial charging observed for a given value of surface electric field is about a factor of 5 less than that observed in GaAs. To date, the most success has been obtained with deposited dielectric materials since the native oxides tend to be too conductive for use as gate insulators. One of the most successful dielectrics has been SiO_2. On suitably prepared surfaces using this dielectric it has been possible to modulate the position of the Fermi level at the surface over nearly the entire band gap of the semiconductor.

5.1. Native Oxides

5.1.1. Thermally Grown Oxides

The thermally grown oxide on InP as grown in dry oxygen at 450–740°C is too conductive ($\rho = 10^8$–10^9 Ω-cm) for meaningful C–V measurements. The relative dielectric constant of these layers was 10.3 at 1 MHz.[73] Capacitance–voltage measurements at 1 MHz on MIS diodes constructed using these layers exhibited a 4:1 capacitance change, but this response may have been dominated by the surface state charge since measurements on deposited layers grown at 500°C have indicated that high-temperature processing steps can produce surface states on InP which respond at frequencies greater than 1 MHz.[74] The resistivity of such layers can be increased to the $\rho = 10^{11}$–10^{12} Ω-cm range by performing the oxidation in a P_2O_5 vapor at a pressure of 0.01–0.2 atm.[75] Capacitance–voltage curves obtained at 1 MHz on such layers yield an oxide dielectric constant of 7, approximately 3-V clockwise hysteresis, and bias-dependent instability. Again, because of the high growth temperature, analysis of the 1-MHz C–V data would be speculative unless it could be shown that the surface states do not respond at 1 MHz.

5.1.2. Anodically Formed Oxides

Anodic layers have been grown on InP using both aqueous and nonaqueous electrolytes. Wilmsen[76] prepared both single- and double-layer dielectric structures using 0.1 N aqueous KOH for the electrolyte and sputtered SiO_2 for the second layer. The resistivity of the anodized layer was low (4×10^{10} Ω-cm) and this motivated the use of the second dielectric layer to limit the dc current flow. C–V measurements at 1 MHz on the anodic layer exhibited a 5:1 capacitance change, however, no C–V modulation whatsoever was seen with the sputtered dielectric only. In the two-layer structures a gradual reduction in the surface modulation was seen as the thickness of the anodized layer was reduced and the thickness of the sputtered layer was increased. Lile and Collins[77] reported a somewhat improved anodized oxide on InP using a nonaqueous electrolyte consisting of 0.1 N sodium salicylate in ethanol. This electrolyte gave layers with resistivities of $\rho = 10^{12}$ Ω-cm and relative dielectric constants which varied from 11 at 100 Hz to 4 at 1 MHz. The electrical properties of these layers were very sensitive to the absorption of water vapor. Similar behavior has been observed by other authors on both single[78,79] and double-layer[80] anodized structures. Analysis of such data to obtain the surface potential and surface state density of the interfaces must be considered speculative due to the high leakage currents and frequency-dependent dielectric constants observed in these layers. The leakage current prevents the taking of quasistatic C–V data which is perhaps the most unambiguous method of obtaining the variation of surface potential with gate voltage. Application of the Terman[6] method of analysis to the high-frequency data is frustrated by the lack of a well-defined value of dielectric capacitance. Since the dielectric constant of the oxide layer monotonically decreases as the frequency is raised, it cannot be claimed with certainty that a high-frequency C–V curve has been obtained.

5.2. Deposited Dielectrics

Probably the most benign technique reported for forming dielectric layers on InP is that of depositing Langmuir films consisting of successive monolayers of cadmium stearate or cadmium arachidate. The deposition technique consists of spreading a monolayer of the amphipathic molecules on a water surface and then transferring it to the substrate wich is dipped and then raised through the interface.[81] The C–V measurements shown in Fig. 17 on MIS devices constructed using $n = 10^{15}$–10^{16} cm^{-3} (100)-oriented substrates indicated qualitatively that both accumulation and inversion are achieved with $V_g = \pm 3$ V. However, these layers suffer some of the same problems as the anodized oxides in that the leakage currents prevent the taking of quasistatic C–V data and make the calculation of surface state densities speculative. Capacitance–voltage measurements reported by Sykes et al.[82] are shown in Fig. 18 on MIS devices employing Langmuir films on p-type InP substrates which have a frequency dispersion with negative bias which is qualitatively very reminiscent of data on p-GaAs MIS structures[11] and indicates that the p-InP surface probably cannot be accumulated with this insulator. This would argue against the formation of an inversion layer on the n-type substrates. Field-effect transistor structures have also been fabricated using such layers and depletion-type devices employing thin $n = 6 \times 10^{14}$ cm^{-3} epilayers have exhibited effective mobilities of 2250 cm^2 V^{-1} s^{-1}.[83]

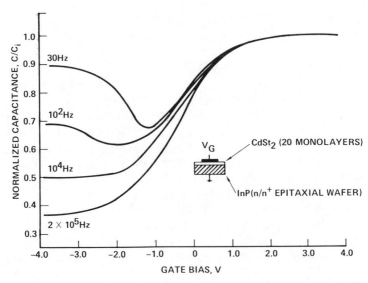

Fig. 17. C–V characteristic of n-type InP coated with an organic Langmuir film (after Roberts et al.[81]).

Deposited SiO_2 layers on InP have been reported by a number of authors. MIS devices constructed with 1100-Å CVD layers deposited on $n = 2 \times 10^{16}$ cm^{-3} InP by the reaction of SiH_4 with O_2 at 330°C as reported by Messick[84] exhibited a 2:1 change in the 1-MHz capacitance with ±4-V gate bias and an apparent surface state density of 2×10^{11} cm^{-2} eV^{-1} near the conduction-band edge as determined from a Terman analysis. Fritzsche[85] reported similar layers which had been doped with HCl during dielectric growth. When HCl was incorporated into the layers significantly less hysteresis was seen in the 1-HMz C–V curves obtained with a 33-Hz seep rate as shown in Fig. 19. Berglund[7] and Terman[6] analyses of the C–V data of Fig. 20 indicate surface potential modulation from −0.4 to ≃0.05 V. This implies the surface can be modulated from accumulation to somewhat less than midgap. From parallel conductance measurements a surface state density of 5×10^{11} cm^{-2} eV^{-1} was calculated. CVD SiO_2 layers deposited in the temperature range 320–370°C using the SiH_4–O_2 process were reported by Stannard[86] and Meiners.[87] As shown in Fig. 21 layers deposited at 320°C[87] exhibited frequency dispersion in the capacitance possibly due to a large contact resistance which made analysis of the C–V measurements somewhat speculative; however, saturated photovoltage measurements of Fig. 22 indicated surface potential modulation from −0.5 V in depletion to near-flat band. C–V measurements on layers deposited at 370°C continued to display charge trapping and hysteretic behavior at 77 K, whereas layers deposited

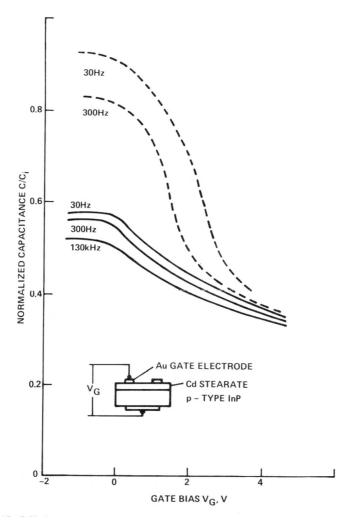

Figure 18. C-V characteristics of p-type InP Langmuir-film MIS diode (after Sykes et al.[82]).

at 340°C gave ideal-appearing deep-depletion curves as shown in Fig. 23.[86] At room temperature hysteresis was also present in the C-V curves of the layers deposited at 340°C.

The first conclusive evidence that inversion layers could be induced on p-type substrates was presented by Lile et al.[88] and Meiners et al.[89] Capacitance–voltage measurements on MIS devices incorporating SiO_2 layers pyrolytically deposited at 320°C in the frequency range from quasistatic to 1 MHz, as shown in Fig. 24, qualitatively indicated the presence of a surface which could be modulated from inversion to

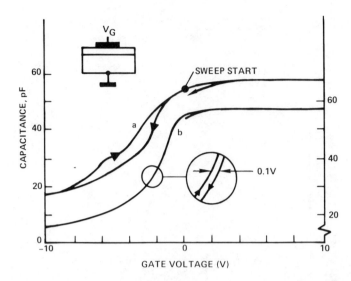

Figure 19. $C-V$ characteristics of SiO_2 layer on n-type InP showing the effects of the incorporation of HCl into the dielectric layer: curve a without HCl etching; curve b with HCl etching (after Fritsche[85]).

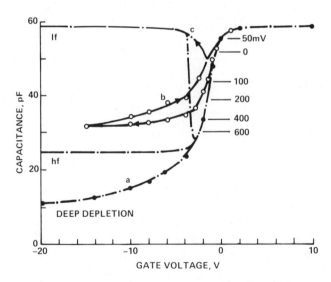

Figure 20. Extended $C-V$ measurements on HCl-doped SiO_2 layer on InP (after Fritzsche[85]).

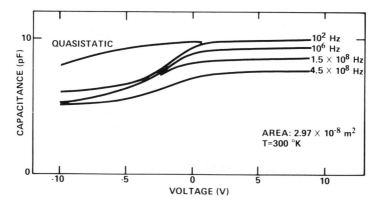

Figure 21. $C-V$ characteristics of pyrolytically deposited (320°C) SiO_2 on n-type InP employing the SiH_4-O_2 process.

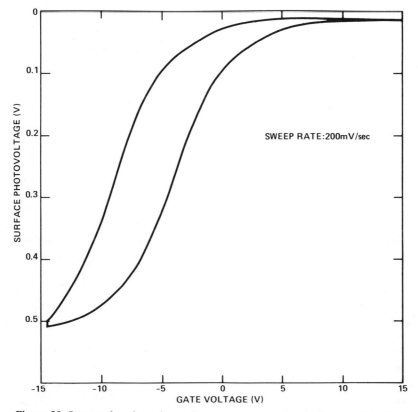

Figure 22. Saturated surface photovoltage measurement for the device of Fig. 21.

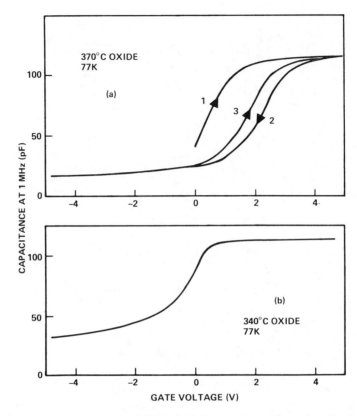

Figure 23. Room-temperature and 77 K $C-V$ characteristics of pyrolytically deposited (340°C) SiO_2 on InP (after Stannard[86]).

depletion. Gate-modulated transport measurements on three-terminal structures indicated a normally off transistor-like behavior with an effective mobility of 400 cm^2 V^{-1} s^{-1} and essentially flat, small signal frequency response from 10^{-2} to 10^6 Hz. The properties of inversion-layer transistor structures formed on $p = 6 \times 10^{16}$ cm^{-3} (111) and (100) substrates have been investigated by Fritzsche[90] and by von Klitzing et al.[91] Pyrolytically deposited SiO_2 Layers formed at 340°C using the SiH_4–O_2 process with 0.2 N HCl and 0.05 N PH_3 added to the gas stream were used in each investigation. Fritzsche[90] investigated the transient response of the source–drain current (I_{SD}) in these structures at 300 and 77 K. At room temperature, he observed a reduction in I_{SD} by as much as a factor of 6 between 10^{-6} and 1 s. At 77 K no drift of I_{SD} was observed. Field-effect mobilities as high as 1000 cm^2 V^{-1} s^{-1} were calculated by curve fitting the drain current-voltage characteristics.

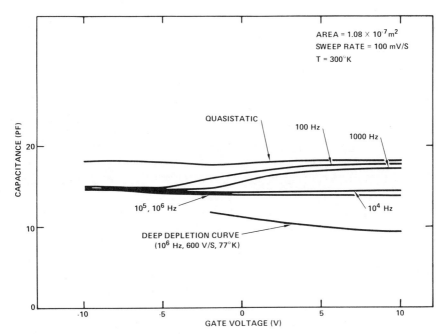

Figure 24. Frequency dispersion of $C-V$ measurements on p-type InP with a pyrolytically deposited (320°C) SiO_2 dielectric layer.

Inversion-layer carrier densities for these devices were determined by von Klitzing et al. from the oscillatory behavior of the magnetoconductivity at 4.2 K. They observed a field-effect mobility of 1000 cm^2 V^{-1} s^{-1} ($\mu_{FE} = d\sigma/C_d V_g$) and an effective mobility of 5000 cm^2 V^{-1} s^{-1} ($\mu_{eff} = \sigma/C(V_g - V_{th})$), and confirmed the observation that no electric field induced changes occurred in the amount of trapped charge at 4.2 K. A maximum inversion-layer charge of 1.66×10^{12} cm^{-2} was observed.

It has also been possible to electrostatically induce electron channels on semi-insulating Fe-doped InP substrates.[92,93] Silicon dioxide layers pyrolytically deposited at 320°C were used to form FET structures with 4-μm gates[92] that had low-frequency behavior similar to devices on p-type substrates[88] and which also had 2-dB gain at 2.5 GHz. Van der Pauw measurements employing five-terminal gated symmetric cross structures fabricated on (100) Fe–InP substrates were used to determine the density and mobility of carriers in the n-type conducting channel. For these devices the SiO_2 was deposited in a plasma-enhanced chemical-vapor-deposition system (PECVD) which employed silane (SiH_4) and nitrous (N_2O) as the reactant gases. It was shown that the conducting channel is most properly thought of as an accumulation layer since the substrate was n-type ($n = 2.2 \times 10^8 \text{ cm}^{-3}$, $\mu_e = 3040 \text{ cm}^2 \text{ V}^{-1} \text{ s}^{-1}$) with a zero-bias surface potential of 0.39 V. A maximum surface carrier density of $2 \times 10^{11} \text{ cm}^{-2}$ was observed, as well as a maximum Hall mobility of $900 \text{ cm}^2 \text{ V}^{-1} \text{ s}^{-1}$. Typically, the field-effect mobility as determined by curve fitting transistor curve-tracer data was one-half of the Hall mobility which indicates that during the measurement of the three-terminal transistor characteristics at a 120-Hz rate roughly one-half of the induced charge is trapped. The surface parameters were highly dependent on the sample preparation before oxide growth.

Plasma-enhanced CVD SiO_2 and Si_3N_4 layers were deposited on $n = 3 \times 10^{16} \text{ cm}^{-3}$ (111) InP substrates using a variety of predeposition surface treatments.[74] Capacitance–voltage measurements on MIS diodes were used to evaluate the electrical properties of the layers. The electrical properties of the SiO_2 layers deposited using a SiH_4–N_2O process at 300°C were sensitive to the predeposition treatment that the surfaces received. Typically used etches, such as Br–methanol, and aqueous solutions containing HCl and/or HF gave surfaces with $\simeq 0.4$-V surface potential modulation and accumulation with positive gate bias. Surface treatments which might be expected to reduce the amount of native oxide present at the interface, such as etching in 95% aqueous hydrazine or predeposition annealing in H_2, tended to increase the total amount of modulation. For such samples, as shown in Fig. 25, the surface potential

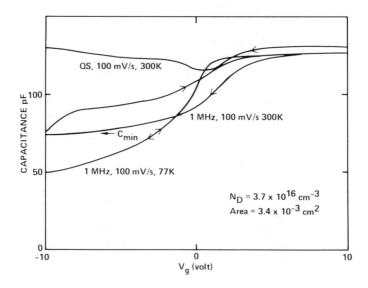

Figure 25. C–V characteristics of SiO_2 layer on n-type InP showing the effect of hydrazine etching of the surface prior to dielectric dsposition.

modulation was as large as 1.1 V. The surface modulation observed with Si_3N_4 layers deposited using a SiH_4–NH_3 process at 300°C was limited to 0.6 V regardless of the surface preparation and in addition large hysteresis effects were present even at low temperature as illustrated in Fig. 26. This was thought to be due to the damage which can be produced on InP substrates by a hydrogen or ammonia plasma. The excited hydrogen liberates phosphine (PH_3) and leaves an indium-rich surface. Similar results have been reported by Cameron et al.[94] for Si_3N_4 layers deposited at temperatures between 350 and 400°C on p-type InP using a PECVD process incorporating NH_3 and SiH_4 as the reactant gases. They observed the very low effective mobility of 7 cm^2 V^{-1} s^{-1} in MIS transistor structures employing these layers. They also deposited SiO_2 at 300°C in a PECVD process which used tetraethoxysilane and oxygen. Capacitance–voltage measurements on MIS diodes indicated strong inversion of the p-type surfaces but no accumulation and I–V measurements on transistor structures indicated an effective mobility of 220 cm^2 V^{-1} s^{-1}. The observed resistivity of 10^{14} Ω-cm was less than that of thermally grown SiO_2. Also aluminum oxide layers were deposited by the pyrolytic decomposition of aluminum isopropoxide in a low-pressure nitrogen atmosphere at 300–400°C. A resistivity of 5×10^{12} cm was observed, as well as a relative dielectric constant of 8.5 at 1 MHz which increased by 25 at 1 Hz. Based on C–V measurements, it was concluded that both accumulation and inversion could be achieved on the p-type ($p = 1 \times 10^{15}$–5×10^{16} cm^{-3}) substrates. This result was calculated from the high-frequency C–V data only, since the low resistivity prevented taking the quasistatic data. An effective mobility of 2100 cm^2 V^{-1} s^{-1} was obtained from I–V measurements.

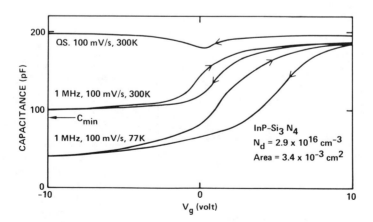

Figure 26. C–V characteristics of plasma-deposited Si_3N_4 layer on n-type InP.

The use of aluminum oxide layers deposited by the pyrolysis of aluminum isopropoxide $Al(OC_3H_7)_3$ at 350–450°C has been reported by a number of investigators. Kawakami and Okamura[95] have reported an effective mobility of 2500 cm^2 V^{-1} s^{-1} in an inversion layer as determined from 100-Hz I–V measurements on MISFET structures constructed on $p = 1 \times 10^{16}$ cm^{-3} ($\mu_h = 100$ cm^2 V^{-1} s^{-1} at 300 K) substrates. They have also reported similar mobilities for FETs constructed on (100)-oriented Fe-doped InP ($p = 10^7$–10^8 cm) substrate material. Okamura and Kobayashi[96,97] have reported an interface with improved electrical properties which they obtained by using an HCl-vapor-etching step at 200°C to remove 1500 Å of InP before the deposition of Al_2O_3 by the pyrolysis of aluminum isopropoxide in a hydrogen atmosphere. MIS capacitors fabricated on (100)-oriented $p = 1 \times 10^{17}$ cm^{-3} and $n = 1 \times 10^{16}$ cm^{-3}

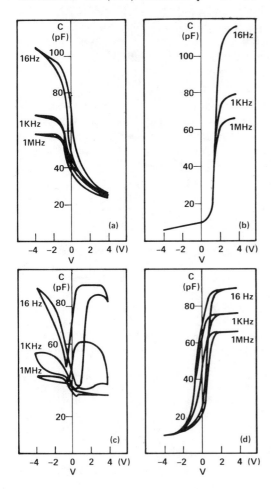

Figure 27. C–V characteristics of Al_2O_3 layers on InP showing the effects of *in situ* predeposition etching in HCl vapor: (a) p-InP and (b) n-InP using HCl etching; (c) p-InP and (d) n-InP without HCl etching (after Okamura and Kobayashi[97]).

substrates exhibited reduced hysteresis and an approximately 2:1 change in the accumulation capacitance between 16 Hz and 1 MHz as shown in Fig. 27. Characteristic of all these layers is a low resistivity ($\rho = 10^{12}$ Ω-cm), which does not allow the quasistatic $C-V$ data to be obtained. MISFET structures constructed on (100)-oriented $n = 1 \times 10^{16}$ cm^{-3} substrate material exhibited *inversion-type* behavior with an effective carrier mobility of 16 cm^2 V^{-1} s^{-1}—as yet, the only reported evidence for inversion on n-type material from a transport measurement. The same authors also report[98] improved drift behavior of n-channel inversion-type MISFETs using the HCl etching technique in combination with an aluminum oxide dielectric. They attribute the improved performance to elimination of the native oxide layer between the InP and Al$_2$O$_3$ by the HCl etching, immediately followed by dielectric growth in a hydrogen-rich atmosphere. The effective mobility of n-channel inversion-type devices fabricated using this technique varied between 100 and 500 cm^2 V^{-1} s^{-1} depending on the substrate temperature during deposition.[99] The maximum mobility was observed at a deposition temperature of 350°C.

The use of germanium nitride (Ge$_3$N$_4$) layers deposited at 400°C on (100)-oriented $n = 10^{16}$ cm^{-3} InP substrates by the pyrolytic decomposition of germane (GeH$_4$) in hydrazine (N$_2$H$_4$) has been investigated by Pande and Pourdavoud.[100] The resistivity of these layers at an electric field of 10^6 V/cm was 10^{12} Ω-cm. The observed no capacitance dispersion in the device in the frequency range 10–10^6 Hz with 4-V gate bias. A

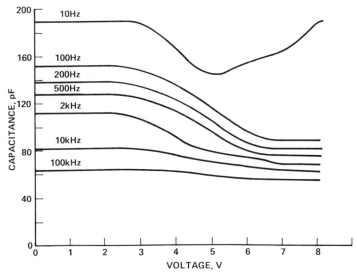

Figure 28. $C-V$ characteristic of p-type InP coated with sputtered InP oxide (after Al-Refaie and Carroll[101]).

novel dielectric material formed by the reactive sputtering of InP in oxygen was been reported by Al-Refaie and Carroll.[101] MIS capacitors formed on (100)- and (111)-oriented $p = 7$–10×10^{17} cm^{-3} InP substrates exhibited an approximate 3:1 change in capacitance in the frequency range from 10 Hz to 100 kHz with negative gate bias as shown in Fig. 28. In this sense the data look quantitatively like those obtained when SiO$_2$ layers are formed on p-type InP and suggest that inversion is obtained, but not accumulation. Auger electron spectroscopy and electron microprobe analyses of the deposited layers indicated that they contained about 46% oxygen.

The electrical properties of MIS diodes fabricated with thermal nitride layers formed by ammoniation of (100) $n \simeq 10^{16}$ cm^{-3} InP have been described by Yamaguchi[102] and by Hirota et al.[103] The layers tend to be rather conductive ($\rho = 10^{11}$–10^{12} Ω-cm) and minimum surface state densities of 5×10^{11} cm^{-2} eV^{-1} have been reported.

6. Indium Arsenide

Kawaji and Kawaguchi[104] were the first to report measurements on the electrical properties of the InAs surface. They constructed Corbino disc structures on $p = 2.6 \times 10^{16}$ cm^{-3} single-crystal substrates which had been oriented so that the surfaces were parallel to the (111) planes. Sheet mylar spacers were used on each side of the sample to provide dielectric isolation of the field electrodes. Magnetoresistance measurements were performed in the temperature range 4.2–77 K. From the low-field magnetoresistance they calculated the mobility and from the zero-field conductance they calculated the carrier density. The surface was normally inverted. An equilibrium concentration of charged surface states of 6×10^{11} cm^2 was estimated from the equilibrium surface carrier concentration of 1.5×10^{11} cm^2 and from the calculated number of ionized acceptors in the space-charge region. From transport measurements on thin-film transistor structures employing thermally evaporated and recrystallized n-type InAs and a silicon monoxide gate insulator, Kunig[105] concluded that a surface accumulation layer of 3.2×10^{12} cm^{-2} existed. Schwartz et al.[106] performed C–V measurements on $n = 2 \times 10^{16}$ cm^{-3} (110) and (100) crystals which were etched in bromine–methanol solutions before the pyrolytic deposition of SiO$_2$ at 400°C using a SiH$_4$–O$_2$–N$_2$ gas mixture. Measurements were performed in the dark at 77 and 295 K. They explain their C–V data in terms of a model which includes a positive fixed interfacial charge density of $0.875\ C_{ox} V_{ox}$ (directly proportional to oxide voltage) and a fast surface state density of 2×10^{11} eV^{-1} cm^{-2} at 77 K and 2.5×10^{12} cm^{-2} eV^{-1} at 295 K. Approximately 6-V hysteresis was observed with ±40-V gate bias on the roughly 1000-Å thick insulating layers.

Terao et al.[107] fabricated MIS diodes employing Al_2O_3 deposited by the pyrolytic decomposition of $Al(OC_3H_7)_3$ at a substrate temperature of 250–450°C. Substrates were (111)B oriented with $n = 2.5 \times 10^{16}$ cm^{-3} and $p = 1.5 \times 10^{17}$ cm^{-3}. Before deposition the surfaces were etched in bromine–methanol solution. The C–V measurements indicated that both the n- and p-type substrates had n-type surfaces at 77 and 300 K. Wilmsen et al.[108] investigated the properties of MIS diodes constructed with both anodized dielectrics and with reactively evaporated SiO_2. The most stable anodized layer was obtained with an electrolyte consisting of 3% tartaric acid (aqueous) buffered to a pH of 5.5 with NH_4OH. The fixed interfacial charge density varied from 0.7 to 4×10^{12} cm^{-2} and was lowest for single-layer SiO_2. The fast interface state density varied between 0.7 and 1.7×10^{12} cm^{-2} eV^{-1}. All surfaces were accumulated with zero gate bias.

Baglee et al.[109] observed inversion-layer transport in (111)B-oriented $p = 2.5 \times 10^{17}$ cm^{-3} substrates with dielectric layers formed by first anodizing the surface and then depositing SiO_2 by sputter deposition. From transport measurements on van der Pauw structures they conclude that the n-type surface wtih zero gate bias has a carrier density of $1-4 \times 10^{11}$ cm^{-2} and that Hall mobilities of 5000–12,000 cm^2 V^{-1} s^{-1} can be obtained at 77 K.

7. Gallium Phosphide

Spitzer et al.[110] and Ikoma and Yokomizo[111] have grown native oxides on n-type GaP substrates by boiling the samples in concentrated H_2O_2 for several hours. The grown oxide is composed of an amorphous mixture of Ga_2O_3 and P_2O_5 which has a breakdown voltage of 0.5–2 V/cm. After annealing in dry N_2 at 250°C for 2 h, the breakdown voltage rises to $6-8 \times 10^6$ V/cm. Both groups report C–V data which look like a nonequilibrium high-frequency curve on GaAs. The reported dielectric constant of the native oxide of 5.7 as obtained from the maximum value of capacitance on the C–V plots seems anomalously low when compared to values for the native oxides of GaAs and InP of 7–8 and 11, respectively. The value of dielectric constant reported was calculated by assuming that $C_{max} = \varepsilon_{ox}/d$. However, if the surface is always depleted, $C_{max} < C_{ox}$ and the calculated value of ε_{ox} would be less than the actual value. This suggests the possibility that with positive bias the surface is pinned in depletion similar to GaAs and that with negative bias the surface deep depletes in a nonequilibrium condition.

8. Gallium Arsenide Phosphide

Phillips et al.[112] oxidized $n = 2 \times 10^{15}$ cm^{-3} (100)-oriented GaAs$_{0.5}$P$_{0.5}$ substrates in dry O$_2$ at atmospheric pressure for 600 min at 700°C. Devices were subsequently annealed for 285 min at 700°C in argon prior to the deposition of chromium gate electrodes and measurement of the C–V characteristics at 500 kHz. They obtained deep-depletion-type curves similar to that of Fig. 12. Forbes et al.[113] deposited Si$_3$N$_4$ on $n = 5 \times 10^{16}$ cm^{-3} GaAs$_{0.62}$P$_{0.38}$ substrates and measured the 1-MHz C–V characteristics after aluminum dots had been deposited on the surface. The C–V characteristics saturate with both positive and negative gate bias and deep-depletion-type C–V curves are obtained unless very slow gate bias ramp rates are used. Ahrenkiel et al.[114] formed anodic native oxide layers on $n = 5 \times 10^{15}$ cm^{-3} epitaxial GaAs$_{0.6}$P$_{0.4}$ substrates using an electrolyte consisting of tartaric acid in ethylene glycol buffered to a pH of 6. The breakdown field of these devices was 1–8 $\times 10^5$ V/cm. The 1-MHz C–V measurements gave results similar to those of Phillips et al.[112] in that deep-depletion-type behavior was obtained at room temperature.

9. Whither Surface States

Our concepts of surface charging on semiconductors have been guided by the early work of Tamm[115] who showed from fundamental quantum-mechanical considerations that the abrupt termination of a semi-infinite lattice creates localized energy states at the surface of a crystal. The energy of these states is referenced to the energy states of the electrons in the bulk of the crystal and the occupational probability of a particular surface state is then expected to be related to the energy of that state with reference to the Fermi level of the crystal bulk. If the amount of energy-band bending at a crystal surface can be modulated (such as by chemical absorption, electric field modulation, or optical modulation), one is then encouraged to consider a type of spectroscopy in which the density of surface states is determined as a function of energy within the fundamental band gap of the semiconductor. At room temperature or below, the Fermi function is sharply defined and as the band bending is modulated the position of the Fermi level at the surface is swept through the band gap) The occupational probability of any particular state abruptly changes as the Fermi level passes through it. This measurement requires the simultaneous determination of the surface band bending or surface potential and the amount of interfacial charge. Typically, the surface potential and interfacial charge are calculated from surface capacitance or surface conductance measurements or from a combination of optical and electrical measurements and the surface state density is evaluated by numerically differentiating the surface charge with respect to surface potential. Typically, the result of this calculation is not a curve with well-defined peaks indicating discrete energy states, but rather a smooth, usually U-shaped curve. The inference made from this result is that the surface states form a continuous distribution within the fundamental band gap of the semiconductor.

There are, however, compelling reasons for plotting the data relating interfacial charge to surface potential in a somewhat different fashion. One of the observations made in the course of developing an understanding of the GaAs–insulator interface was that (for GaAs) Q_{ss} depends linearly on the insulator electric field, E_{ox}. It was thought, at the time, that this is a specific characteristic of GaAs due entirely to the extremely high surface state densities observed on GaAs MIS structures. However, when the values of Q_{ss} for InP, calculated from Figs. 21 and 25, are plotted vs. E_{ox}, as shown in Fig. 29, a similar linear relationship is observed where the slope of the curve is proportional to the rate of charge trapping. A change in slope appears as the electric field passes through zero, however, on either side of this break; the data were strictly linear. The electric field is positive when directed toward the interface (positive voltage on gate).

Langan[116] has reported extensive C–V measurements on SiO_2 layers pyrolytically deposited on (111) $n = 4.5 \times 10^{14}$ cm^{-3} InSb at 200°C employing the chemical reaction between O_2 and SiH_4. The relationship between N_{ss} and E_{ox} obtained from his measurement is plotted in Fig. 29; it is a linear relationship with a slope nearly the same as that for the SiO_2–InP devices. Schwartz et al.[117] have reported C–V measurements on SiO_2 layers pyrolytically deposited on (100) and (110) $n = 2.2 \times 10^{16}$ cm^{-3} InAs. Their SiO_2 layers were deposited at 400°C, again using the reaction between SiH_4 and O_2. A large positive trapped charge in the interfacial region is characteristic of InAs MIS structures. This necessitates that the scales of Fig. 29 be offset in order to plot the values of Q_{ss} vs. E_{ox} obtained from their measurements in which case the same linear relationship is obtained; it is interesting to note that the same slope is exhibited as for the SiO_2–InP capacitors. Ziegler and Klausmann[118] describe static C–V measurements on MOS capacitors constructed from (111) $n = 4.9 \times 10^{15}$ cm^{-3} silicon wafers oxidized in dry oxygen at 1140°C and annealed for 30 min at 350°C in dry hydrogen after deposition of aluminum dots. From their data, Q_{ss} vs. E_{ox} was determined and is plotted in Fig. 29. The striking feature is that on the (111) interface of Si (high surface state density orientation) the electrons for negative gate bias are ejected from the interface linearly with the electric field and with a

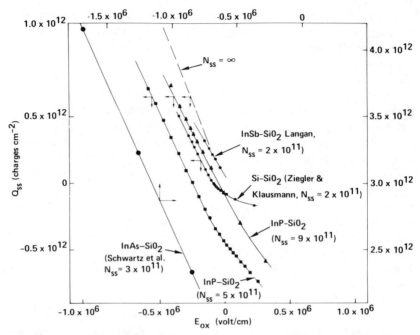

Figure 29. Total interfacial charge Q_{ss} as a function of electric field in the oxide layer for MOS capacitors on n-type InAs, InP, InSb, and Si. Also indicated is the minimum value in units of $cm^{-2}\,eV^{-1}$ of the N_{ss} vs. the energy plot for each set of data. The origin of the axes for the plot of the InAs–SiO$_2$ data has been displaced as indicated by the scales on the top and right-hand edges of the figure. (*Note:* Q_{ss} is given here in units of electronic charges cm^{-2}; the ordinate should be labeled Q_{ss}/q to be consistent with the text.)

characteristic of exactly the same slope as observed for the deposited oxides on the III–V materials. It can be seen from Fig. 29 that electron trapping, when the field is positive, occurs with a smaller slope in this n-type Si–SiO$_2$ device. Quasistatic C-V measurements on Si–SiO$_2$ capacitors constructed on (100) $p = 5 \times 10^{15}$ cm^{-3} silicon wafers have been reported by Johnson *et al.*[119] and Fig. 30 shows Q_{ss} vs. E_{ox} curves calculated from their data. The first sample was oxidized at 1000°C in O$_2$. The second sample was annealed in argon at 1000°C for 30 min and in H$_2$/N$_2$ at 450°C for 60 min after oxidation. Their published data allowed the determination of Q_{ss} only to within an arbitrary constant; therefore, Q_{ss} at $E_{ox} = 0$ has been set equal to zero. These data have a somewhat different behavior than that of the n-type semiconductors. The linear portion of the curve occurs for that region in which electrons are trapped by the interface (holes ejected). The annealed specimen also exhibits a linear Q_{ss} vs. E_{ox} characteristic but with a reduced slope. Because $\varepsilon_{ox}E_{ox} = Q_{sc} + Q_{ss}$, a linear relationship between Q_{ss} and E_{ox} implies that

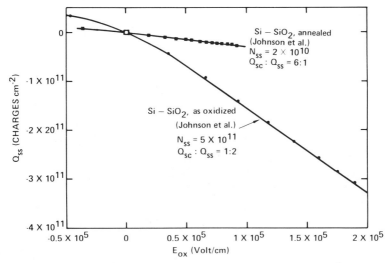

Figure 30. Total interfacial charge Q_{ss} as a function of electric field in the oxide layer for MOS capacitors on p-type Si. Also indicated is the minimum value in units of $cm^{-2}\,eV^{-1}$ of the N_{ss} vs. energy plot for each set of data. (*Note:* Q_{ss} is given here in units of electronic charges cm^{-2}; the ordinate should be labeled Q_{ss}/q to be consistent with the text.)

the space-charge capacitance Q_{sc} of the semiconductor is linearly related to Q_{ss}. It is interesting to compare these ratios on the Si–SiO$_2$ interface for the as-oxidized and for the annealed samples. In the as-oxidized device, two charges are induced in interfacial states for every charge induced in the semiconductor. In the annealed device one charge is induced in interfacial states for every six charges induced in the semiconductor, these ratios being maintained over the entire V_s range for which the Q_{ss}/E_{ox} plot is linear.

Calculated plots are presented in Figs. 31 and 32 in order to illustrate the appearance of a trapped interfacial charge related linearly to the insulator electric field when plotted, as is more usually the case, on the N_{ss} vs. energy curve. A relationship $Q_{ss} = -0.77 C_{ox} V_{ox}$ has been assumed which is approximately that observed for InP, where V_{ox} is the potential drop across the dielectric.

It can be seen that the typically reported U-shaped "density-of-states" distribution is obtained when the linear data of Fig. 29 are plotted in this manner due to the nonlinear dependence of V_s on V_g. Curves for several different doping densities of InP are shown in Fig. 31. The apparent minimum of the surface state density decreases with doping as $[N/\ln(N/n_i)]^{1/2}$ in the same manner as does the space-charge capacitance C_{sc}. Furthermore, the range of surface potential displacement which is accessible for a given range of applied gate voltages decreases as the

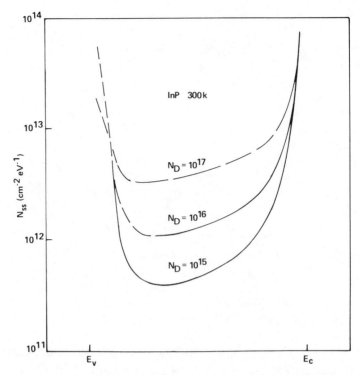

Figure 31. Apparent distribution of interface states on n-type InP at 300 K if the interfacial charge depends linearly on the electric field in the oxide according to the formula $Q_{ss} = -0.77 C_{ox} V_{ox}$ with $E_{ox} = 4$ and $d = 1000$ Å. For p-type material the position of E_v and E_c would be interchanged. The solid lines delineate the range of V_s which can be accessed for $V_g = \pm 10$ V.

doping level increases. For a semiconductor such as InP ($E_g = 1.35$ eV), this effect becomes noticeable at $N_D = 10^{16}$ cm^{-3} for the Q_{ss} vs. E_{ox} relation given above; about three-quarters of the band gap is accessible with ± 10-V gate bias. At $N_D = 10^{17}$ cm^{-3}, less than one-fifth of the band gap is accessible for the same gate bias range. For InSb, as shown in Fig. 32 assuming exactly the same oxide-charging behavior, the entire band gap is accessible ($E_g = 0.22$ eV) where fewer carriers need to be induced in the semiconductor in order to swing the surface over the full range from accumulation to inversion.

The data compiled in Fig. 29 and 30 fall into two groups. One group contains the deposited silicon dioxide layers on the III–V semiconductors InSb, InAs, and InP, hydrogen-annealed thermal oxides on (111) Si, and as-oxidized layers on (100) Si. The quasistatic charging behavior of all interfaces contained in this group of devices is essentially identical. The other group contains the hydrogen-annealed thermal oxide on (100) Si.

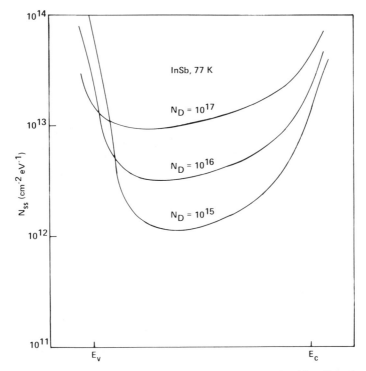

Figure 32. Apparent distribution of interface state on n-type InSb at 77 K if the interfacial charge depends linearly on the electric field in the oxide according to the formula $Q_{ss} = -0.77 C_{ox} V_{ox}$ with $E_{ox} = 4$ and $d = 1000$ Å. For p-type material the position of E_v and E_c would be interchanged. The solid lines delineate the range of V_s which can be accessed for $V_g = \pm 10$ V.

The linear Q_{ss} vs. E_{ox} relationship is observed for this interface also but the total amount of interfacial charging is smaller by a factor of 12.

A number of inferences might be drawn from these investigations. The first, and most important of these, is that the equilibrium value of trapped interfacial charge depends, to first order, not on the position of the surface Fermi level but rather on the electric field or the effective force that charged particles are subjected to in the surface region. This implies that the occupancy of the charge-trapping centers is not determined by the equilibrium statistics of the semiconductor and suggests the possibility that what have traditionally been called semiconductor surface states may be more properly associated with the oxide layer. This would perhaps explain why *ad hoc* curve-fitting techniques (such as surface potential patch variation) are necessary to make the current potential models of surface state behavior agree with experiment.

There is ample evidence to suggest that the processes commonly used to form insulating layers on the III–V semiconductors produce dielectrics which are nonstoichiometric and which contain large numbers of charge-trapping centers. For example, Wilmsen *et al.*[20] have presented Auger data which imply that In diffuses into pyrolytically deposited SiO_2 on InP. It is commonly argued[121] that the (111) surface of Si is the high interface state orientation because the density of available bonds is twice that of the (100) orientation. This argument is weakened, however, by the fact that the measured surface state densities for the two orientations differ by more than a factor of 10. An argument consistent with conditions existing on the III–V interfaces is that the reaction kinetics may differ for the (111) surface as compared to the (100) surface with a resulting large number of Si inclusions in thermally grown oxides on (111) surfaces. Indeed, the role of hydrazine in providing an improved surface of InP is a complex one. The results have not been entirely reproducible; however, improved surface properties seem to be obtained most consistently if hydrocarbon solvents are used immediately after hydrazine etching. While the hydrazine treatment undoubtedly reduces the quantity of native oxide at the InP surface, it may be that the combination of a strong reducing agent, in addition to a carbon-containing solvent, acts to stabilize the InP surface and reduce In diffusion into the oxide. Our present understanding of the electrical properties of such impurities and flaws in insulators is meager, at best. We known, for example, that such flaws can give rise to hysteresis or memory effects in interfacial charging phenomena. The present work suggests that insulator defects may play an even greater role in the electrical behavior of insulator–semiconductor interfaces.

A change in slope is observed in all of the Q_{ss} vs. E_{ox} plots as the electrical field passes through zero. It is consistently observed that electrons are ejected from the interfacial region of oxides on n-type semiconductors more readily than extra electrons are trapped. Conversely, holes are ejected from the interfacial region on a p-type semiconductor more readily than extra holes are trapped. This seems to be suggesting a model for the charging behavior based on a multivalent defect in the insulator.

Figures 31 and 32 illustrate the somewhat arbitrary manner in which low values of surface state densities can be obtained by using a substrate material with a low impurity level. Figure 29 suggests the use of a somewhat more universal quality factor to characterize the interfaces of a variety of semiconductors. If the ratio $\Delta Q_{ss}/\Delta Q_{sc}$ in the region in which the semiconductor is depleted is defined as the interfacial quality factor, a parameter for assessing interfacial behavior is obtained which is independent of the semiconductor band gap or impurity level. This ratio can vary from zero for an ideal interface to infinity for a pinned surface.

This interfacial quality factor is equal to $\frac{1}{6}$ or less for a "state-of-the-art" silicon (100) surface; it is roughly equal to 2 for the unannealed thermal silicon oxide on (100) silicon; it is approximately 6 for the annealed thermal oxide on (111) silicon and for deposited SiO_2 on InSb, InAs, and InP; and it is equal to 30 for anodic oxides and deposited dielectrics on GaAs.[10]

It would be premature to say whether the above arguments either prove or disprove the existence of true surface states at MIS interfaces. However, they do suggest that the typical interfacial charging behavior observed in MIS devices is analytically simpler to describe than that predicted by a surface state model. At the same time a first-principles model which gives a charging behavior which varies linearly with the electric field has not been forthcoming. It is tempting to speculate that such behavior may be due to the collective effects of surface charges at interfaces. However, at this point it is unclear what the coupling mechanism would be since the ordinary electrostatic forces acting on a uniformly spaced distribution of interfacial charge seem to be too small to give the observed effect.

10. Low-Temperature Deposition of Dielectric Layers

Dielectric isolation of the gates of metal–insulator–semiconductor field-effect transistors (MISFETs) requires the synthesis of compatible thin insulating layers on semiconductor surfaces. In contrast to the native oxide on silicon, the native oxide layers that can be grown on many III–V semiconductors are either too conductive or produce an unacceptably large interface state density for such applications. Consequently, various types of deposited layers are currently under investigation for use in passivating these materials. A suitable deposition technique should maintain the substrate temperature below the point at which incongruent evaporation at the surface occurs. Evaporation of III–V semiconductors invariably leads to surface depletion of one of the constituents and consequent degradation of the electrical properties of the surface.

A suitable process for the deposition of insulating layers on III–V substrates must therefore provide not only for the preservation of the stoichiometry of the semiconductor surface but also must provide conditions for the formation of a refractory material with a very high melting temperature. Typical thin-film processes for the deposition of dielectrics, such as sputtering or electron-beam evaporation, have not been very successful for this application because of the problems associated with the incorporation of charges into the deposited layers.

Thin insulating layers of silicon dioxide can be deposited by the oxidation of silane (SiH_4). This is normally accomplished by passing a combination of silane, an oxidizing agent, and a nonreactive diluent gas through a heated region in which the substrates have been placed. By passing combinations of oxidizer and silane through a quartz-chip-filled reactor and measuring the infrared absorption spectrm of the effluent Strater[122] determined that SiH_4–O_2 mixtures begin to react at 120–270°C and silane–nitrous oxide (SiH_4–N_2O) mixtures begin to react at 370–525°C. Silane–carbon dioxide (SiH_4–CO_2) mixtures did not react at temperatures up to 520°C. In practical silicon oxide deposition systems the temperature required for reasonable deposition rates is generally somewhat higher than those observed by Strater.[22] Chemical-vapor-deposition (CVD) systems employing oxygen generally require 325–450°C,[123] those employing N_2O require 840–900°C,[124] and those employing CO_2 require 700–1000°C.[125] If this substrate temperature is too high for practical applications, then it is feasible to deposit SiO_2 layers at lower temperatures or even at room temperature by exciting the gas mixture in a radio-frequency (rf) discharge. Sterling and Swann[126] have obtained growth of SiO_2 via the reaction between N_2O and SiH_4 by reducing the pressure in the reaction vessel to the 100 μm Hg range and coupling an rf field into the substrate region. The low-pressure plasma discharge thus created contains excited species which combine to form silicon dioxide and water vapor. Nitrous oxide is to be preferred over oxygen for the plasma-enhanced chemical vapor deposition (PECVD) of SiO_2 because large amounts of silica dust form in the discharge zone when an oxygen–silane mixture is used.

MIS devices employing the thermally grown native oxide on silicon are extremely sensitive to ultraviolet radiation,[127–129] ion bombardment,[130] and electron-beam exposure.[131–136] In a conventional PECVD system the substrates are exposed to a rather harsh chemical environment which contains a nonthermal (~5 eV) distribution of electrons and ionized atoms and molecules, as well as rather energetic photons and excited metastable neutral species. Ephrath and DiMaria[137] have observed that trapping centers can be induced in thermally grown SiO_2 layers by the exposure of the oxides to either oxygen or argon low-energy plasmas. These traps were located within 10 nm of the exposed surface and were not completely removed by annealing at 400°C for 20 min in forming gas. It seems reasonable, therefore, to assume that typical PECVD processes produce traps throughout the deposited layers which can only be removed by high-temperature annealing. Since the annealing treatment may affect adversely the surface properties of III–V materials, damage may exist in PECVD layers which cannot be removed by annealing.

We have developed an indirect PECVD which attempts to improve the chemical environment of III–V compound surfaces during SiO_2 deposition.[138] This was accomplished by exciting the oxidizer (oxygen) in a separate chamber and mixing the effluent from this chamber with silane in the substrate region. Silica films were observed to grow downstream of the gas-mixing region even though the substrate and the ionization chamber were separate by a path length of nearly 1 m. The reactor configuration is shown schematically in Fig. 33. The oxygen entered a pyrex ionization chamber 20 cm long and 6.3-cm ID which was inductively coupled to a 10-W, 13.56-MHz rf source. A small amount of nitrogen was added to the oxygen flow to catalyze the formation of atomic oxygen.[139] A mixture of silane and nitrogen was fed into the deposition zone via a 3-mm-ID quartz tube. The added nitrogen served only to improve the uniformity of the deposition by slowing down the silane–oxygen reaction in the gas-mixing region. The substrates were heated with a muffle furnace supported on rails which could be positioned over the entire gas-mixing–substrate deposition region. One effect of keeping the chamber walls hot throughout the gas-mixing and substrate regions was to completely eliminate the dust buildup which is often a problem with insulator deposition systems. Cleanup of the system was required only when the deposited layers became so thick that they began to crack and peel off—normally after many runs.

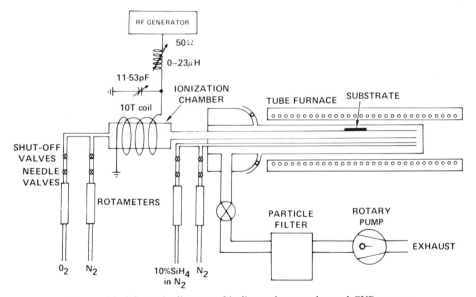

Figure 33. Schematic diagram of indirect plasma-enhanced CVD reactor.

Although the indirect PECVD system described herein was developed for the purpose of depositing SiO_2 layers on III–V semiconducting substrates, specifically InP, layers have also been deposited on silicon substrates as a means of evaluating the dielectric properties. From C–V measurements on MIS devices constructed on both silicon and InP substrates we conclude that the indirect PECVD technique gives lower interface state densities, less hysteresis, and better reproducibility as compared to results obtained in our own conventional PECVD system.

Layers grew in the indirect PECVD system at temperatures from ambient upwards suggesting that oxidation of the silane is at least precipitated by an excited species in the oxidant stream. Charged-particle recombination is expected to occur with a time constant of 1 m or less for oxygen plasma of this density.[140,141] The calculated transit time from the ionization chamber to the substrate of 7 m would seem to rule out the possibility that ionized particles exist in significant quantities in the substrate region.

The dissociation of molecular oxygen occurs as a result of excitation from the ground state to either the $\sum_u^{3+}$ or $\sum_u^{3-}$ excited state by electron collision and subsequent decay to either the $O(^3P)$ or $O(^1D)$ atomic state.[142] The most abundant excited species to survive the transit from the ionization chamber to the deposition zone is likely to be atomic oxygen.

The action of the atomic oxygen which flows from the discharge chamber seems, in some respects, to be only that of initiating the decomposition of silane. It appears unlikely that the electrical properties of the deposited layers would vary with substrate position if the silane were to react with atomic oxygen in a single step:

$$SiH_4 + 4[0] \rightarrow SiO_2 + 2H_2O$$

More plausible is the model for silane oxidation proposed by Emeleus and Stewart.[143] They postulate a branching-chain reaction which is initiated when an excited oxygen reacts with silane to form a silyl radical as follows:

$$SiH_4 + [0] \rightarrow [SiH_2] + H_2O$$

This is followed by the branching reactions:

$$[SiH_2] + O_2 \rightarrow [SiH_2O] + [0]$$
$$[SiH_2O] + O_2 \rightarrow [SiH_2O_2] + [0]$$

and by the regenerating and terminating reaction:

$$[SiH_2O_2] + O_2 \rightarrow SiO_2 + H_2O + [0]$$

The reaction proceeds until all of the hydrogen is extracted from the silane or until terminated by collision with unreactive molecules or one of the surfaces in the reaction zone. If a SiH_xO_y radical is absorbed on a surface, the hydrogen may remain imbedded in the deposit or may combine with oxygen. The water thus formed may remain in the layer or at high substrate temperatures will be driven off the surface. The composition of the deposited material SiH_xO_y would be expected to vary with distance along the flow direction of the reactants. It therefore is reasonable to suppose that the electrical properties of layers deposited in different regions of the reaction zone might vary significantly.

Nonuniformity in the deposition rate seemed to be controlled primarily by variation in the gas flow velocity due to boundary layer effects on the surfaces of the deposition tube. For the cylindrical deposition geometry that was used, maximum deposition was observed to occur on the axis of the tube at the point of maximum flow velocity. The samples were placed approximately on the tube axis on a horizontal quartz plate which to some degree disrupted the raidial symmetry of the flow pattern. In the 50-mm-diam. deposition tube that was used, a thickness variation of approximately 5% was observed across a sample width of 15 mm. A rectangular deposition tube, as well as improved nozzle design, would be required in order to obtain greater deposition uniformity.

Another interesting technique for the deposition of dielectric materials at low temperatures is the photo-CVD or PHOTOX process reported by Peters.[144] Photogenerated atomic oxygen is combined with silane or other metal hydrides, chlorides, or organic compounds to form a variety of oxide layers. The atomic oxygen is formed by the dissociation of nitrous oxide as caused by the emission from a low-pressure mercury arc lamp. Direct photolysis can be achieved by absorption in the 1750–1950 Å band according to the reaction

$$N_2O + h\nu \,(1750\text{--}1950 \text{ Å}) \rightarrow N_2 + O(^1D)$$

or alternatively by a more efficient mercury-catalyzed process which proceeds according to the reactions shown below:

$$Hg_0 + h\nu \,(2537 \text{ Å}) \rightarrow Hg^* \quad \text{(photoexcited)}$$
$$Hg^* + N_2O \rightarrow N_2 + O(^3P) + Hg_0$$

Atomic oxygen is considerably more reactive than molecular oxygen; the enthalpy of formation $H_f \simeq 59$ kcal/mol is one indication of this increased reactivity. Not surprisingly, atomic oxygen will react with a number of materials at room temperature to form oxide layers although the physical properties of layers deposited at such low temperatures may suffer. In the PHOTOX system reported by Peters[144] photoexcitation of N_2O, reaction of the atomic oxygen with silane, and deposition of the SiO_2 on the substrate all take place within the same volume. It seems likely that intermediate products of the silane–oxygen reaction may be incorporated into the SiO_2 layers in the same manner as discussed for the indirect PECVD system unless precautions are taken to fractionate the intermediate products. Yet to be reported is the effect of the presence of mercury vapor in the deposition chamber on the electrical behavior of MOS devices.

11. Conclusion

Dielectric–semiconductor interfaces on III–V semiconductors are in general characterized by larger amounts of charge trapping and hysteresis than observed in high-quality SiO_2–Si interfaces. At the present time the thermally grown native oxide on (100) silicon is unique in its nearly ideal electrical behavior. The reasons for this nearly ideal behavior are not understood on a microscopic level but surely are related to the atomic structure of the interfaces. The properties of the structure, such as interface abruptness, stoichiometry, and formation of intermediate layers, that can be achieved in a particular material are related to the thermodynamic properties of the insulator and the semiconductor and may be severely constrained regardless of the details of the insulator formation process. One interesting example in which the electrical behavior of the interface *can* be significantly altered by the process is the (111) face of silicon. A thermally grown oxide on the (111) orientation is much inferior to a thermally grown oxide on (100) silicon and is in fact very similar to the better interfaces on the III–V compounds. However, a SiO_2 layer can be grown on (111) silicon by chemical vapor deposition and if care is taken to inhibit growth of the native oxide by introducing hydrogen into the growth chamber, interfaces may be produced which are very nearly as good as the thermally grown oxide on the (100) interface.[145] This strongly suggests that SiO_2 somehow grows differently on the (111) face of silicon with perhaps more silicon in the oxide layer immediately adjacent to the semiconductor.

Similar considerations are surely at work in the III–V materials. The electrical charging behavior at gallium arsenide interfaces occurs with a very short response time ($<10^{-6}$ s) and the charging centers can be considered as true surface states. At the same time there is now general agreement that an arsenic-rich layer forms immediately adjacent to the GaAs when it is exposed to oxygen. Deposited arsenic layers go through an amorphous to polycrystalline transition at a temperature between 300 and 400°C. This is accompanied by a transition in the electrical properties from semiconducting to semimetallic behavior. It is interesting to note that this is the same temperature range in which GaAs MIS devices, when annealed, undergo dramatic changes in their electrical behavior. As grown, anodic layers display $C-V$ characteristics with very large hysteresis loops and anomalous shape. When the devices are annealed at temperatures above 300°C, most of the hysteresis disappears and $C-V$ behavior similar to that of Fig. 10 is observed, indicating a very large density of fast interface states. A qualitative argument that something like this might occur can be made if the semiconductor interface is thought of as a heterojunction. The ability of an interfacial layer to act as a reservoir for charge could be greater if the interlayer were metallic than if it were semiconducting. Similarly, an insulating interlayer would be expected to hold less charge than a semiconducting interlayer. This picture is consistent with observations of InP interfaces in which it is thought that the interfacial layer is either In_2O_3 or $InPO_4$ or a combination of the two. Interfacial charges respond at much lower frequencies in InP than in GaAs and the charging behavior is more likely due to traps in the oxide than to surface states. When high-quality InP MIS devices are cooled to 77 K, all interfacial charging effects disappear which also argues strongly against the presence of surface states. In this respect InAs, InSb, and $In_{0.53}Ga_{0.47}As$ seem to be like InP, and GaP and GaAsP seem to be like GaAs.

For III–V semiconductors which appear not to have surface states, such as InP, it is still not clear if insulator–semiconductor interfaces can be prepared which will be sufficiently ideal for the manufacture of large-scale integrated circuits. The outstanding problem is the long-term drift of device characteristics due to charging of traps in the interfacial layer. Very recent work in our laboratory indicates that improvements may be achieved with InP interfaces when insulating layers are deposited with an overpressure of phosphorus present during insulator growth. Presumably this helps to preserve the stoichiometry of the InP surface during growth. Whether such techniques can be successful in producing technologically useful gate insulators on III–V semiconductors will only be known after much further study of the chemical, structural, and electrical properties of the layers.

References

1. S. R. Hofstein and G. Warfield, *Solid-State Electron* 8, 321 (1965).
2. F. P. Heiman, *IEEE Trans. Electron. Devices* ED-14, 781 (1967).
3. A. Goetzberger, E. Klausmann, and M. Schulz, *CRC Crit. Rev. Solid-State Sci.* 6, 1 (1976).
4. S. M. Sze, *Physics of Semiconductor Devices*, Wiley–Interscience, New York, 1969, p. 436.

5. S. Schottky, *Naturwissenschaften*, **26**, 843 (1938).
6. L. M. Terman, *Solid-State Electron.* **5**, 285 (1962).
7. C. N. Berglund, *IEEE Trans. Electron. Devices* ED-13, 701 (1966).
8. C. Hilsum, *Prog. Semicond.* **9**, 137 (1965).
9. J. I. Pankove, *Optical Processes in Semiconductors*, Prentice-Hall, Englewood Cliffs, NJ (1971), p. 53.
10. L. G. Meiners, Electrical properties of the gallium arsenide–insulator interface, *J. Vac. Sci. Technol.* **15**, 1402–1407 (1978).
11. L. G. Meiners, Surface potential of anodized p-GaAs MOS capacitors, *Appl. Phys. Lett.* **33**, 747–748 (1978).
12. E. Kohn and H. L. Hartnagel, On the interpretation of electrical measurements on the GaAs-MOS system, *Solid-State Electron.* **21**, 409–416 (1978).
13. D. Gerlich, Beat frequency bridge for large signal field effect, *J. Appl. Phys.* **33**, 1815–1816 (1962).
14. M. H. Pilkuhn, Study of gallium arsenide surfaces, *J. Phys. Chem. Solids* **25**, 141–146 (1964).
15. I. Flinn and M. Briggs, Surface measurements on gallium arsenide, *Surf. Sci.* **2**, 136–145 (1964).
16. D. N. Butcher and B. J. Sealy, Electrical properties of thermal oxides on GaAs, *Electron. Lett.* **13**, 558–559 (1977).
17. H. Hasegawa, K. E. Forward, and H. L. Hartnagel, New anodic native oxide of GaAs with improved dielectric and interface properties, *Appl. Phys. Lett.* **12**, 567–569 (1975).
18. G. Weimann and W. Schlapp, Anodic oxidation of gallium arsenide, *Thin Solid Films* **38**, L5–L7 (1976).
19. G. Weimann, Oxide and interface properties of anodic oxide MOS structures on III–V compound semiconductors, *Thin Solid Films* **56**, 173–182 (1979).
20. S. Varadarajan, M. A. Littlejohn, and J. R. Hauser, Inversion and accumulation layer formation at elevated temperatures in n-type GaAs–anodic oxide MIS devices, *Thin Solid Films* **56**, 235–242 (1979).
21. T. Sawada and H. Hasegawa, Interface state band between GaAs and its anodic native oxide, *Thin Solid Films* **56**, 183–200 (1979).
22. G. Sixt, K. H. Ziegler, and W. R. Fahrner, Properties of anodic oxide films on n-type GaAs, $GaAs_{0.6}P_{0.4}$ and GaP, *Thin Solid Films* **56**, 107–116 (1979).
23. C. R. Zeisse, L. J. Messick, and D. L. Lile, Electrical properties of anodic and pyrolytic dielectrics on gallium arsenide, *J. Vac. Sci. Technol.* **14**, 957–960 (1977).
24. A. Shimano, A. Moritani, and J. Nakai, GaAs-MOS capacitor with native oxide film anodized in nonaqueous elecltrolyte, *Solid-State Electron.* **21**, 1149–1152 (1978).
25. B. M. Arora and A. M. Narsale, Electrical instabilities of Al–anodic oxide–n-GaAs MOS structures and the effect of annealing, *Thin Solid Films* **56**, 153–161 (1979).
26. R. P. F. Chang and A. K. Sinha, Plasma oxidation of GaAs, *Appl. Phys. Lett.* **29**, 56–58 (1976).
27. L. A. Chesler and G. Y. Robinson, DC plasma anodization of GaAs, *Appl. Phys. Lett.* **32**, 60–62 (1978).
28. L. A. Chesler and G. Y. Robinson, Plasma anodization of GaAs in a dc discharge, *J. Vac. Sci. Technol.* **15**, 1525–1529 (1978).
29. K. Yamasaki and T. Sugano, Determination of the interface states in GaAs MOS diodes by deep-level transient spectroscopy, *Appl. Phys. Lett.* **35**, 932–934 (1979).
30. Y. Hirayama, F. Koshiga, and T. Sugano, Capacitance–voltage characteristics of Al-plasma anodic Al_2O_3–GaAs diodes, *J. Appl. Phys.* **52**, 4697–4699 (1981).
31. N. Yokoyama, T. Mimura, K. Odani, and M. Fukuta, Low-temperature plasma oxidation of GaAs, *Appl. Phys. Lett.* **32**, 58–60 (1978).

32. R. P. H. Chang and J. J. Coleman, A new method of fabricating gallium arsenide MOS devices, *Appl. Phys. Lett. 32*, 332–333 (1978).
33. F. Koshiga and T. Sugano, The anodic oxidation of GaAs in an oxygen plasma generated by a D.C. electrical discharge, *Thin Solid Films 56*, 39–49 (1979).
34. R. Hall and J. P. White, Surface capacity of oxide coated semiconductors, *Solid-State Electron 8*, 211–226 (1965).
35. H. Becke, R. Hall, and J. White, Gallium arsenide MOS transistors, *Solid-State Electron. 8*, 813–823 (1965).
36. H. W. Becke and J. P. White, Gallium arsenide FETs outperform conventional silicon MOS devices, *Electronics 40*(12), 82–89 (1967).
37. W. Kern and J. P. White, Interface properties of chemically vapor deposited silica films on gallium arsenide, *RCA Rev. 31*, 771–783 (1970).
38. J. E. Foster and J. M. Swartz, Electrical characteristics of the silicon nitride–gallium aresnide interface, *J. Electrochem. Soc. 117*, 1410–1417 (1970).
39. H. Klose, Y. E. Maronchuk, and O. V. Senoshenko, On the photo-capacitance of the MIS structure Al–Si_3N_4–n-GaAs, *Phys. Status Solidi A 21*, 659–664 (1974).
40. T. Ito and Y. Sakai, The GaAs inversion-type MIS transistors, *Solid-State Electron. 17*, 751–759 (1974).
41. W. Y. Lum and H. H. Wieder, Thermally converted surface layers in semi-insulating GaAs, *Appl. Phys. Lett. 31*, 213–215 (1977).
42. T. Mimura, K. Odani, N. Yokoyama, Y. Nakayama, and M. Fukuta, GaAs microwave MOSFETs, *IEEE Trans. Electron. Devices ED-25*, 573–579 (1978).
43. S. Yokoyama, K. Yukitomo, M. Hirose, Y. Osaka, A. Fischer, and K. Ploog, GaAs MOS structures with Al_2O_3 growth by molecular beam reaction, *Surf. Sci. 86*, 835–848 (1979).
44. M. Hirose, S. Yokoyama, and Y. Osaka, Surface states in GaAs tunnel MIS structures, *Phys. Status Solidi A 42*, 483–488 (1977).
45. M. Hirose, A. Fischer, and K. Ploog, Growth of Al_2O_3 layer on MBE GaAs, *Phys. Status Solidi A, 45*, K175–K177 (1978).
46. M. Hirose, S. Hirose, and Y. Osaka, Surface states in GaAs tunnel MIS structures, *Phys. Status Solidi A, 42*, 483–488 (1977).
47. S. Yokoyama, K. Yukitomo, M. Hirose, and Y. Osaka, GaAs MOS structures with Al_2O_3 grown by molecular beam reaction under UV excitation, *Thin Solid Films 56*, 81–88 (1979).
48. K. Kamimura and Y. Sakai, The properties of GaAs–Al_2O_3 and InP–Al_2O_3 interfaces and the fabrication of MIS field effect transistors, *Thin Solid Flms 56*, 215–223 (1979).
49. R. L. Streever, J. T. Breslin, and E. H. Ahlstron, Surface states at the n-GaAs–SiO_2 interface from conductance and capacitance measurements, *Solid-State Electron. 23*, 863–868 (1980).
50. N. Suzuki, T. Hariu, and Y. Shibata, Effect of native oxide on the interface property of GaAs MIS structures, *Appl. Phys. Lett. 33*, 761–762 (1978).
51. B. Bayraktaroglu, R. L. Johnson, D. W. Langer, and M. G. Mier, Germanium (oxy)nitride based surface passivation techniques as applied to GaAs and InP, *Physics of MOS Insulators* (G. Lucvosky, S. T. Pantelides and F. L. Galeener, eds.), pp. 207–211, Pergamon Press, Oxford (1980).
52. G. D. Bagratishvili, R. B. Dzhanelidze, N. I. Kurdiani, and O. V. Saksagenskii, MIS structure GaAs–Ge_3N_4–Al, *Phys. Status Solidi A 36*, 73–79 (1976).
53. K. P. Pande, M. L. Chen, M. Yousuf, and B. Laleric, Interface characteristics of Ge_3N_4–(n-type)GaAs MIS devices, *Solid-State Electron. 24*, 1107–1109 (1981).
54. G. K. Eaton, R. E. J. King, F. D. Morten, A. T. Partridge, and J. G. Smith, Surface conductance on p-type InSb at 77 K, *J. Phys. Chem. Solids 23*, 1473–1477 (1962).

55. J. L. Davis, Surface states on the (111) surface of indium antimonide, *Surf. Sci. 2*, 33–39 (1964).
56. J. F. Dewald, The kinetics and mechanism of formation of anode films on single-crystal InSb, *J. Electrochem. Soc. 104*, 224–251 (1957).
57. R. K. Mueller and R. L. Jacobson, Photo-controlled surface conductance in anodized InSb, *J. Appl. Phys. 35*, 1524–1529 (1964).
58. L. L. Chang and W. E. Howard, Surface inversion and accumulation of anodized InSb, *Appl. Phys. Lett. 7*, 210–212 (1965).
59. H. Huff, S. Kawaji, and H. C. Gates, Field-effect measurements on the A and B (111) surfaces of indium antimonide, *Surf. Sci. 5*, 399–409 (1966).
60. L. L. Chang, Orientation dependence of surface charge on anodized InSb, *Solid-State Electron. 10*, 69–70 (1967).
61. D. L. Lile and J. C. Anderson, Electrical surface properties of polycrystalline layers of PbTe and InSb, *Brit. J. Appl. Phys. (J. Phys. D) 2*, 839–853 (1969).
62. K. F. Komatsubara, H. Kamioka, and Y. Katayama, Electrical conductivity in an n-type surface inversion layer of InSb at low temperature, *J. Appl. Phys. 40*, 2940–2944 (1969).
63. K. F. Komatsubara, Y. Katayama, N. Kotera, and T. Kobayashi, Transport properties of electrons in inverted InSb surface, *J. Vac. Sci. Technol. 6*, 572–575 (1969).
64. R. Y. Hung and E. T. Yon, Surface study of anodized indium antimonide, *J. Appl. Phys. 41*, 2185–2189 (1970).
65. M. L. Korwin-Pawlowski and E. L. Heasell, Characteristics of MOS capacitors formed on p-type InSb, *Phys. Status Soldi A 24*, 649–652 (1974).
66. H. L. Henneke, Comment on "Polarity effects in InSb-alloyed p–n junctions," *J. Appl. Phys. 36*, 2967–2968 (1965).
67. J. C. Kim, Interface properties of InSb MIS structures, *IEEE Trans. Parts, Hybrids, Packaging PHP-10*, 200–207 (1974).
68. A. Etchels and C. W. Fischer, Interface-state density and oxide charge measurements on the metal–anodix oxide–InSb system, *J. Appl. Phys. 47*, 4605–4610 (1976).
69. H. Fufiyasu, M. Suzuki, K. Nakao, S. Itho, and O. Ohtsuki, Properties of metal–borosilicate glass–InSb oxide–p-type InSb structures, *Japan. J. Appl. Phys. 16*, 1473–1474 (1977).
70. T. Nakagawa and H. Fujisada, Method of reporting by hysteresis effects from MIS capacitance measurements, *Appl. Phys. Lett. 31*, 348–350 (1977).
71. A. Heime and H. Pagnia, Influence of the semiconductor–oxide interlayer on the ac-behavior of InSb MOS-capacitors, *Appl. Phys. 15*, 79–84 (1977).
72. J. D. Langan and C. R. Viswanthan, Characterization of improved InSb interfaces, *J. Vac. Sci. Technol. 16*, 1474–1477 (1974).
73. M. Yamaguchi and K. Ando, Thermal oxidation of InP and properties of oxide film, *J. Appl. Phys. 51*, 5007–5012 (1980).
74. L. G. Meiners, Electrical properties of SiO_2 and Si_3N_4 dielectric layers on InP, *J. Vac. Sci. Technol. 19*, 373–379 (1981).
75. M. Yamaguchi, Thermal oxidation of InP in phosphorus pentoxide vapor, *J. Appl. Phys. 52*, 4885–4887 (1981).
76. C. W. Wilmsen, The MOS InP interface, *CRC Crit. Rev. Solid-State Sci. 5*, 313–317 (1975).
77. D. L. Lile and D. A. Collins, An InP MIS diode, *Appl. Phys. Lett. 28*, 554–556 (1976).
78. K. P. Pande and G. G. Roberts, Interface characteristics of InP MOS capacitors, *J. Vac. Sci. Technol. 16*, 1470–1473 (1979).
79. T. Ota and Y. Horikoshi, InP MIS diodes prepared by anodic oxidation, *Japan. J. Appl. Phys. 18*, 989–990 (1979).

80. S. Hannah and B. Livingstone, Composite Al_2O_3 and native oxide on GaAs and InP incorporating enhanced group III oxides for surface passivation, *Inst. Phys. Conf. Ser. No. 50*, 271–279 (1980).
81. G. G. Roberts, K. P. Pande, and W. A. Barlow, InP–Langmuir film M.I.S. structures, *Electron. Lett. 13*, 581–583 (1977).
82. R. W. Sykes, G. G. Roberts, T. Fok and D. T. Clark, p-type InP/Langmuir film M.I.S. diodes, *IEEE Proc. Part I*, 137–139 (1980).
83. G. G. Roberts, K. P. Pande, and W. A. Barlow, InP–Langmuir-film M.I.S.F.E.T., *Solid-State Electron. Devices 2*, 169–175 (1978).
84. L. Messick, InP/SiO_2 MIOS structure, *J. Appl. Phys. 47*, 4949–4951 (1976).
85. D. Fritzsche, $InP–SiO_2$, M.I.S. structure with reduced interface state density near conduction band, *Electron. Lett. 14*, 51–52 (1978).
86. J. Stannard, Carrier generation and trapping in n-InP/SiO_2 capacitors, *J. Vac. Sci. Technol. 16*, 1462–1465 (1979).
87. L. G. Meiners, Capacitance–voltage and surface photovoltage measurements of pyrolytically-deposited SiO_2 and InP, *Thin Solid Films 56*, 201–207 (1979).
88. D. L. Lile, D. A. Collins, L. G. Meiners, and L. J. Messick, N-channel inversion-mode InP MISFET, *Electron. Lett. 14*, 657–659 (1978).
89. L. G. Meiners, D. L. Lile, and D. A. Collins, Inversion layers on InP, *J. Vac. Sci. Technol. 16*, 1458–1461 (1979).
90. D. Fritzsche, Interface studies on InP inversion FETs with SiO_2 gate insulation, *Inst. Phys. Conf. Ser. No. 50*, 258–265 (1980).
91. K. Von. Klitzing, T. Englert, and D. Fritsche, Transport measurements on InP inversion MOS transistors, *J. Appl. Phys. 51*, 5893–5897 (1980).
92. L. G. Meiners, D. L. Lile, and D. A. Collins, Microwave gain from an n-channel enhancement-mode InP M.I.S.F.E.T., *Electron Lett. 15*, 578 (1979).
93. L. G. Meiners and H. H. Wieder, in: *Semi-Insulating III–V Materials* (G. J. Ress, ed.), pp. 198–205, Shiva Press, Orpington (1980).
94. D. C. Cameron, L. D. Irving, G. R. Jones, and J. Woodward, MISFET and MIS diode behavior of some insulator–InP systems, presented at INFOS 1981, Erlangen, West Germany.
95. T. Kawakami and M. Okamura, InP/Al_2O_3 n-channel inversion-mode M.I.S.F.E.T.s using sulfur-diffused source and drain, *Electron Lett. 15*, 502–504 (1979).
96. M. Okamura and T. Kobayashi, Reduction of interface states and fabrication of p-channel inversion-type InP-MISFET, *Japan. J. Appl. Phys. 19*, L599–L602 (1980).
97. M. Okamura and T. Kobayashi, Improved interface in inversion-type InP-MISFET by vapor etching technique, *Japan. J. Appl. Phys. 19*, 2151–2156 (1980).
98. M. Okamura and T. Kobayashi, Slow current-drift mechanism in n-channel inversion type InP-MISFET, *Japan, J. Appl. Phys. 19*, 2143–2150 (1980).
99. T. Kobayashi, M. Okamura, E. Yamaguchi, Y. Shinoda, and Y. Hirota, Effect of pyrolytic Al_2O_3 deposition temperature on inversion-mode InP metal–insulator–semiconductor field-effect transistor, *J. Appl. Phys. 52*, 6434–6436 (1981).
100. K. P. Pande and S. Pourdavoud, Ge_3N_4–InP MIS structures, *IEEE Electron. Devices Lett. EDL-2*, 182–184 (1981).
101. S. N. Al-Refaie and J. E. Carroll, Indium phosphide oxide on InP for MOSFET applications, *IEE Proc. 128*, 207–210 (1981).
102. M. Yamaguchi, Thermal nitridation on InP, *Japan. J. Appl. Phys. 19*, L401–L404 (1980).
103. Y. Hirota, M. Okamura, and T. Kobayashi, The effects of annealing metal–insulator–semiconductor diodes employing a thermal nitride–InP interface, *J. Appl. Phys. 53*, 536–640 (1982).

104. S. Kawaji and Y. Kawaguchi, Galvanomagnetic properties of surface layers in indium arsenide, Proceedings of the Conference on Physics of Semiconductors, Kyoto, 1966, *J. Phys. Soc. Japan Suppl. 21*, 336–340 (1966).
100. H. E. Kunig, Analysis of an InAs thin film transistor, *Solid-State Electron. 110*, 335–342 (1968).
106. R. J. Schwartz, R. C. Dockerty, and H. W. Thompson, Capacitance–voltage measurements on n-type InAs MOS diodes, *Solid-State Electron. 14*, 115–124 (1971).
107. H. Terao, T. Ito, and Y. Saki, Interface properties of InAs-MIS structures and their application to FET, *Elec. Eng. Japan 94*, 127–132 (1974).
108. C. W. Wilmsen, L. G. Meiners, and D. A. Collins, Single- and double-layer insulator metal–oxide semiconductor capacitors on indium arsenide, *Thin Solid Films 46*, 331–337 (1977).
109. D. A. Baglee, D. K. Ferry, C. W. Wilmsen, and H. H. Wieder, Inversion layer transport and properties on InAs, *J. Vac. Sci. Technol. 17*, 1032–1036 (1980).
110. S. M. Spitzer, B. Schwartz, and M. Kuhn, Electrical properties of a native oxide on gallium phosphide, *J. Electrochem. Soc. 120*, 669–672 (1973).
111. T. Ikoma and H. Yokomizo, C–V characteristics of GaP MOS diode with anodic oxide film, *IEEE Trans. Electron. Devices ED-23*, 521–523 (1976).
112. D. H. Phillips, W. W. Grannermann, L. E. Coerver, and G. J. Kuhlmann, Fabrication of GaAsP MIS capacitors using a thermal-oxidation dielectric-growth process, *J. Electrochem. Soc. 120*, 1087–1091 (1973).
113. L. Forbes, J. R. Yeargan, D. L. Keune, and M. G. Craford, Characteristics and potential applications of $GaAs_{1-x}P_x$ MIS structures, *Solid-State Electron. 17*, 25–29 (1974).
114. R. K. Ahrenkiel, F. Moser, S. L. Lyu, and T. J. Coburn, Electronic properties of anodic oxides grown on $GaAs_{0.6}P_{0.4}$, *Thin Solid Films 56*, 117–128 (1979).
115. I. Tamm, *Physik. Z. Sowjetunion 1*, 733 (1933).
116. L. D. Langan, *Study and Characterization of Semiconductor Surfaces and Interfaces*, Ph.D. Thesis, University of California, Santa Barbara (1979).
117. R. J. Schwartz, R. C. Dockerty, and H. W. Thompson, Capacitance voltage measurements on N-type InAs MOS diodes, *Solid-State Electron. 14*, 115 (1971).
118. R. Zeigler and E. Klausmann, Static technique for precise measurements of surface potential and interface state density in MOS structures, *Appl. Phys. Lett. 26*, 400 (1975).
119. N. M. Johnson, D. K. Biegelsen, and M. D. Moyer, Characteristic defects at the $Si–SiO_2$ interface, *Physics of MOS Insulators* (G. Lucovsky, S. T. Pantelides, and F. L. Galeenev, eds.), p. 311, Pergamon Press, New York (1980).
120. C. W. Wilmsen, J. F. Wager, and J. Stannard, Chemical vapour deposited SiO_2–InP interface, *Inst. Phys. Conf. Ser. No. 50*, 251 (1980).
121. A. Goetzberger, E. Klausmann, and M. J. Schulz, Interface states on semiconductor/insulator interfaces, *CRC Crit. Rev. Solid-State Sci. 6*, 1 (1976).
122. K. Strater, Controlled oxidation of silane, *RCA Rev. 29*, 618–629 (1968).
123. N. Goldsmith and W. Kern, The deposition of vitreous silicon dioxide films from silane, *RCA Rev. 28*, 153–165 (1967).
124. R. S. Rosler, Low pressure CUD production processes for poly, nitride and oxide, *Solid-State Technol. 20*, 63–70 (1977).
125. R. C. G. Swann and A. E. Pyne, The preparation and properties of silica films deposited from silane and carbon dioxide, *J. Electrochem. Soc. 116*, 1014–1017 (1969).
126. H. F. Sterling and R. C. G. Swann, Chemical vapor deposition promoted by r.f. discharge, *Solid-State Electron.*, 8, 653–654 (1965).
127. K. Saminadayer and J. C. Pfister, Surface state generation on MOS capacitors irradiated with UV light and electrons, *Phys. Status Solidi A 36*, 679–686 (1976).

128. R. J. Powell, Vacuum-ultraviolet-induced space charge in Al_2O_3 films, *Appl. Phys. Lett.* **28**, 643–645 (1976).
129. G. W. Hughes, R. J. Powell, and M. H. Woods, Oxide thickness dependence of high energy electron, VUV and corona-induced charge in MOS capacitors, *Appl. Phys. Lett.* **29**, 377–379 (1976).
130. R. J. Powell, Hole photocurrents and electron tunnel injection induced by trapped holes in SiO_2 films, *J. Appl. Phys.* **46**, 4557–4563 (1975).
131. N. M. Johnson, W. C. Johnson, and M. A. Lampert, Electron trapping in aluminum implanted silicon dioxide films on silicon, *J. Appl. Phys.* **46**, 1216–1222 (1975).
132. J. M. Aitken, D. R. Young, and K. Pan, Electron trapping in electron beam irradiated SiO_2, *J. Appl. Phys.* **49**, 3386–3391 (1978).
133. T. H. Ning, Electron trapping in SiO_2 due to electron beam deposition of aluminum, *J. Appl. Phys.* **49**, 4077–4082 (1978).
134. J. R. Szedon and J. E. Sandor, The effect of low energy electron irradiation of metal oxide semiconductor structures, *Appl. Phys. Lett.* **6**, 181–182 (1965).
135. A. J. Spetn and F. F. Fang, Effect of low-energy electron irradiation on Si-insulated gate FETs, *Appl. Phys. Lett.* **7**, 145–146 (1965).
136. J. M. Fanet and R. Poirier, Charge storage in SiO_2 under low energy electron bombardment, *Appl. Phys. Lett.* **25**, 183–185 (1974).
137. L. M. Ephrath and D. J. DiMaria, Review of RIE induced radiation damage in silicon dioxide, *Solid-State Technol.* **24**, 182–188 (1981).
138. L. G. Meiners, Indirect plasma deposition of silicon dioxide, *J. Vac. Sci. Technol.* **21**, 655–658 (1982).
139. F. Kaufman and J. R. Kelso, Catalytic effects in the dissociation of oxygen in microwave discharges, *J. Chem. Phys.* **32**, 301–302 (1960).
140. W. L. Fite, in: *Chemical Reactions in Electrical Discharges* (B. D. Blaustein, ed.), p. 9, American Chemical Society, Washington, D.C. (1969).
141. A. T. Bell, *Techniques and Applications of Plasma Chemistry* (J. R. Hollahan and A. T. Bell, eds.), p. 28, Wiley, New York (1974).
142. A. T. Bell and K. Kwong, Dissociation of oxygen in a radiofrequency electrical discharge, *J. Am. Inst. Chem. Engrs.* **18**, 990–998 (1972).
143. H. J. Emeleus and K. Stewart, The oxidation of the silicon hydrides, Part I, *J. Chem. Soc. 1935*, 1182–1189.
144. J. W. Peters, Low temperature photo-CVD oxide processing for semiconductor device applications, International Electron Devices Meeting, December 7–9, Washington, D.C., 1981.
145. L. A. Kasprzak nd A. K. Gaind, Near-ideal Si–SiO_2 interfaces, *IBM J. Res. Develop.* **24**, 348–352 (1980).

5

III–V Inversion-Layer Transport

S. M. Goodnick and D. K. Ferry

1. Introduction

Application of an electric field normal to the surface of a semiconductor, such as in a metal–oxide–semiconductor (MOS) device, results in band bending and, for sufficiently strong fields, the formation of an inversion or accumulation layer. It has long been recognized[1] that the strong potential necessary to invert or accumulate the surface, shown schematically in Fig. 1, can quantize the motion of carriers normal to the surface and thus give rise to quasi-two-dimensional behavior for the parallel motion. The rationale for this is easily understood in the following context. Classically, the inversion-layer charge density falls to $1/e$ of its surface value in a distance over which the surface potential varies by an amount kT. Thus, if the total inversion density is 10^{12} cm^{-2}, the effective classical width of the inversion layer is only about 5 Å. Clearly, when the electron wavelength is closer to 100 Å, one must expect quantization of the motion perpendicular to the oxide–semiconductor interface. Here the carriers are trapped in a potential well in which the discrete energy eigenvalues for the motion normal to the surface form the minima of a set of quasicontinuous two-dimensional (parallel to the surface) energy bands referred to as subbands. In many cases, these subbands are separated by energies on the order of 50 meV which is significant when compared to the thermal energy kT, even at room temperature. This comparison suggests that the quantized nature of surface carriers can be an important consideration in normal MOS device applications. Indeed, it is generally thought that the combination of the quantization and the

---

S. M. Goodnick and D. K. Ferry ● Department of Electrical Engineering, Colorado State University, Fort Collins, CO 80523.

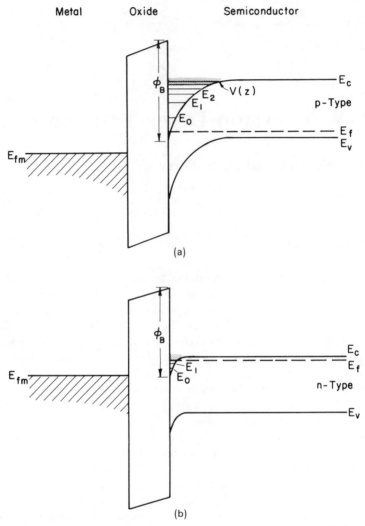

Figure 1. (a) Potential diagram for inversion of p-type semiconductor. ϕ_B is the oxide–semiconductor barrier, $V(z)$ is the surface potential, and E_0, E_1, and E_2 are the first three subband levels. (b) Potential diagram for accumulation on n-type semiconductor.

presence of interface scattering combine to cause the large mobility reduction at room temperature seen in MOS transistors compared to the bulk.

A variety of interesting and unusual phenomena have been observed in connection with quasi-two-dimensional behavior in these systems, which recently have been the subject of extensive review by Ando et al.[2]

In the context of the present review, we consider only the transport properties of electrons in inversion (or accumulation) layers of III–V compounds and the role that surface quantization plays in determining these properties. While the main emphasis here is on transport at an oxide–semiconductor interface, it is noteworthy to mention recent advances in heterojunctions and superlattices in III–V compounds where high-mobility, gate-controlled structures have been fabricated.[3,4] Qualitatively, the theoretical discussion for transport in heterostructures is the same as that of MOS devices.[5,6] Practically, however, scattering mechanisms which appear dominant in MOS structures (interface charge and interface roughness scattering) are relatively unimportant in heterolayers. Thus, the mobility in these latter structures provides an interesting contrast to the case of MOS surface mobility.

The room-temperature mobility of electrons in an inversion layer is invariably less than the corresponding bulk value, typically by a factor of 2 to 4 in Si and more in III–V compounds. While part of this decrease may be attributable to two-dimensional confinement, the main contributions are most likely due to the additional scattering mechanisms associated with the surface. Ionized impurities, defects, and the termination of the crystal lattice create chargeable surface states which result in Coulomb scattering of inversion electrons. Short-range disorder due to "surface roughness" at the oxide–semiconductor interface distorts the surface potential and causes fluctuations in the subband energies, which lead to scattering. Additional phonon modes due to the interface with the oxide contribute to the scattering as well. On the other hand, screening plays a much larger role for inversion-layer electrons[7] than in the bulk. This reduces the strength of some of these additional scattering mechanisms, especially at low temperature, as evidenced by the large increases in mobility measured for Si inversion-layer electrons as liquid-helium temperature is approached.[8]

The main problem in understanding the transport in III–V inversion and accumulation layers is the lack of detailed mobility measurements over a wide range of carrier densities and temperatures. Transport studies in III–V's have been hampered by the difficulty in obtaining clean interfaces in a gate-controlled structure where clear inversion or accumulation layers may be formed. Some of these difficulties are associated with the poor quality of native oxides on III–V's[9] for use as gate insulators in contrast to the fortuitous case of SiO_2 on Si. At present, one can only extrapolate from the limited transport data available on III–V surface layers what the dominant scattering mechanisms are based on the relative wealth of data concerning Si inversion transport.

Electronic transport at III–V surfaces differs from Si in a number of fundamental aspects. The most conspicuous difference is the smaller effective mass of the III–V's compared to Si, which usually results in higher bulk mobilities for the III–V compounds. For two-dimensional systems, however, a smaller mass results in a lower density of states and leads to the possibility of several occupied subbands, even at low temperatures. The possibility of multisubband occupancy thus makes the possibility of intersubband transitions an important consideration in the III–V's. The lower density of states also leads to decreased carrier screening for two-dimensional transport in III–V's compared to that of Si, which influences the relative strength of the various scattering mechanisms. While multivalley effects need not be considered for low-field transport in III–V's, nonparabolicity plays an important role in determining the surface subband structure and transport in narrow-gap semiconductors. Here, the admixture of s-like conduction-band states and p-like valence-band states (an effect which is reinforced by the surface potential) is expected to contribute to the scattering matrix element, in analogy to the bulk case.[10,11]

Finally, one must consider the relatively complex structure of the III–V oxide–semiconductor interface in determining the electronic properties at the surface. Recent advances in interface studies,[12,13] as well as theoretical surface electronic structure calculations,[14,15] have given a qualitative model for the electronic structure of these interfaces. It appears that defects due either to missing anions or cations[16] at the surface or antisite defects[17] are responsible for the large densities of observed interface states and subsequent Fermi-level pinning in many compound semiconductors. Such a defect structure may form during oxide growth on III–V's through preferential oxidation of one element over another and/or selective dissolution or dissociation of the surface atoms leaving a nonstoichiometric interface. Even in the case of deposited oxides on III–V's, intermediate native oxides appear[18,19] which may give rise to a layered structure at the interface. Such a nonstoichiometric interface structure will certainly give rise to additional and more complex scattering mechanisms than have previously been considered for the Si/SiO_2 system.

A brief overview is given in Section 2 of the formation of quantized states in inversion and accumulation layers. For more detail of quantization concepts and two-dimensional behavior, the interested reader is referred to the recent review of Ando *et al.*[2] In Sections 3 and 4, the important scattering mechanisms expected for III–V compound interfaces arising from the surface (Section 3) and the lattice (Section 4) are discussed. Finally, in Section 5, a review is given of the current status of experimental transport measurements on III–V materials as well as the prospects for the near future.

2. Quantization

2.1. Surface Subbands

A description of surface quantization from first principles is formally difficult, requiring a detailed model of the chemical bonding and local atomic configuration at the interface. While significant progress has been made in the understanding of the interfacial electronic structure of III–V compounds on a microscopic level, for most applications the effective-mass approximation (EMA) has proved sufficient in describing the effect of the surface potential in the formation of quantized states.

In the EMA for a single band, the wave function of an inversion-layer electron is written as a product of the Bloch function of that band, $e^{i\mathbf{K}\cdot\mathbf{R}}u_{\mathbf{K}}(\mathbf{R})$, and an envelope function $\Psi(\mathbf{R})$, where $\Psi(\mathbf{R})$ satisfies the effective-mass equation

$$\left(\frac{\hbar^2}{2m^*}\nabla^2 + V(z) - E\right)\Psi(\mathbf{R}) = 0 \tag{1}$$

Here m^* is the appropriate isotropic effective mass for the band in question and $V(z)$ is the surface potential, shown in Fig. 1 for both inversion [Fig. 1(a)] and accumulation [Fig. 1(b)]. For the III–V compounds of interest, the lowest conduction-band minima and the valence-band maximum occur at the Γ point in the Brillouin zone, and are characterized by an isotropic (albeit nonparabolic) effective mass. Thus the complications of an effective mass tensor, as occur in connection with Si and Ge, are not considered here. For now, we restrict the discussion to the lowest conduction band of wide-band-gap III–V compounds and defer the problem of nonparabolicity to Section 2.3.

The potential in Eq. (1) is assumed to depend solely on the z direction and hence the solution is separable as

$$\Psi(\mathbf{R}) = \zeta_i(z)\, e^{i\mathbf{k}\cdot\mathbf{r}} \tag{2}$$

Here $\mathbf{k}$ and $\mathbf{r}$ represent the two-dimensional wave vector and position vector in the plane parallel to the surface, and $\zeta_i(z)$ is an envelope function which satisfies

$$\left(\frac{\hbar^2}{2m^*}\frac{d^2}{dz^2} + V(z)\right)\zeta_i(z) = E_i\zeta_i(z) \tag{3}$$

The eigenvalues of Eq. (3) form a ladder of parabolic two-dimensional subbands which satisfy the dispersion relation

$$E(k) = E_i + \frac{\hbar^2 k^2}{2m^*} \tag{4}$$

The density of states for a two-dimensional electron gas (for parabolic bands) is constant in energy as opposed to the three-dimensional case. Each subband contributes a constant density of states, such that the total density of states may be written

$$D(E) = \sum_{i=0} \frac{n_s m^*}{2\pi\hbar^2} \theta(E - E_i) \tag{5}$$

where n_s is the spin degeneracy and $\theta(x)$ is the unit step function. Here a constant density of states for each subband is added as successive minima in energy are passed.

The surface potential $V(z)$ represents contributions from the space-charge layer, the inversion electrons, and the image potential which arises from the differing dielectric constants between the oxide and the semiconductor. The electronic contribution to $V(z)$ depends on the spatial distribution of the inversion electrons through Poisson's equation. In the Hartree approximation, the average potential due to the electron distribution depends on the magnitude of the envelope function, $\zeta_i(z)$, and thus (3) must be solved self-consistently. Numerical self-consistent solutions of (3) have been performed in connection with Si accumulation[20] and inversion layers,[21] as well as for accumulation layers on InAs[22,23] and InP.[24] In addition, one must usually consider the electronic contributions to the subband energies due to exchange and correlation. While for Si these effects are substantial, it has been suggested that the III–V compounds are adequately described in the Hartree approximation,[24,25] due to the smaller effective density r_s for electrons in III–V compound surface layers. However, the magnitude of r_s depends on the value of the effective mass m^*, which varies substantially between the different III–V compounds. Thus it is not clear that neglect of exchange and correlation is a valid assumption in all cases.

2.2. Approximate Solutions

By making a suitable approximation for the self-consistent electronic potential, Eq. (3) may be uncoupled from Poisson's equation and analytic solutions found. This neglect of the detailed form of the electronic potential is usually only valid at low inversion or accumulation densities. In connection with inversion and accumulation layers, two model potentials have appeared in the literature which give exact solutions for (3), the triangular well potential and the exponential potential.

The triangular well potential arises from neglect of quadratic terms which appear in the usual expressions for the space-charge potential. This is not a bad assumption in moderately doped, wide-band-gap semiconductors, since the inversion electrons are usually confined to within

100 Å of the surface, much less than the extent of the space-charge region. The model potential in this case is of the form $V(z) = eF_s z$ for $z > 0$, where F_s is the effective surface field in the semiconductor. For a linear potential, (3) yields Airy-function solutions[26]

$$\zeta_i(z) = A_i \left[\left(\frac{2m^* eF_s}{\hbar^2} \right)^{1/3} \left(z - \frac{E_i}{eF_s} \right) \right] \tag{6}$$

Here F_s can be taken as the average surface field $= (e/\bar{\varepsilon})(N_{\mathrm{depl}} + \tfrac{1}{2}N_s)$, where N_s is the inversion density, $N_{\mathrm{depl}} = (N_d - N_A)d$ is the total depletion charge density with d the depletion region width, and $\bar{\varepsilon}$ is the average dielectric constant between the semiconductor and the oxide. This choice for F_s is not unique, however. The eigenvalues E_i are determined by the condition that the wave function vanish at the interface $z = 0$, which is valid for an infinite interfacial barrier. The solutions are given by

$$E_i \cong \left(\frac{\hbar^2}{2m^*} \right)^{1/3} [\tfrac{3}{2} \pi e F_s (i + \tfrac{3}{4})]^{2/3} \tag{7}$$

The actual eigenvalues are the zeros of the Airy functions and differ very slightly from the value given in Eq. (7). Properties and asymptotic solutions for the Airy functions may be found in Abramowitz and Stegun,[26] while normalization integrals are given in Appendix B of Stern.[21]

In Fig. 2, the fraction of carriers in the lowest subband and the average width of the inversion layer are shown as a function of inversion density for several different temperatures. For the Airy-function solutions, Eq. (6), the average width of the ith level is $\tfrac{2}{3} E_i / e F_s$, and thus the levels become narrower as the surface field increases. For $N_s \gg N_{\mathrm{depl}}$, the curves are amenable to scaling as

$$T = aT_1, \qquad N_s = a^2 N_1, \qquad z_{\mathrm{avg}} = z_1/a \tag{8}$$

where T_1, N_1, and z_1 are, respectively, the temperature, the surface charge density, and the average width shown in Fig. 2 for InP and $a = m^*/m_1$ ($m_1 = 0.077 m_0$ is the appropriate value for InP). From Fig. 2(a) it is apparent that for the range of densities considered, as N_s is increased the surface field F_s increases, causing a deeper potential well. Thus the subbands are moved farther apart, causing the higher-lying subbands to be emptied of carriers.

The above results considered the case of an infinite oxide–semiconductor barrier. In real systems, however, this barrier is finite and electrons can penetrate the oxide region. In this case, the energy eigenvalues are found from the logarithmic derivative at the surface $z = 0$. Care must be taken, however, in matching boundary conditions at the interface between

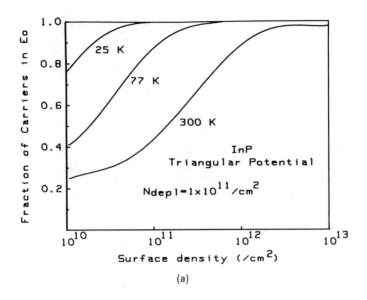

(a)

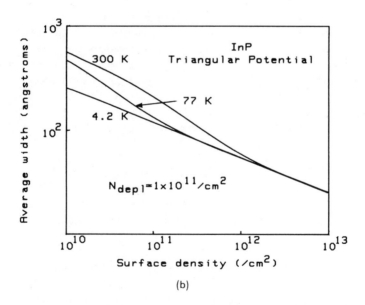

(b)

Figure 2. (a) Fraction of carriers residing in the lowest subband as a function of channel density for the triangular well potential. Parameters are for InP with a fixed effective depletion charge $N_{\text{depl}} = 1 \times 10^{11}/\text{cm}^{-2}$. (b) Average width of the inversion layer as a function of density using the parameters of Fig. 2(a). (c) Energy eigenvalues and penetration of electrons for InP with $m_{0x} = 0.5 m_0$ as a function of the barrier height ϕ_B. E_{00} is the subband energy with an infinite oxide barrier.

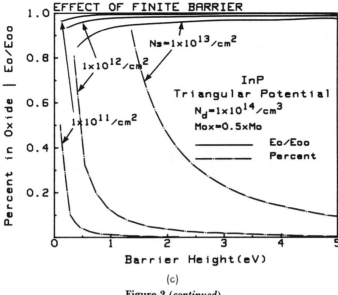

Figure 2 (*continued*)

two materials of differing effective masses. In order to conserve probability current, the eigenfunctions on either side can be scaled by the approximate effective mass.[27] The barrier between InP and SiO_2 is on the order of 3 eV which is large. Recently, however, experimental measurements by Wager *et al.*[28] have implied a barrier height of around 1 eV for the InP–native oxide interface using a combined ELS (Electron Loss Spectroscopy)–XPS (X-ray Photoemission Spectroscopy) technique. In Fig. 2(c), a plot is made of the energy eigenvalues and barrier penetration for InP for various values of the barrier height and an effective mass in the oxide of $0.5m_0$. This is calculated in the extreme quantum limit (EQL), where only one subband is occupied, using the Airy-function solution in the semiconductor and decaying exponential solutions for the oxide. From Fig. 2(c) it is seen that the effect of finite barrier height on the energy eigenvalues and barrier penetration is very small except at large surface densities where the Airy-function solution is probably inappropriate in any case. Thus it appears that the assumption of an infinite oxide–semiconductor barrier is justified for this model, even for a 1 eV barrier.

In accumulation layers, where the depletion region is small, the band-edge curvature is sufficiently appreciable that the triangular well model leads to noticeable errors. In this case one may assume an exponential such as $V(z) = -V_0 e^{-\lambda z}$. Here the zero of energy is chosen as the bulk conduction-band edge. V_0 represents the depth of the potential well and λ represents a characteristic decay which is related to V_0 via Gauss's law

$$V_0\lambda = \frac{e^2 N_s}{\varepsilon_{sc}} \qquad (9)$$

The solution of (3) for this potential are fractional-order Bessel functions

$$\zeta_i(z) = J_{P_i}\left[8\left(\frac{m^*V_0}{\hbar^2\lambda^2}\right)^{1/2} e^{-\lambda z/2}\right] \qquad (10)$$

$$P_i^2 = -\frac{8m^*E_i}{\hbar^2\lambda^2}, \qquad E_i < 0 \qquad (11)$$

The allowed values of P_i are determined by the vanishing of the wave functions at $z = 0$,

$$a_i P_i + b_i = \left(8\frac{m^*V_0}{\hbar^2\lambda^2}\right)^{1/2} \qquad (12)$$

Equation (10) is a parameterized equation for the zeros of the fractional-order Bessel function which are tabulated in Jahnke and Emde.[29] Only positive P_i are considered for (9) to vanish in the bulk. Thus, in contrast to the triangular well model, only a finite number of levels can exist in the experimental well, and at a given critical density, a new level may appear.

One may assume that all the surface carriers are trapped in the potential well and that the density is determined by the bulk Fermi level. Then the potential V_0 must be varied to give the correct amount of surface charge N_s. In Fig. 3(a), the self-consistent potential well determined in this manner in InAs for several different accumulation densities is shown. The potential well becomes deeper and narrower as the surface density increases as one would expect. At higher surface densities, new levels are observed to appear as the potential becomes deeper. In Fig. 3(b) the fraction of charge in the lowest subband is shown for various accumulation surface densities for several temperatures. In contrast to the triangular well model, the EQL is not approached as density increases. As in Fig. 2, when the depletion charge is small the parameters can be scaled via Eq. (8), except that here $a = (m^*/m_1)^{1/2}$ and the parameters used in Fig. 3(b) are appropriate for electrons in InAs.

An alternate approach to obtain analytic solutions is through the use of variational wave functions. Such solutions are useful due to their functional simplicity in which the self-consistent effects neglected in the prior treatment may be incorporated. The usual form of the variational solution is due to Fang and Howard[30] who considered an exponential

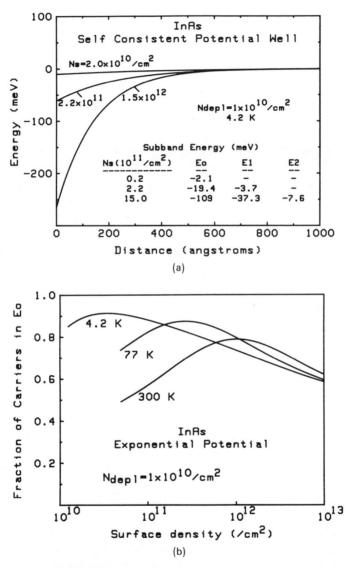

Figure 3. (a) Potential well $V(z) = -V_0 e^{-\lambda z}$ for several surface carrier densities including the calculated subband energies using Eq. (12). Parameters are for InAs at 4.2 K with a small depletion charge, $N_{\text{depl}} = 1 \times 10^{10}$ cm^{-2}. (b) Fraction of carriers in the lowest subband vs. carrier density for $T = 4.2$, 77, and 300 K using the parameters of Fig. 3(b) and a constant depletion charge.

function for the lowest subband,

$$\zeta_0(z) = \left(\frac{b^3}{2}\right)^{1/2} z\, e^{-bz/2} \tag{13}$$

The parameter b is determined by minimizing the one-electron energy and has been calculated, neglecting the image potential, as

$$b = \frac{12 m^* e^2 (N_{\text{depl}} + \tfrac{11}{32} N_s)^{1/3}}{\varepsilon_{\text{sc}} \hbar^2} \tag{14}$$

Das Sarma et al.[31] have considered a general form of (13) to include higher subbands. However, such solutions become increasingly unwieldy as more subbands are considered. Kawamoto et al.[32] have used such an approach to investigate the subband structure in III–V semiconductors in the presence of polar LO phonon-mediated electron–electron interactions. This latter effect was found to have a substantial effect in the subband structure for InP and GaAs. More recently, Yi and Ferry[33] have performed a variational calculation including the effects of exchange and correlation for InP which shows that many body effects do in fact give significant contributions to the subband energies, though not as large as the Si case.

2.3. Effects of Nonparabolicity

The discussion so far has been relevant to the larger band-gap semiconductors such as InP and GaAs where nonparabolic effects are not serious. However, for materials such as InAs and InSb, strong nonparabolicity of the bulk dispersion relation tends to mix the motion of electrons parallel to the surface with the normal motion. If the surface potential is on the order of the semiconductor band gap, additional mixing of valence- and conduction-band wave functions is expected. This may give rise to splitting of the spin degeneracy due to reduction of the inversion symmetry caused by the surface potential.

The bulk band structure for narrow band-gap III–V semiconductors at the Γ point may be described in terms of the six-band $\mathbf{k} \cdot \mathbf{p}$ Hamiltonian due to Kane,[34] which includes the s-like Γ_6 conduction band and the two p-like Γ_8 valence bands. The Γ_7 spin–split-off band is usually far enough below the Γ_8 that it may be neglected. Rather than solving a single-band effective-mass equation as was done in the previous section, a matrix equation involving the 6×6 bulk Hamiltonian results which yields six coupled EMA equations.

Various schemes have been applied to decouple these equations. Ohkawa and Uemura[35] considered only states arising from doubly degenerate light-mass Γ_8 and the Γ_6 band which results in only two coupled equations. These results were used to analyze Shubnikov–de Haas (SdH) oscillations on $Hg_{1-x}Cd_xTe$ surfaces with reasonable success. However, their work was later criticized by Därr et al.[36] as overestimating the surface potential mixing. Takada et al.[37] improved the calculation by diagonalizing the effective kinetic energy term in the EMA equation. This was found to approximately decouple the conduction- and valence-band equations and hence facilitate the solution. Their results for subband occupation and effective mass in InSb compared quite well to cyclotron resonance and magnetotransport data of Därr et al.[38,39] More recently, Marques and Sham[40] have applied the surface scattering theory of Sham and Nakayama[41] to the problem of subbands on narrow-gap semiconductors. The effects of exchange and correlation were included and found to be small in contrast to the case with Si. The calculated subbands showed strong nonparabolicity as well as noticeable spin splitting which was suggested as explaining doublets observed by Weisinger et al.[42] using intersubband spectroscopy. However, the latter[43] attribute this to independent excitations by the perpendicular and parallel components of the incident field due to the admixture of s- and p-like states occurring in nonparabolic bands rather than spin splitting. The fact that similar doublets are observed in Ge(111) subbands[44] may cast some doubt on this latter explanation.

3. Surface Scattering Mechanisms

As discussed earlier, additional scattering mechanisms are present at the surface which strongly influence the channel mobility, especially at low temperatures. Here topological imperfections at the interface, as well as charged interface states and impurities, perturb the energies of quantized electrons at the surface and thus give rise to scattering. Additional phonon scattering mechanisms due to interface and oxide modes are also present and are discussed in more detail in Section 4.

In many cases, the scattering cross section may be reduced to a two-dimensional problem through suitable averaging of the perturbing potential over the envelope function (3). However, nonvanishing matrix elements between different subbands result in intersubband transitions which couple the motion of electrons in different subbands. This may be especially problematic in materials like InAs and InSb where typically three subbands are occupied for carrier densities as low as 5×10^{11} cm^{-2} due to the low density of states in these compounds (see Section 5.1).

The problem of multisubband transport has formally been approached by Siggia and Kwok[45] where the solution is found by simultaneously solving a coupled set of Boltzmann equations, one for each subband. Although somewhat tedious, calculations for multisubband transport have been carried out in some cases.[46–48] In the following discussion, however, we restrict ourselves to the extreme quantum limit (EQL) in which only the ground subband is occupied, and only intrasubband scattering is important. Hence the mobility is determined in the usual manner from an average over the distribution function calculated in the relaxation time approximation. In this way, the dependence of the channel mobility on physical parameters becomes much more obvious at the cost of quantitative accuracy.

3.1. Coulomb Scattering

Disorder at the oxide–compound semiconductor interface, which as discussed earlier is associated with defects[12,13,15–17] in the vicinity of the interface, contributes to the high density of surface states observed[49,50] for MOS devices on III–V compounds. The charge of these states, usually either neutral or singly ionized, depends on their type (i.e., either donor or acceptor) as well as their energy relative to the Fermi level. In addition to surface states, ionized impurities incorporated in the oxide (i.e., Na$^+$ ions) and in the depletion region are present. Charge states give rise to Coulomb scattering of the inversion or accumulation electrons, which differs from the case of bulk impurity scattering[51] due to the reduced dimensionality of these carriers.

Coulomb scattering in the context of surface quantization was first described by Stern and Howard[7] for electrons confined to the lowest subband. Since that time, similar treatments have appeared[45,52–55] which treat the screened interaction of the inversion electrons with charged impurities to various degrees of approximation. In most cases, the oxide–semiconductor interface is treated as abrupt with an infinite potential barrier, so that problems associated with interfacial nonstoichiometry, as may be the case with III–V interfaces, are neglected.

In the discussion of Coulomb and surface roughness scattering, it is convenient to use the electrostatic Green's function for charges in the presence of a dielectric interface. The scattering matrix element involves integration over plane-wave states parallel to the surface, and thus we consider the two-dimensional Fourier transform of the Green's function

$$G(\mathbf{q}, z - z') = \frac{1}{2q\varepsilon_{sc}} (e^{-q|z-z'|} + \alpha\, e^{-q(z+z')}), \qquad z' > 0$$
$$= \frac{1}{2q\bar{\varepsilon}} e^{-q(z-z')}, \qquad z' < 0 \quad (15)$$

where $\mathbf{q}$ is the two-dimensional wave vector and α is given for an abrupt interface as $(\varepsilon_{sc} - \varepsilon_i)/(\varepsilon_{sc} + \varepsilon_i)$. Equation (15) is obtained by solving the classical image problem for a charge located at z' and its image at $-z'$ (the interface is located at $z' = 0$). For a nonabrupt interface, α above becomes in general a complex function of **q**. This wave-vector dependence has been calculated for several simple models by Stern.[56]

For two-dimensional scattering in the Born approximation, the scattering cross section is determined by the matrix element over the bare Coulomb potential of a charge located at z_i (in the presence of a dielectric interface located at $z = 0$):

$$\langle \mathbf{k}| V_c |\mathbf{k}'\rangle = e^2 \int_0^\infty dz |\zeta_0(z)|^2 G(\mathbf{q}, z - z_i) \qquad (16)$$

where **q** is the difference between the incident and scattered wave vector, $\mathbf{q} = |\mathbf{k} - \mathbf{k}'|$.

In the linear screening approximation, the scattering potential is simply the Fourier transform of the bare potential (16) divided by the static dielectric function for the inversion electrons, which in the random-phase approximation for a single subband is written[57]

$$\varepsilon(q) = 1 + \frac{q_0}{q} F(q) \Pi(q) \qquad (17)$$

Here q_0 is the inverse screening length, and $F(q)$ is a slowly varying function on the order of unity accounting for the broadening of the inversion layer from an ideal two-dimensional electron gas. $\Pi(q)$ is the wave-vector dependence of the static polarizability given at absolute zero as

$$\Pi(q) = 1 - \Theta(q - 2k_f)\left[1 - \left(\frac{2k_f}{q}\right)^2\right]^{1/2} \qquad (18)$$

where k_f is the Fermi wave vector and Θ is the unit step function. Extension of (18) to finite temperatures has been given by Maldague.[58]

The inverse screening length q_0 depends on the density of states and is given for a singly occupied subband as

$$q_0 = \frac{n_s m^*}{2\pi\hbar^2}\left(\frac{e^2}{2\varepsilon}\right) \qquad (19)$$

Here the term outside the parentheses is merely the density of states (5) for the lowest subband. The relative magnitude of q_0 compared to the wave vector q is a measure of the effectiveness of screening in the inversion layer.

It is not a bad assumption to consider only scattering from charges located at the interface due to the high density of defects there. In this idealization, charges are assumed to be uniformly distributed in the plane forming the interface ($z_i = 0$). The scattering rate in the Born approximation for the ground subband gives[7]

$$\frac{1}{\tau_i(k)} = \frac{e^4 m^* N_i}{8\pi\hbar^3 \bar{\varepsilon}^2} \int_0^{2\pi} d\theta \frac{(1 - \cos\theta)A(q)^2}{[q + q_0 F(q)\Pi(q)]^2} \tag{20}$$

where N_i is the density of charge centers, and $q = 2k\sin(\theta/2)$ where k is the wave vector and θ the scattering angle. The function $A(q)$ is given by

$$A(q) = \int_0^\infty dz\, e^{-qz} |\zeta_0(z)|^2 \tag{21}$$

which follows from (15) and (16) and is essentially a spatial average of the Coulomb potential over the electron distribution.

Two limiting cases may be observed. For $q_0 \ll q$ we have essentially the unscreened case[54] where the scattering rate varies as $1/k^2 \propto 1/N_s$. Here the mobility, $(e/m^*)\langle 1/\tau\rangle^{-1}$, is proportional to the surface density, reflecting the decrease in scattering cross section with the increase in average energy of the incident electrons. In the other extreme, $q_0 \gg q$, scattering is totally dominated by screening. Here the wave-vector dependence is determined by the ratio $A(q)/F(q)$, which is not a strong function of inversion density. Even in the totally screened case, however, the scattering rate for Coulomb charge centers tends to decrease as the inversion density increases, and thus this mechanism tends to dominate at low carrier densities.

Typical values of the screening length $1/q_0$ range from less than 5 Å in Si to greater than 180 Å for InSb. Thus, for the appropriate density range, the two extremes discussed above may be realized for these two materials, with the other III–V's representing intermediate cases. Of course the dielectric function, Eq. (17), and the scattering rate, Eq. (3.6), are appropriate only for a parabolic band structure, and thus are not truly applicable to narrow-gap materials like InSb. However, two-dimensional forms for screening and scattering in nonparabolic bands are not available at present.

The temperature dependence for these two extremes is quite different. In the unscreened case, the mobility is proportional to the average electron energy and hence kT for high temperatures. In the other extreme, however, the temperature dependence of the screening dominates. Qualitatively, the effect of increasing temperature is to randomize to correlated motion of electrons and hence reduce screening. Stern[59] has shown using the temperature dependence of the dielectric function[58] that the mobility dominated by impurity scattering will decrease with increasing temperature between 4.2 and 77 K due to the reduction in screening, an effect seen in experimental Si transport data.[8]

Experimentally, the observed density dependence of the mobility in Si MOSFETs appears much stronger than predicted by (20).[60] This has led to the consideration of additional scattering mechanisms as well as modifications to (20). Cheng[55] has considered impurities which are randomly distributed with a Gaussian correlation rather than the uniform distribution assumed in the previous derivation. This was found to fit the experimental data much better, however, at the cost of introducing additional fit parameters which describe the statistical properties of the scatterers. Yagi et al.[61,62] have alternately suggested that tailing in the band extremum reduces the density of states and hence increases the screening length (19) at very low densities. This results in increased scattering and strong density dependence at low carrier densities. Other authors have suggested dipole scattering[63,64] resulting from impurity substrate complexes at the interface as well as scattering from neutral impurities or bound states[47] to explain the observed N_s dependence. For III-V compounds, certainly many of these considerations for Si are applicable. However, as yet there are not sufficient data for comparison between theory and experiment.

3.2. Surface Roughness Scattering

In addition to Coulombic scattering due to surface states and ionized impurities, short-range scattering associated with interfacial disorder limits the mobility of electrons at the surface. On a microscopic level, the oxide–semiconductor interface is never truly abrupt, although evidence seems to suggest this is nearly true for the case of Si–SiO_2.[65,66] Rather, the interface will be distorted to accommodate lattice mismatch and nonstoichiometry between the two dissimilar materials. The local atomic environment along the surface varies in a more or less random fashion which, coupled to the surface potential, gives rise to fluctuations of the subband energy and hence a finite lifetime for electrons in a given momentum state.

At present, a calculation of the scattering rate based on the microscopic detail of the interface does not exist. The usual models instead rely on a semiclassical description in terms of a phenomenological surface roughness. Early models treated the diffuse and specular scattering of electrons from the crystal surface, the degree of diffuse scattering indicative of the magnitude of "roughness."[67,68] However, for quantized surface electrons, such a description is inadequate. Surface scattering within the framework of quantum subbands was first described in connection with magnetically induced surface states,[69,70] and later extended to electrons confined to narrow semiconductor layers.[68,71,72]

In present surface roughness models, the displacement of the interface from a perfect plane is assumed to be described by a random function $\Delta(\mathbf{r})$, where $\mathbf{r}$ is the two-dimensional position vector parallel to the surface. This model assumes that $\Delta(\mathbf{r})$ changes slowly over atomic dimensions such that the boundary may be treated as continuous and abrupt. This assumption is obviously in error when surface fluctuations occur on the order of atomic dimensions.

The scattering potential may be obtained by expanding the surface potential in terms of $\Delta(\mathbf{r})$[68,73]:

$$V^0_{SR}(\mathbf{r}, z) = V(z + \Delta(\mathbf{r})) - V(z) \cong \Delta(\mathbf{r})eF(z) + \cdots \qquad (22)$$

where $F(z)$ is the electric field in the channel. Other authors have considered alternatively expanding the wave function at the surface[60,69,71] or finding a unitary transformation which removes $\Delta(\mathbf{r})$ from the potential term.[70] All three methods have been shown by Cheng[74] to be equivalent to lowest order in $\Delta(\mathbf{r})$.

The scattering rate for the potential (22) is usually treated in the Born approximation, although it is not clear on what grounds this is valid. The scattering matrix element using (22) gives[60]

$$\langle \mathbf{k} | V^0_{SR} | \mathbf{k}' \rangle = e\Delta_q \int_0^\infty dz F(z) |\zeta_0(z)|^2 = \frac{e^2 \Delta_q (N_{depl} + \tfrac{1}{2} N_s)}{\varepsilon_{sc}},$$
$$\mathbf{q} = |\mathbf{k} - \mathbf{k}'| \qquad (23)$$

where Δ_q is the transform of $\Delta(\mathbf{r})$. The factor of $\tfrac{1}{2}$ multiplying N_s arises from (23) depending on the average surface field in the channel.

Additional corrections to the roughness potential (22) arise from the normal displacement of charge carriers to the surface as well as electric field modification caused by a rough interface.[75] These corrections give rise to an additional roughness potential

$$V^1_{SR}(\mathbf{r}, z) = e^2 \int d^2\mathbf{r}' \, dz' n(z' + \Delta(\mathbf{r})) G(\mathbf{r}, \mathbf{r}'; z, z')$$
$$+ \int_{S'} d^2\mathbf{r}' \left(G \frac{\partial V(z')}{\partial z'} - V(z') \frac{\partial G}{\partial z'} \right) - V(z) \qquad (24)$$

where $n(z)$ is the carrier density $N_s \zeta_0^*(z) \zeta_0(z)$, $V(z)$ is the unperturbed surface potential in the semiconductor, G is the electrostatic Green's function, the transform of (15), and the surface integral is taken over the perturbed surface described by $\Delta(\mathbf{r})$. The volume integral represents the potential arising from the fluctuation normal to the surface of the inversion-layer electrons, while the surface integral gives rise to first-order corrections to the electric field due to the perturbed dielectric boundary.

The scattering matrix element can be calculated keeping only first-order terms in Δ_q and making use of convolution

$$\langle \mathbf{k}| V'_{SR}|\mathbf{k}'\rangle = \Delta_q \int_0^\infty dz\, \zeta_0(z)^2 \left[e^2 N_s \int_0^\infty dz'\, G(\mathbf{q}, z-z') \frac{\partial}{\partial z'} \zeta_0(z')^2 \right.$$
$$\left. + eF_s \left(\frac{\partial G(\mathbf{q}, z-z')}{\partial z'} - \frac{\partial G(\mathbf{q}, z-z')}{\partial z} \right) \right] \quad z' = 0 \quad (25)$$

where $G(\mathbf{q}, z-z')$ is given by (15) for $z' > 0$. When (25) is evaluated using the variational wave function (14), the result of Ando[75] is obtained. Saitah[76] has considered additional corrections to the above due to image-potential modification.

In the Born approximation, only the statistical properties of $\Delta(\mathbf{r})$ need be considered. Thus, the assumption is usually made of a Gaussian distribution for the autocorrelation[69]

$$\langle \Delta(\mathbf{r})\Delta(\mathbf{r}-\mathbf{r})\rangle \cong \Delta^2 e^{-r^2/L^2} \quad (26)$$

where Δ represents the rms height of the surface roughness and L is the correlation length, which is roughly equivalent to the average distance between "bumps" on the surface. By convolution (26) is related to the Fourier transform of $\Delta(\mathbf{r})$ which gives

$$|\Delta_q|^2 = \pi \Delta^2 L^2 e^{-q^2 L^2/4} \quad (27)$$

Combining (23), (26), and (27), the scattering rate for a single subband in the Born approximation may be written

$$\frac{1}{\tau_{SR}} = \frac{\Delta^2 L^2 e^4 (N_{depl} + \tfrac{1}{2} N_s)^2 m^*}{2\bar{\varepsilon}^2 \hbar^3}$$
$$\times \int_0^{2\pi} d\theta\, (1-\cos\theta) \exp(-k^2 L^2 \sin^2 \theta/2) \left(\frac{\Gamma(2k \sin \theta/2)}{\varepsilon(2k \sin \theta/2)} \right)^2 \quad (28)$$

Here $\varepsilon(2k \sin \theta/2)$ is the dielectric function (17) which incorporates screening, while Γ contains the corrections due to Ando[75] given in (25) and Saitah[76] (with the prefactors removed) which are on the order of unity. The explicit dependence of the scattering rate on the square of the average surface field results in decreasing mobility with increasing surface field which agrees with the trend observed in the experimental mobility data of all materials. This decrease in mobility with surface density qualitatively arises from the increased electric field dispersion around surface discontinuities at higher surface fields which in turn gives rise to a larger scattering potential as reflected in (23).

The roughness parameters, Δ and L, are usually found by fitting experimental mobility data. Typical values obtained in the manner are

2–6 Å for Δ and 15–40 Å for L for Si,[75] and more for III–V's,[77,78] depending on whether a screened or unscreened model is used. Direct measurement of Δ and L have been attempted for Si–SiO$_2$ by Sugano[79] and Goodnick et al.,[80,81] who find values for Δ and L and the same order of magnitude as given by mobility fits. As yet no similar comparisons have been reported on III–V's. In Fig. 4(a), the calculated low-temperature mobility using (20) and (28)* is shown along with experimental inversion mobility data of von Klitzing et al.[78] for InP measured at liquid-He temperature. Although the mobility is rather low and vanishes at nonzero N_s indicating some degree of trapping, a qualitative fit has been made (assuming that carriers are trapped for $\mu \cong 0$) which shows the range of carrier density over which the respective scattering mechanisms should dominate. As expected, impurity scattering dominates at low carrier density ($< 1 \times 10^{12}$ cm^{-2}), while surface roughness is dominant in the high-density regime. The tradeoff between impurity and surface roughness usually leads to the peaked behavior in the mobility vs. carrier density at low temperatures as seen in Fig. 4(a).

As is evident from (28), the scattering rate depends strongly on the rms height of the roughness Δ. The dependence on the correlation length L is somewhat more complicated due to the competing effects of the quadratic preintegral term and the exponential dependence inside the integral. Figure 4(b) shows a theoretical plot of the combined effect of Coulomb and surface roughness scattering vs. density in InP for various values of L at low temperature. Here the preintegral (ΔL) product has been held constant, while L is varied to illustrate the mobility dependence on this parameter. For L on the order of $\frac{1}{2}k_f$, the exponential inside the integral becomes sizable and thus the scattering rate due to surface roughness may actually decrease at high-enough densities. In the unscreened case, an electron of incident wave vector k is most strongly scattered when the surface roughness is coherent with k, that is, $L = 1/k$. The effect of screening, however, is to reduce the long-wavelength fluctuations of $\Delta(\mathbf{r})$ and hence shift the above value to a shorter wavelength.

In the high-temperature limit, where screening is greatly reduced, the scattering rate may be averaged over a Maxwellian to yield[77]

$$\left\langle \frac{1}{\tau} \right\rangle = \frac{m^* \Delta^2 L^2}{\hbar^3 \bar{\varepsilon}^2} e^4 (N_{\text{depl}} + \tfrac{1}{2} N_s)^2 / (1 + 2m^* L^2 k_B T/\hbar^2)^{3/2} \qquad (29)$$

Here the scattering rate decreases with temperature, the magnitude of the decrease depending on the relative size of the correlation length L. For usual values of L (< 40 Å), the denominator of (29) is close to one and the temperature dependence is weak, even at room temperature.

---

*Excluding the image corrections of Saitah.[76]

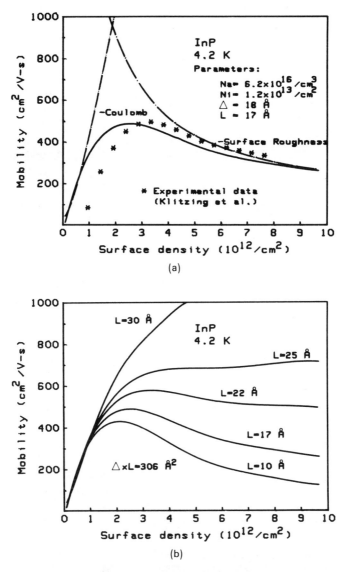

Figure 4. (a) Plot of the theoretical low-temperature mobility versus total surface density in InP dominated by impurity and surface roughness scattering. Included are the data of von Klitzing et al.[78] which have been fitted (assuming the data may be shifted) using the parameters shown. (b) Theoretical low-temperature mobility holding the ΔL product constant while varying L. Parameters are the same as Fig. 4(a) with $\Delta L = 306$ Å.

The present formalism for roughness scattering is weak in several respects. The expansion of the potential in terms of the quasicontinuous roughness function assumes that one may treat the fields classically at the surface. Yet the typical fit parameters from (28) are less than 20 Å, which is on the order of interatomic distances. This suggests that atomic fluctuations are contributing to the scattering and perhaps it is more appropriate to treat the problem similarly to that of a disordered alloy, where multiple scattering is important. In compound semiconductors, interface nonstoichiometry and broadening are present, which are ignored in the derivation of (28). In such cases, while (28) probably serves to fit experimental mobility data, Δ and L lose their physical significance for a complicated interface structure, and then are merely indicative of the degree of disorder present.

Some attempt has been made to address the effects of interface grading on surface roughness scattering by Stern[82] and Price and Stern.[83] Here, a simple model was used between the oxide and the semiconductor in which the bulk properties (i.e., m^*, ε, etc.) were assumed to vary continuously across the interface. The calculated corrections due to grading do not appear to be too substantial until the interface width exceeds 10 Å. Stern[56] has also calculated the image potential in the presence of a graded interface. This can be included in the Green's function, Eq. (15). When included in the formalism of (25) and (28), the net effect of grading tends to reduce the image potential which decreases the scattering rate somewhat for both Coulomb and surface roughness.

4. Phonon Scattering

Numerous papers have been devoted to the scattering of inversion-layer charge carriers by interface and bulk phonons. To obtain a rough understanding, one generally only has to consider two limiting cases. While the electrons or holes form a quasi-two-dimensional system because of size quantization, this does not imply that the phonons are interface modes. The possible "two-dimensionality" of the phonons depends on the elastic constants of the two neighboring media, which are only remotely related to the interface potential confining the charge carriers. Therefore, we treat the charge carriers as two-dimensional, but for the phonons we have to consider: (1) that the elastic (dielectric) constants are close enough and bulk modes exist, or (2) that the elastic (dielectric) constants are sufficiently different that interface modes exist. This distinction has not always been made in the literature and some of the formulas given are rather meaningless mixtures of these cases. The basic interaction mechanisms have been applied unchanged to scattering at interfaces, which may not be a valid approximation. The deformation-potential concept is based on the assumption that the interaction of electrons with a long-wavelength acoustic mode is equivalent to the effect of a locally homogeneous strain. This conjecture (deformation potential theorem) is certainly violated if the width of the inversion layer is on the order of the phonon wavelength, which is the case over a wide range of temperatures. It is trivial to show that in a square well the vibration of the walls contributes to scattering that is not included in the ordinary deformation potential theory. However, it is not known how this differs from scattering by surface interface wave modes.

While the formulas that follow represent the best interpretation of current understanding, they should be carefully applied, especially in certain temperature ranges. The polar interaction mechanism has been considered here in more detail due to the possibility of scattering by remote polar phonon-generated interface modes.

The III–V materials differ considerably from Si, for which the majority of work has been carried out. In the latter material, acoustic scattering makes a major contribution, while the dominant optical interaction is the equivalent intervalley scattering process.[84] In the III–V materials, however, acoustic scattering is relatively weak due to the low effective mass of the electrons. In addition, the dominant optical interaction arises from the polar (Fröhlich) interaction. Here we review the importance of these scattering processes and then discuss the relative strengths of the various scatterers. For a more extensive review, readers are referred to Ferry et al.[85]

4.1. The Acoustic Interaction

A first-principles calculation of the electron–phonon interaction for inversion-layer electrons at the semiconductor–insulator interface has not yet been performed, and the formulas found in the literature are based on the well-known phenomenological approaches. At the beginning of this section we discussed the conjecture that the equivalence of phonons and homogeneous strain breaks down when the wavelength is on the order of the inversion-layer width. Over some ranges of temperature, we therefore have to take the deformation potential concept with a "grain of salt." However, for small phonon wave vectors, it is probably not a bad approximation. Kawaji,[86] Ezawa et al.,[87] and Sah et al.[54] have calculated the acoustic scattering rate in this approximation to be

$$\frac{1}{\tau_{ac}} = \frac{\pi}{\hbar}\left(\frac{k_B T \Xi^2}{v_s^2 \rho_m V}\right) \sum_{j,q,q_z} B_{ij}^2 \delta(E_{k'} - E_k \pm \hbar\omega_q), \qquad q = |\mathbf{k} - \mathbf{k}| \qquad (30)$$

where Ξ is the deformation potential; the upper and lower signs stand for phonon absorption and emission of energy $\hbar\omega_q$, respectively; v_s is the velocity of sound; ρ_m is the mass density; and

$$B_{ij} = \int_0^\infty g_{ij}(z) \, e^{iq_z z} \, dz \tag{31}$$

Here, $g_{ij}(z) = \zeta_i^*(z)\zeta_j(z)$ is the product of the initial- and final-state wave functions to account for bulk scattering, both between different subbands (elastic) and within a subband, and surface scattering. For intrasubband scattering, (31) yields a factor on the order of W_j^{-1}, where W_j is the inversion-layer thickness of the jth subband, and

$$\frac{1}{\tau_{ac}} = \frac{\lambda m^* \Xi^2 k_B T}{\hbar^3 \rho_m v_s^2 W_j} \tag{32}$$

where λ is an overlap integral of order unity and the other variables have their normal meanings. The value of λ has slightly different values for scattering from bulk phonons or from interface Rayleigh waves, but the form (32) is unchanged. Equation (32) contains no dependence on the kinetic energy and so is expected to show no field-dependent behavior. However, (32) is calculated in the equipartition limit, which is not expected to be valid at low temperatures. At these low temperatures, where equipartition is invalid, different temperature dependences and some kinetic energy dependence is expected. It should also be pointed out that at low temperatures the inversion layers are degenerate and that more importantly the Fermi energy is of the order of phonon energies of the system. In this case, interesting coupled electron–phonon interactions can be expected.

4.2. Scattering by Polar Modes

In covalent semiconductors with two or more nonequivalent atoms per unit cell, optical modes, corresponding to movements of these atoms relative to one another, are allowed. In compound semiconductors, where the bonding is partially ionic, a polarization field arises due to the Coulomb interaction between the different effective charges sitting on the nonequivalent lattice sites. The polarization field arising from the typical longitudinal–optical mode causes a polar–optical interaction with the electrons. The matrix element for this interaction is found by first treating the lattice as a continuum, characterized by its polarization $P(\mathbf{r})$ at the point $\mathbf{r}$. This gives rise to a scattering Hamiltonian from which the scattering rate can be calculated[88,89] as

$$\frac{1}{\tau} = \sum_j \frac{\pi e E_0}{m^* \omega_0 W_j} N\{U_0(E + \hbar\omega_0 - E_j)[1 - f_0(E + \hbar\omega_0)] \\ + \exp(\hbar\omega_0/k_B T) U_0(E - \hbar\omega_0 - E_j)[1 - f_0(E - \hbar\omega_0)]\} \tag{33}$$

where E_j is the bottom energy of the jth subband, $U_0(x)$ is the unit step function, and

$$E_0 = \frac{m^* e \omega_0}{4\pi \hbar \varepsilon_0}\left(\frac{1}{\varepsilon_\infty} - \frac{1}{\varepsilon_{st}}\right) \tag{34}$$

is an effective interaction field. Here ω_0 is the frequency of the relevant phonon mode which is usually taken to be independent of the phonon wave vector for the polar–optical interaction, N is the Bose–Einstein occupation factor for the phonons, and ε_∞ and ε_{st} are the high- and low-frequency relative permittivities, respectively. The first term in (34) refers to the absorption of a phonon by the electron, while the second term refers to the emission of a phonon by the electron. An overlap integral relates the initial and final states of the electron which also gives rise to nonparabolicity corrections to the matrix element. We have taken the overlap integral as unity in this work, however.

In Fig. 5, the variation of the quasi-two-dimensional momentum relaxation rate for electrons at the surface of GaAs at 77 K is shown as a function of the inversion-layer thickness. For comparison, the three-dimensional polar–optical phonon result taken from Conwell[51] is also shown. In this case, the quantization effects will begin to affect the

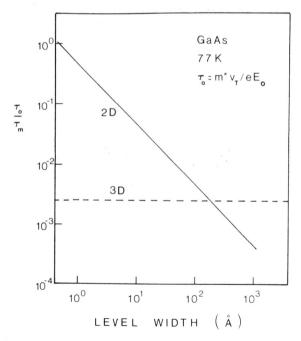

Figure 5. The scattering rate for the polar-optical interaction in two- and three-dimensional systems normalized to $1/\tau_0 = eE_0/m^* v_T$, where $v_T = (\omega_0 k_B T/m^*)^{1/2}$.

polar-limited mobility when W_j, the subband effective width, becomes smaller than about 200 Å. Only a single subband is considered, but additional effects on the mobility can be expected to occur if more than one subband becomes important.

The polar–optical phonon interaction represents an important scattering mechanism in compound semiconductors. At low temperatures, however, it often is ineffective. In quasi-two-dimensional semiconductors, though, the quantization of the motion perpendicular to the surface enhances this interaction so that this scattering mechanism can be important even at low temperatures, although the exact level of importance can only be ascertained in individual semiconductors by detailed consideration of all the residual scattering mechanisms.

4.3. Remote Optical Phonons

In the preceding discussion we mentioned the main lattice scattering mechanisms one would expect. We have to consider, however, that the charge carriers are at a distance of 10^{-6} cm or less from the strongly polar oxide. Therefore, the "fringing fields" of the polar interface modes can also scatter the charge carriers. For a free surface, the macroscopic electric fields outside the crystal and the scattering rate of electrons due to the polar surface modes were calculated by Wang and Mahan.[90] Their treatment was generalized to the case of two adjacent dielectrics and size quantization of inversion-layer carriers by Hess and Vogl.[91] This interface scattering rate is

$$\frac{1}{\tau_{\text{int}}} = \frac{e^2 \chi}{4\pi\varepsilon_0} \int_0^\infty \int_0^{2\pi} (N + \tfrac{1}{2} \pm \tfrac{1}{2}) \delta(E_{k\pm q} - E_k \pm \hbar\omega_i) \, dq \, d\phi \qquad (35)$$

For the case of two polar neighboring media, the term χ is given by

$$\chi = \frac{1}{\omega_\pm} \left(\frac{\omega_{T0,1}^2 (\varepsilon_{\text{st},1} - \varepsilon_{\infty,1})}{(\omega_\pm^2 - \omega_{T0,1}^2)^2} - \frac{\omega_{T0,2}^2 (\varepsilon_{\text{st},2} - \varepsilon_{\infty,2})}{(\omega_\pm^2 - \omega_{T0,2}^2)^2} \right)^{-1} \qquad (36)$$

where $\omega_\pm$ are solutions of

$$(\omega^2 - \omega_{T0,1}^2)(\omega^2 - \omega_{L0,2}^2) = -\frac{\varepsilon_{\infty,1}}{\varepsilon_{\infty,2}} (\omega^2 - \omega_{T0,2}^2)(\omega^2 - \omega_{L0,2}^2) \qquad (37)$$

Here $\omega_{T0,i}$ ($i = 1, 2$) is the angular frequency of the transverse polar–optical mode, $\omega_{L0,i}$ the longitudinal frequency, and $\varepsilon_{\text{st},i}$, $\varepsilon_{\infty,i}$ the static and optic dielectric constants.

An example of the calculated contributions of the various scattering mechanisms considered thus far is shown in Fig. 6(a) for InAs at 77 K. The fit parameters are relevant to the experimental mobility data of

Baglee et al.[92] and Kawaji and Kawaguchi[93] shown in Fig. 6(b). As can be seen in Fig. 6(a), surface roughness is dominant at all densities studied, with contributions at low densities from interface charge scattering (unscreened) and at high densities from the bulk L0 phonon. The fit parameters used are $\Delta = 15$ Å, $L = 29$ Å using (29), and $N_i = 1.3 \times 10^{11}$ cm^{-2}. The interface phonon scattering is found to be very

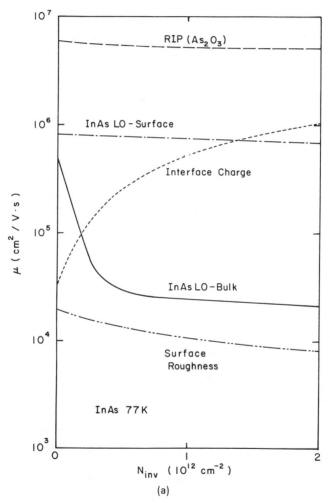

Figure 6. (a) Relative contributions to the mobility of the various phonon mechanisms, impurity charge, and surface roughness for InAs at 77 K. Here the roughness parameters are $\Delta = 15$ Å and $L = 20$ Å and the interface charge density is $N_{int} = 1.3 \times 10^{11}$ cm^{-2}. (b) Expanded plot of Fig. 6(a) including the data of Baglee et al.[92] and Kawaji and Kawaguchi[93] (represented by Δ).

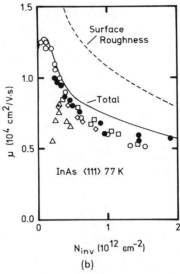

Figure 6 (*continued*)

weak, contributing less than 1% to the total scattering. The calculation shown is for the surface phonon generated by the low-frequency oxide phonon present in As_2O_3 (bulk). The parameters used for this phonon are $\hbar\omega_1/k_B = 533$ K, $\varepsilon_{st} = 4.04$, and $\varepsilon_\infty = 3.69$. The higher-energy phonons in As_2O_3 and In_2O_3 make an even smaller contribution since their occupation numbers are much smaller.[77]

4.4. High Fields

The treatment of transport at moderate and high electric fields along the surface has received much extensive review recently.[94] Most of this work deals primarily with inversion layers in silicon, however. For very high fields, where the kinetic temperature of the carriers becomes large, the essential two-dimensional behavior is lost and the carriers respond as if they were in a three-dimensional system but with a lower initial mobility. One exception to this occurs, however, in many-valleyed semiconductors where repopulation among the subbands can occur, much as in the Gunn effect.

At modest fields and low temperatures, however, the changes in mobility, scattering rates, and energy-loss mechanisms in electric fields can be utilized as a sensitive probe of the details of the various scattering mechanisms. This has been most successful in the treatment of warm carrier magnetotransport but has illuminated a number of other problems, notably in magnetotransport.

5. Experimental Results

As discussed earlier, extensive study of the transport properties of III–V surface layers has been hampered by the development of MOS technology comparable to that of Si–SiO$_2$. Investigations of actual MOS inversion and accumulation layers in III–V's have thus far been restricted to the indium compounds: InSb, InAs, InP, and ternary or quaternary compounds of In$_{1-x}$Ga$_x$As$_y$P$_{1-y}$. The viability of the In compounds for MOS applications has been attributed to a favorable defect pinning level,[13] which for In vacancies lies close to or above the bulk conduction-band edge. Such defects may form at the oxide–semiconductor interface due to preferential oxidation or dissolution of the constituent elements composing the binary-compound substrate. Even for the In compounds, high densities of surface states are still present, at least an order of magnitude greater than found in Si, which degrade the carrier mobility and complicate interpretation of conventional transport measurements.

Such considerations do not seem as important in heterojunction and superlattices of III–V compounds, which are usually fabricated using molecular-beam epitaxy (MBE)[95] between approximately lattice-matched semiconductors. Investigations in this field have concentrated on the alloy combination of Ga$_{1-x}$Al$_x$As–GaAs and the heterostructure InAs–GaSb, both of which appear to form atomically smooth junctions as opposed to the MOS case. This is reflected in the relatively weak influence of surface scattering for these structures.

Historically, the earliest studies of III–V surface layers began in the mid-1960s with InSb and InAs, both of which form naturally inverted surfaces. Later interest in nonparabolic subband structure, as well as the development of suitable dielectric insulators, spurred continued studies of these two materials through the 1970s. More recently, attention has focused on InP and quaternary compounds of In$_{1-x}$Ga$_x$As$_y$P$_{1-y}$ for possible use in high-speed logic applications. This has led to a number of device-related papers although relatively few serious transport studies.

In the following sections, we separate the current literature into studies relating to subband structure and studies which deal primarily with carrier mobility at the surface. Of course, these two are related and thus the distinction is not always clear. However, for the present review it is useful to differentiate between the two.

5.1. Subband Structure

At liquid-He temperature, thermal broadening is sufficiently negligible compared to the subband energies that actual measurement of the subband separation and occupation may be made. In particular, magnetotransport experiments such as Shubnikov–de Haas (SdH) and cyclotron resonance have been widely employed in studies of both Si and III–V compound inversion and accumulation layers. Also, intersubband spectroscopy using far-infrared radiation has proved to be a sensitive probe of subband separation.

Qualitatively, the SdH effect for two-dimensional carriers is the same as the three-dimensional effect with a maximum in the conductivity occurring when successive Landau levels pass through the Fermi perimeter. In the purely two-dimensional case, however, the effect depends only on the component of the magnetic field perpendicular to the surface. Also, the periodic oscillations in $1/H$ are independent of the carrier effective mass which allows a direct measurement of the surface carrier density independent of the mass.

It is well known from studies on Si inversion layers that the intersubband transition energies for optical absorption are modified by depolarization and final-state many-body interactions, and thus a direct measure of subband separation is not possible.[25] However, advantage may be taken of the admixture of s- and p-like states in the nonparabolic III–V compounds to excite intersubband transitions using parallel (to the surface) polarized light and hence avoid depolarization effects.[42] As mentioned earlier, it is also argued that many-body effects are greatly reduced in the III–V's,[25] and thus one can neglect the final-state interaction, although it is not clear that this is always true. In cases where this assumption is valid, however, direct measurement of intersubband energies should be possible using far-infrared spectroscopy.

Experiments by Kawaji et al.[96,97] in the 1960s, which investigated the effects of a normal magnetic field on conduction in InSb accumulation and InAs[93,98] inversion layers, represents some of the earliest work on III–V's in which the two-dimensional nature of surface space-charge layers were revealed. The observation of weak Shubnikov–de Haas (SdH) oscillations in InAs,[93] paralleled the more definitive studies of quantum oscillation in Si inversion layers by Fowler et al.[99]

Direct measurement of subband energy levels on naturally accumulated InAs surfaces was facilitated by the tunneling measurements of Tsui.[100–102] Here, the tunneling current normal to the surface is thought to provide a direct measure of the density of occupied states at the semiconductor surface, reflecting discontinuities in the two-dimensional density of states (5) caused by successive subbands. Magneto-oscillations in the tunnel current were analyzed to determine the nonparabolic effective mass of surface subbands. Subtle effects due to bulk free-carrier screening, which plays an important role for accumulation layers on degenerate semiconductors, were also investigated[103] in the context of the self-consistent theory of Baraff and Appelbaum.[23]

SdH oscillations were later studied in gate-controlled InAs accumulation layers[104–107] formed from depositing SiO_2 onto nondegenerate InAs epilayers. Multiple periodicity in the transconductance data was resolved to reveal three partially occupied subbands for surface densities as low as 1×10^{12} cm^{-2} at 4.2 K.[104,105] The ground subband effective mass for various surface densities was determined from the temperature dependence of the oscillation amplitude,[106,107] where the energy dependence of the mass was found to agree with the two-band Kane model[34] including a crude correction for surface potential mixing.[108]

More recent investigations of InAs accumulation layers have employed far-infrared optical absorption,[109] cyclotron resonance,[110] and SdH oscillations for both normal and tilted[111,112] magnetic fields. Intersubband transitions are found to be generated by both perpendicular and parallel excitation in the surface plane, similar to InSb[43,109] (see Section 2.4). For nondegenerate samples, line fits to the far-infrared surface cyclotron resonance coupled with SdH data were employed to determine the effective mass vs. carrier density in each of the four occupied subbands[110] at densities as high as 7×10^{12} cm^{-2}. Ground-state values of the mass differed somewhat from the previous work of Washburn et al.,[106,107] probably reflecting differences in the experimental techniques used.

Photoconductivity and SdH oscillations in inversion layers of p-type InSb were reported by Katayama et al.[113,114] using evaporated SiO_2 for the gate insulator. Here, two occupied subbands were identified from the periodicity of the SdH oscillations. In later studies (see Section 2.4) intersubband spectroscopy,[42,115,116] cyclotron resonance,[38,117] and SdH oscillations[39] were measured on laquer-coated p-type InSb surface layers. Here, three occupied subbands were observed at densities as low as 5×10^{11} cm^{-2}, consistent with the low density of states for this material. Experimental values for the nonparabolic mass and subband energies agreed quite well with theoretical values of Takada et al.[37] and unpublished calculations of Arai,[118] although the latter calculation has been shown to be inadequate for high surface densities.[119]

InP has only recently received attention as a candidate MOS system, and hence subband measurements are somewhat less extensive than those on InAs and InSb. SdH oscillations at 4.2 K on inversion layers of p-type InP were first reported by von Klitzing et al.,[78,120] although due to the low mobility of the samples, rather large magnetic fields were required (14T) to observe oscillations. Cheng and Koch[121–123] later investigated intersubband spectroscopy and SdH oscillations on n-type surface layers of both n- and p-type substrates. Surprisingly, identical results were obtained regardless of bulk doping, which could imply freeze-out of impurities at 4.2 K. The results showed two occupied subbands at densities above 1.5×10^{12} cm^{-2}, which essentially agreed with the previous studies.[78,120] Anomalous cyclotron resonance was also observed,[123] which was interpreted as a trapping effect for conduction-band carriers. Such an interpretation would not be inconsistent with the large density of measured surface states near the InP conduction-band edge from capacitance measurements.[50]

Briefly we mention that similar probes of two-dimensional subbands have been applied to heterojunctions and superlattices of $Ga_{1-x}Al_xAs$–$GaAs$[124,125] and InAs–GaSb.[126] Through the use of modulation doping,[3] high mobility has been achieved for these structures, particularly at low temperatures. Thus, due to the low scattering rates, well-defined cyclotron resonance peaks and SdH oscillations are observed. This structure has proved particularly useful for investigation of the quantized Hall effect in heterojunctions of $Ga_{1-x}Al_xAs$–$GaAs$.[127]

5.2. Transport Measurements

The carrier mobility in inversion and accumulation layers may be determined in several ways. The most direct method is to measure the channel conductance and divide by the induced charge density. The inversion-layer density is usually assumed to be given by the gate capacitance

$$N_s = \frac{C_{0x}}{e}(V_g - V_T) \tag{38}$$

where C_{0x} is the gate capacitance per unit area, V_g the gate voltage, and V_T the threshold or turn-on voltage. Use of (38) to derive the effective mobility assumes that no significant trapping of surface carriers occurs past threshold which may be a poor assumption in devices with a high surface state density. There is also some ambiguity in the definition of V_T as well, which may change anomalously at low temperatures[8] due to localization effects. Also, for many of the dielectrics that are employed in III–V MOS devices, frequency dispersion in the dielectric constant is present which makes the determination of C_{0x} uncertain.

The threshold voltage may be eliminated by measuring the so-called field-effect mobility derived from the transconductance

$$g_m = \frac{\partial I_D}{\partial V_g} = \frac{W}{L}\mu_{FE}C_{0x}V_D, \tag{39}$$

where W and L are the channel width and length, while I_D and V_D are the source–drain current and voltage. The field-effect mobility is also influenced by trapping of carriers past threshold. In addition, if the mobility depends strongly on the gate voltage, then additional terms appear in (39) which causes μ_{FE} to differ from the true channel mobility.

Low-field magnetotransport measurements, such as the Hall and gated van der Pauw, are often used to avoid the problems discussed above. Such measurements usually involve fabrication of special-geometry MOS devices which can be a disadvantage in some cases. In the usual Hall measurement, the Hall field arising from a normal magnetic field applied to the surface layer is related to the channel density through

$$R_H = \frac{E_H}{J_x B} \quad (40)$$

$$R_H = -\frac{r}{N_s e} \quad (41)$$

where R_H is the Hall coefficient, E_H is the Hall field, B the magnetic induction, and J_x the channel current density. The Hall factor r is dependent on the thermal average of the scattering rate,

$$r = \frac{\langle \tau^2 \rangle}{\langle \tau \rangle^2} \quad (42)$$

which usually varies between 1 and 2. Uncertainty in r is the main drawback in the Hall technique. Also, (40) and (41) are based on a single-band model, whereas several subbands may be occupied. In this case, the expressions (40) and (41) may become complicated, especially if intersubband transitions are important.

Early field-effect studies on InSb and InAs in the mid-1960s relied on mica insulators, mylar sheets, or an absorbed external ambient to modulate the surface conductance. Other than from an historical perspective, the only noteworthy early transport study here was by Kawaji and Kawaguchi,[93] who investigated the density and temperature dependence of InAs inversion electrons from 4.2 to 77 K. Overall, the measured mobility was not especially dependent on temperature, with the low-temperature mobility found to be consistent with ionized impurity scattering[7] over the limited density range of the study ($N_s < 5 \times 10^{11}$ cm^{-2}).

Transport studies on accumulated surfaces are complicated by the contribution of bulk free carriers to the total conductance. In Hall measurements on naturally accumulated InAs epilayers, Sites and Wieder[128] separated the bulk and surface contribution to the Hall coefficient R_H by analyzing the magnetic field dependence of R_H in a two-layer model. In contrast to the bulk mobility, the surface mobility was found to weakly depend on temperatures from 4.2 to 300 K, which is suggestive of surface-dominated scattering. Later measurements by Hall and van der Pauw on gated InAs accumulation layers[129,107] were interpreted similarly in terms of two- and three-layer models. As in Si, the mobility of the accumulation-layer electrons decreased with increasing surface field, which is consistent with the presence of surface roughness scattering (Section 3.2).

Relatively high mobilities have been implied for InAs inversion layers from MOSFET field-effect measurements[130] ($\mu_{FE} \sim 5000$ cm^2/V-300 K). However, no extensive transport measurements on these devices have been reported. Baglee et al.[92,131] have studied Hall mobilities of InAs inversion layers at 77 K for various gate insulators. These data, along with that of Kawaji and Kawaguchi[93] are plotted in Fig. 6(b) along with the theoretical fit by Moore and Ferry[77] discussed in Section 4.3. As can be seen, surface roughness scattering dominates over the entire density range. The values for Δ and L required to fit these data ($\Delta = 15$ Å, $L = 29$ Å) are substantially larger than the values found by Hartstein et al.[132] ($\Delta = 3$ Å, $L = 6$ Å) for Si using essentially the same unscreened model, suggesting a much more disordered interface for these InAs samples than for the case of Si–SiO$_2$.

Line fits to cyclotron resonance data on InAs[110] have been used to imply the scattering rates in higher occupied subbands. Here, the widths of the resonance peaks associated with each occupied subband were fitted using a phenomenological relaxation model. The scattering rate was always lower for higher-lying subbands, although for increasing carrier density the scattering rate in a given subband increased, with the greatest increase occurring in the lowest subband. This behavior is consistent with surface-dominated scattering, which qualitatively is stronger for more tightly bound electrons. A recent study of the low-temperature conductance in InAs MOSFETS[133] has shown structure in the channel mobility vs. gate voltage corresponding to the onset of occupation of a second subband as ascertained from SdH measurements. This correlation seems to demonstrate a contribution to the mobility arising from intersubband scattering.

No extensive transport studies of InSb inversion layers have been undertaken to supplement the exhaustive theoretical and experimental investigations of subband structure. Usually the inversion mobility of samples used for SdH and spectroscopic measurements are on the order of 10^4 cm^2/V-s.[114,39] Field-effect measurements on InSb have been performed at 4.2 K by Komatsubara et al.[134] which showed anomalous discontinuities in the mobility vs. density curves. This was interpreted in terms of trap states in the conduction band. Later measurements[114] on higher-mobility samples by the same group appeared more like Si low-temperature mobility curves.

InP has recently received considerable attention for high-speed device applications following the demonstration of an InP MOSFET by Lile et al.[135] Since that time, favorable results have been reported on devices using CVD (chemical vapor deposition) SiO$_2$,[135–142] Al$_2$O$_3$,[143–151]

Si_3N_4,[50] and Ge_3N_4[152] as gate insulators. Field-effect mobilities for these devices typically range from[400 to 2000 cm^2/V-s at 300°C. Some of this variation may be associated with the quality of the bulk material,[153] although as discussed in Section 3, the low-field mobility of inversion electrons is more sensitive to surface scattering than to bulk scattering mechanisms (other than phonons). Hence, much of the dispersion in reported device mobilities reflects a sensitivity to surface preparation associated with the surface scattering. Some success has recently been achieved as well using anodically grown[154,155] native oxides in InP. For these devices, mobilities as high as 3000 cm^2/V-s have been reported.

Care must be taken in comparing reported field-effect mobilities depending on the measurement technique employed. It is well documented[50,141,146–150] that slow source–drain current drift is present in InP enhancement devices associated with long-time-constant surface states at the interface or in the oxide. For this reason, pulsed or ac measurements are usually employed to remove the effect of these long-time-constant states. However, as discussed previously, the field-effect mobility in the presence of surface states may differ from the transport mobility roughly by a factor proportional to the fraction of free carriers in the inversion layer,[8,156] and hence is sensitive to the measurement frequency or pulse width. Thus, high-frequency field-effect mobilities will invariably be greater than the corresponding dc mobility.

At present, only limited low-temperature mobility measurements have been performed on InP surfaces. Field-effect mobilities up to carrier densities of 7×10^{12} cm^{-2} have been reported by von Klitzing et al.[78,120] at 4.2 K, some of which were shown in Fig. 4(a). The peak mobility in these experiments was always less than 1000 cm^2/V-s and strongly dependent on substrate bias. These results were fit by von Klitzing et al.[78] using unscreened versions of impurity and surface roughness, with the corresponding (ΔL) product for roughness five times greater than that of Si.[132] Some field-effect measurements were reported by Cheng and Koch,[123] who observed negative magnetoresistance in their laquer-coated samples at low densities. This was believed to be related to similar observations on Si[157] associated with localization phenomena. Dramatic shifts in mobility were also observed with electrical prestressing at room temperature, probably related to shifts in the charge state at the interface.

Enhancement-mode devices have recently been reported on ternary[158–161] and quaternary[156,162,163] compounds of $In_{1-x}Ga_xAs_yP_{1-y}$. Usually, these alloy semiconductors are lattice matched to InP substrates on which they are grown. Thus far, field-effect mobilities reported for the ternary $In_{0.53}Ga_{0.47}As$ have been less than 1700 cm^2/V-s, although this should improve. In studies of quaternary compounds, Shinoda and Kobayashi[156] observed an increase in mobility with increasing Ga and

As concentration. The highest electron mobilities in the study were on the order of 2000 cm^2/V-s, which was twice the value of InP control devices made simultaneously, although the mobility in the control samples was somewhat low. This implies that refinement of their surface preparation technique should lead to higher channel mobilities.

Recent novel approaches to quantum well devices like GaAlAs–GaAs have resulted in high-mobility, gate-controlled structures.[3,4,164,165] Through the use of modulation doping[3] (MD), carriers in the GaAs channel are separated from their donor impurities, which to this date has led to Hall mobilities upwards from 1×10^5 cm^2/V-s at 77 K.[164] Recently, gated structures on MD heterojunctions have been developed,[3,164] which allow the carrier concentration to be varied over a limited range. Tsui et al.[166,167] have studied the density dependence in these structures at 4.2 and 77 K. They concluded that residual ionized impurity scattering still dominated the mobility, although this is not surprising considering the highest density achieved was 5×10^{11} cm^{-2}. Störmer et al.,[168] using similar structures, have observed a sudden decrease in the Hall mobility with density associated with the onset of intersubband scattering. The sharpness of this transition reflects the lack of significant level broadening for these high-mobility devices.

6. Summary

We have presented an overview of the important scattering mechanisms expected for surface layers of III–V compounds. These include Coulomb and roughness scattering from the interface as well as scattering by bulk optical phonon and surface modes, modified to account for the reduced dimensionality of carriers at the surface. A review of the transport data available on III–V MOS inversion and accumulation layers, while by no means conclusive, seems to indicate that the mobility in present devices is strongly influenced by disorder and charge states at the oxide–semiconductor interface. This is evidenced by the sensitivity to surface preparation and weak temperature dependence reported in transport studies up to now. In fact, the high mobilities reported for modulation-doped quantum well structures, where the influence of surface roughness and interfacial charge appears negligible, is an indirect affirmation of the limiting scattering mechanisms for MOS devices.

Physical studies of III–V compound–oxide interfaces seem to indicate an inherently disordered surface subject to nonstoichiometric interface broadening and defect formation. Advances in surface analysis and high-resolution TEM of the oxide–semiconductor interface should prove interesting in the understanding of surface scattering and electronic structure in III–V MOS devices. Improvements in surface preparation technique seem to be resolving some of the problems associated with surface scattering, as evidenced by the continual increase in reported transconductance and mobilities for InP enhancement devices. Presumably, as device interfaces become better, the room-temperature mobility will become dominated by thermal phonon mechanisms rather than surface scattering. This will be evidenced by a much stronger temperature dependence of the channel mobility than currently is observed.

New materials and novel device structures hold the promise of interesting future studies. Ternary and quaternary compounds of InGaAsP have already demonstrated feasibility for device applications, and with the possibility of a controllable band gap and effective mass (through the relative mole fraction), may provide some interesting fundamental studies. A recent revival of interest in InSb for use in IR CCD imagers[169,170] may also spur renewed research into the transport properties of carriers in nonparabolic band structures. Finally, the recent advances in modulation-doped superlattices and HEMT devices, in which the surface scattering mechanisms limiting conventional MOS structures are virtually absent, have already proved instrumental, if not controversial, in probing low-temperature galvanomagnetic phenomena such as the quantized Hall effect. It is expected that future studies using such high-mobility devices may provide insight into other collective phenomena in two-dimensional systems.

References

1. J. R. Schrieffer, in: *Semiconductor Surface Physics* (R. H. Kingston, ed.), pp. 55–69, Univ. Pennsylvania Press, Philadelphia (1957).
2. T. Ando, A. B. Fowler, and F. Stern, Electronic properties of two-dimensional systems, *Rev. Mod. Phys.* 54, 437–672 (1982).
3. R. Dingle, H. L. Störmer, A. C. Gossard, and W. Wiegmann, Electron mobilities in modulation-doped semiconductor heterojunction superlattices, *Appl. Phys. Lett.* 33, 665–667 (1978).
4. T. Mimura, S. Hiyamizu, T. Fuji, and K. Nanbo, A new field-effect transistor with selectively doped GaAs/n-$Al_xGa_{1-x}As$ heterojunctions, *Japan. J. Appl. Phys.* 19, L225–L227 (1980).
5. S. Mori and T. Ando, Electronic properties of a heavily-doped n-type GaAs–$Ga_{1-x}Al_xAs$ superlattice, *Surf. Sci.* 98, 101–107 (1980).
6. P. J. Price, Electron transport in polar heterolayers, *Surf. Sci.* 113, 199–210 (1982).
7. F. Stern and W. E. Howard, Properties of semiconductor surface inversion layers in the electric quantum limit, *Phys. Rev.* 163, 816–835 (1967).
8. F. Fang and A. B. Fowler, Transport properties of electrons in inverted silicon surfaces, *Phys. Rev.* 169, 619–631 (1968).

9. C. W. Wilmsen, Chemical composition and formation of thermal and anodic oxide/III–V compound semiconductor interfaces, *J. Vac. Sci. Technol.* 19, 279–289 (1981).
10. H. Ehrenreich, Electron scattering in InSb, *J. Phys. Chem. Solids* 2, 131–149 (1957).
11. D. L. Rode, Electron transport in InSb, InAs, and InP, *Phys. Rev. B* 3, 3287–3299 (1971).
12. H. H. Wieder, Perspectives on III–V compound MIS structures, *J. Vac. Sci. Technol.* 15, 1498–1506 (1978).
13. W. E. Spicer, I. Lindau, P. Skeath, and C. Y. Su, Unified detect model and beyond, *J. Vac. Sci. Technol.* 17, 1019–1027 (1980).
14. F. Herman, Electronic structure calculations of interfaces and overlayers in the 1980s, *J. Vac. Sci. Technol.* 16, 1101–1107 (1979).
15. M. S. Daw, P. L. Smith, C. A. Swarts, and T. C. McGill, Surface vacancies in II–VI and III–V zinc blende semiconductors, *J. Vac. Sci. Technol.* 19, 508–512 (1981).
16. W. E. Spicer, P. W. Chye, P. R. Skeath, C. Y. Su, and I. Lindau, New and unified model for Schottky barrier and III–V insulator interface states formation, *J. Vac. Sci. Technol.* 16, 1422–1433 (1979).
17. R. E. Allen and J. D. Dow, Unified theory of point-defect electronic states, core excitons, and intrinsic electronic states at semiconductor surfaces, *J. Vac. Sci. Technol.* 19, 383–387 (1981).
18. J. F. Wager and C. W. Wilmsen, Plasma-enhanced chemical vapor deposited SiO_2/InP interface, *J. Appl. Phys.* 53, 5789–5797 (1982).
19. R. P. Vasquez and F. J. Grunthaner, XPS study of interface formation of CVD SiO_2 on InSb, *J. Vac. Sci. Technol.* 19, 431–436 (1981).
20. C. B. Duke, Optical absorption due to space-charge-localized states, *Phys. Rev.* 159, 632–644 (1967).
21. F. Stern, Self-consistent results for n-type Si inversion layers, *Phys. Rev. B*5, 4891–4899 (1972).
22. M. E. Alfereiff and C. B. Duke, Energy and lifetime of space-charge-induced localized states, *Phys. Rev.* 168, 832–842 (1968).
23. G. A. Baraff and J. A. Appelbaum, Effect of electric and magnetic fields on the self-consistent potential at the surface of a degenerate semiconductor, *Phys. Rev. B* 5, 475–497 (1972).
24. S. Das Sarma, Energy levels of n-channel accumulation layer on InP surface, *Solid State Commun.* 41, 483–485 (1982).
25. T. Ando, Electron–electron interaction and electronic properties of space charge layers on semiconductor surfaces, *Surf. Sci.* 73, 1–18 (1978).
26. M. Abramowitz and I. A. Stegun, *Handbook of Mathematical Functions*, Dover, New York (1972), pp. 446–452.
27. H. Kroemer and Qi-Gao Zhu, On the interface connection rules for effective-mass wave function at an abrupt heterojunction between two semiconductors with different effective mass, *J. Vac. Sci. Technol.* 21, 551–554 (1982).
28. J. F. Wager, K. Geib, C. W. Wilmsen, and L. L. Kazmerski, Native oxide formation and electrical instabilities at the insulator/InP interface, *J. Vac. Sci. Technol. B* 1, 778–781 (1983).
29. E. Jahnke and F. Emde, *Tables of Functions*, Dover, New York (1945), p. 152.
30. F. F. Fang and W. E. Howard, Negative field-effect mobility on (100) Si surfaces, *Phys. Rev. Lett.* 16, 797–799 (1966).
31. S. Das Sarma, R. K. Kalia, M. Nakayama, and J. J. Quinn, Stress and temperature dependence of subband structure in silicon inversion layers, *Phys. Rev. B* 19, 6397–6406 (1979).
32. G. H. Kawamoto, J. J. Quinn, and W. L. Bloss, Subband structure of n-channel inversion layers on polar semiconductors, *Phys. Rev. B* 23, 1875–1886 (1981).

33. K. S. Yi and D. K. Ferry, Many-body effects on the subband structure of n-type surface space-charge layers in InP, *Phys. Rev. B 28*, 1127–1129 (1983).
34. E. O. Kane, Band structure of indium antimonide, *J. Phys. Chem. Solids 1*, 249–261 (1957).
35. F. J. Ohkawa and Y. Uemura, Quantized states of a narrow gap semiconductor, *J. Phys. Soc. Japan 37*, 1325–1333 (1974).
36. A. Därr, J. P. Kotthaus, and T. Ando, in: *Proceedings of the 13th International Conference on the Physics of Semiconductors, Rome* (F. G. Funi, ed.), pp. 774–777, North-Holland, Amsterdam (1976).
37. Y. Takada, K. Arai, N. Uchimura, and Y. Uemura, Theory of the electronic properties of n-channel inversion layers on narrow-gap semiconductors. I Subband structure of InSb, *J. Phys. Soc. Japan 49*, 1851–1858 (1980).
38. A. Därr, J. P. Kotthaus, and J. F. Koch, Surface cyclotron resonance in InSb, *Solid State Commun. 17*, 455–458 (1975).
39. A. Därr and J. P. Kotthaus, Magnetotransport in an inversion layer on p-InSb, *Surf. Sci. 73*, 549–559 (1978).
40. G. E. Marques and L. J. Sham, Theory of space charge layers in narrow-gap semiconductors, *Surf. Sci. 113*, 131–136 (1982).
41. L. J. Sham and M. Nakayama, Effective mass approximation in the presence of an interface, *Phys. Rev. B 20*, 734–747 (1979).
42. K. Weisinger, W. Beinvogel, and J. F. Koch, in: *Proceedings of the 14th International Conference on the Physics of Semiconductors, Edinburgh* (B. L. H. Wilson, ed.); *Inst. Phys. Conf. Ser. 43* (Inst. Phys. London) 1215 (1979).
43. K. Weisinger, H. Reisinger, and F. Koch, The nonparabolicity parallel excitation mechanism and doublet peak problem in subband resonance, *Surf. Sci. 113*, 102–107 (1982).
44. J. Scholz and F. Koch, Spectroscopy of electron subbands on Ge-(111), *Solid State Commun. 34*, 249–251 (1980).
45. E. D. Siggia and P. C. Kwok, Properties of electrons in semiconductor inversion layers with many occupied electric subbands. I. Screening and impurity scattering, *Phys. Rev. B 2*, 1024–1036 (1970).
46. S. Mori and T. Ando, Intersubband scattering effect on the mobility of a Si(100) inversion layer at low temperatures, *Phys. Rev. B 19*, 6433–6441 (1979).
47. Y. Takada, Effects of screening and neutral impurity on mobility in silicon inversion layers under uniaxial stress, *J. Phys. Soc. Japan 46*, 114–122 (1979).
48. H. Ezawa, Inversion layer mobility with intersubband scattering, *Surf. Sci. 58*, 25–32 (1976).
49. J. E. Stannard, T. A. Kennedy, and B. D. McCombe, Properties of surface carriers at GaAs–native oxide interfaces, *J. Vac. Sci. Technol. 13*, 869–872 (1976).
50. L. G. Meiners, Electric properties of SiO_2 and Si_3N_4 dielectric layers on InP, *J. Vac. Sci. Technol. 19*, 373–379 (1981).
51. E. M. Conwell, *High Field Transport in Semiconductors*, Academic Press, New York (1967).
52. F. Berz, Ionized impurity scattering in silicon surface channels, *Solid-State Electron. 13*, 903–906 (1970).
53. J. L. Rutledge and W. E. Armstrong, Effective surface mobility theory, *Solid-State Electron. 15*, 215–219 (1972).
54. C. T. Sah, T. H. Ning, and L. L. Tschopp, The scattering of electrons by surface oxide charges and by lattice vibrations at the silicon–silicon dioxide interface, *Surf. Sci. 32*, 561–575 (1972).
55. Y. C. Cheng, Effect of charge inhomogeneities on silicon surface mobility, *J. Appl. Phys. 44*, 2425–2427 (1973).

56. F. Stern, Image potential near a gradual interface between two dielectrics, *Phys. Rev. B 17*, 5009–5015 (1978).
57. F. Stern, Polarizability of a two-dimensional electron gas, *Phys. Rev. Lett. 18*, 546–548 (1967).
58. P. F. Maldague, Many-body corrections to the polarizibility of the two-dimensional electron gas, *Surf. Sci. 73*, 296–302 (1978).
59. F. Stern, Calculated temperature dependence of mobility in silicon inversion layers, *Phys. Rev. Lett. 44*, 1469–1472 (1980).
60. Y. Matsumoto and Y. Uemura, Proceedings of the Second International Conference on Solid Surfaces, Kyota; *Japan. J. Appl. Phys. Suppl. 2, Part 2*, 367–370 (1974).
61. A. Yagi and M. Nakai, Coulomb scattering in the band tail of n-channel silicon MOSFETs, *Surf. Sci. 98*, 174–180 (1980).
62. A. Yagi and S. Kawaji, Effects of tailing of density of state on the mobility of Si-MOSFETs at low temperatures—A proposal for the characterization of Si–SiO$_2$ interfaces, *Japan. J. Appl. Phys. 20*, 909–915 (1981).
63. K. Hess and C. T. Sah, Dipole scattering at the Si–SiO$_2$ interface, *Surf. Sci. 47*, 650–654 (1975).
64. E. Vass, R. Lassnig, and E. Gornik, Electron mobility analysis of n-Si inversion layers, *Surf. Sci. 113*, 223–227 (1982).
65. S. T. Pantelides and M. Long, in: *The Physics of SiO$_2$ and Its Interfaces* (S. T. Pantelides, ed.), pp. 339–343, Pergamon Press, New York (1978).
66. O. L. Krivanek and J. H. Mazur, The structure of ultrathin oxide on SiO$_2$, *Appl. Phys. Lett. 37*, 392–394 (1980).
67. J. M. Ziman, *Electrons and Phonons* (2nd Ed.), Oxford Univ. Press, Oxford (1979), pp. 456–460.
68. R. F. Greene, in: *Molecular Processes on Solid Surfaces* (E. Draugle, R. D. Gretz, and R. J. Jaffee, eds.), pp. 239–263, McGraw-Hill, New York (1969).
69. R. E. Prange and Tsu-Wei Nee, Quantum spectroscopy of the low-field oscillations in the surface impedance, *Phys. Rev. 168*, 779–786 (1968).
70. J. Mertsching and H. J. Fishbeck, Surface scattering of electrons in magnetic surface states, *Phys. Status Solidi 41*, 45–46 (1970).
71. A. V. Chaplik and M. V. Entin, Energy spectrum and electron mobility in a thin film with non-ideal boundary, *Sov. Phys.—JETP 28*, 514–517 (1969).
72. M. V. Entin, Surface mobility of electron with quantizing band bending, *Sov. Phys.—Solid State 11*, 781–783 (1969).
73. Y. C. Cheng, Electron mobility in an MOS inversion layer, Proceedings of the Third Conference on Solid Devices, Tokyo; *Japan. J. Appl. Phys. Suppl. 41*, 173–180 (1972).
74. Y. C. Cheng, On the scattering of electrons in magnetic and electric surface states by surface roughness, *Surf. Sci. 27*, 663–666 (1971).
75. T. Ando, Screening effect and quantum transport in a silicon inversion layer in strong magnetic fields, *J. Phys. Soc. Japan 43*, 1616–1626 (1977).
76. M. Saitah, Warm electrons on liquid ^{4}He surface, *J. Phys. Soc. Japan 42*, 201–209 (1977).
77. B. T. Moore and D. K. Ferry, Scattering of inversion layer electrons by oxide polar mode generated interface phonons, *J. Vac. Sci. Technol. 17*, 1037–1040 (1980).
78. K. von Klitzing, Th. Englert, and D. Fritzsche, Transport measurements on InP inversion metal–oxide–semiconductor transistors, *J. Appl. Phys. 51*, 5893–5897 (1980).
79. T. Sugano, Physical and chemical properties of Si–SiO$_2$ transition regions, *Surf. Sci. 98*, 145–153 (1980).
80. S. M. Goodnick, R. G. Gann, D. K. Ferry, C. W. Wilmsen, and O. L. Krivanek, Surface roughness induced scattering and band tailing, *Surf. Sci. 113*, 233–238 (1982).

81. S. M. Goodnick, R. G. Gann, J. R. Sites, D. K. Ferry, C. W. Wilmsen, D. Fathy, and O. L. Krivanek, Surface roughness scattering at the Si–SiO$_2$ interface, *J. Vac. Sci. Technol. B 1*, 803–808 (1983).
82. F. Stern, Effect of a thin transition layer at a Si–SiO$_2$ interface on electron mobility and energy levels, *Solid State Commun. 21*, 163–166 (1977).
83. P. J. Price and F. Stern, Carrier confinement effects, *Surf. Sci. 132*, 577–593 (1983).
84. D. K. Ferry, Optical and intervalley scattering in quantized inversion layers in semiconductors, *Surf. Sci. 57*, 218–228 (1976).
85. D. K. Ferry, K. Hess, and P. Vogl, in: *VLSI Electronic Microstructure Science* (N. Einspruch, ed.), Vol. 2, pp. 67–103, Academic Press, New York (1983).
86. S. Kawaji, The two-dimensional lattice scattering mobility in a semiconductor inversion layer, *J. Phys. Soc. Japan 27*, 906–908 (1969).
87. H. Ezawa, T. Kuroda, and K. Nakamura, Surfaces and the electron mobility in a semiconductor inversion layer, *Surf. Sci. 27*, 218–220 (1971).
88. E. Vass and K. Hess, Energy loss of warm and hot carriers in surface inversion layers of polar semiconductors, *Z. Phys. B 25*, 323–325 (1976).
89. D. K. Ferry, Scattering by polar-optical phonons in a quasi-two-dimensional semiconductor, *Surf. Sci. 75*, 86–91 (1978).
90. S. Q. Wang and G. D. Mahan, Electron scattering from surface excitations, *Phys. Rev. B, 6*, 4517–4524 (1972).
91. K. Hess and P. Vogl, Remote polar phonon scattering in silicon inversion layers, *Solid State Commun. 30*, 807–809 (1979).
92. D. A. Baglee, D. K. Ferry, C. W. Wilmsen, and H. H. Wieder, Inversion layer transport and properties of oxides on InAs, *J. Vac. Sci. Technol. 17*, 1032–1036 (1980).
93. S. Kawaji and Y. Kawaguchi, Proceedings of the International Conference on the Physics of Semiconductors, Kyoto; Galvanomagnetic properties of surface layers in indium arsenide, *J. Phys. Soc. Japan 21, Suppl.*, 336–340 (1966).
94. D. K. Ferry, Transport of hot carriers in semiconductor quantized inversion layers, *Solid-State Electron. 21*, 115–121 (1978).
95. L. Esaki and L. L. Chang, Semiconductor superfine structures by computer-controlled molecular beam epitaxy, *Thin Solid Films 36*, 285–298 (1976).
96. S. Kawaji, H. Huff, and H. C. Gates, Field effect on magnetoresistance of n-type indium antimonide, *Surf. Sci. 3*, 234–242 (1965).
97. S. Kawaji and H. C. Gatos, The role of surface treatment in the field effect anomaly of n-type InSb at high magnetic fields, *Surf. Sci. 6*, 362–368 (1967).
98. S. Kawaji and H. C. Gatos, Electric field effect on the magnetoresistance of indium arsenide surfaces in high magnetic fields, *Surf. Sci. 7*, 215–228 (1967).
99. A. B. Fowler, F. F. Fang, W. E. Howard, and P. J. Stiles, Magnetooscillatory conductance in silicon surfaces, *Phys. Rev. Lett. 16*, 901–903 (1966).
100. D. C. Tsui, Observation of surface bound state and two-dimensional energy band by electron tunneling, *Phys. Rev. Lett. 24*, 303–306 (1970).
101. D. C. Tsui, Electron-tunneling studies of a quantized surface accumulation layer, *Phys. Rev. B 4*, 4438–4449 (1971).
102. D. C. Tsui, Electron tunneling and capacitance studies of a quantized surface accumulation layer, *Phys. Rev. B 8*, 2657–2669 (1973).
103. D. C. Tsui, Landau-level spectra of conduction electrons at an InAs surface, *Phys. Rev. B 12*, 5739–5748 (1975).
104. R. J. Wagner, T. A. Kennedy, and H. H. Wieder, in: *Proceedings of the Third International Conference of Narrow Gap Semiconductors* (J. Rauluskiewicz, M. Gorska, and E. Kaczmarek, eds.), pp. 427–432, PWN-Polish Scientific Publishers, Warsaw (1977).

105. R. J. Wagner, T. A. Kennedy, and H. H. Wieder, Magneto-transconductance study of surface accumulation layers in InAs, *Surf. Sci.* **73**, 545 (1978).
106. H. Washburn and J. R. Sites, Oscillatory transport coefficients in InAs surface layers, *Surf. Sci.* **73**, 537–544 (1978).
107. H. Washborn, J. R. Sites, and H. H. Wieder, Electronic profile of n-InAs on semi-insulating GaAs, *J. Appl. Phys.* **50**, 4872–4878 (1979).
108. G. A. Anticliffe, R. T. Bates, and R. A. Reynolds, in: Proceedings of the Conference on the *Physics of Semimetals and Narrow Gap Semiconductors, Dallas* (D. L. Carter and R. T. Bate, eds.) Oxford, New York, Pergamon Press (1971); *J. Phys. Chem. Solids 32, Suppl.* **1**, 499–510 (1970).
109. H. Reisinger and F. Koch, Spectroscopy of InAs subbands, *Solid State Commun.* **37**, 429–431 (1981).
110. H. Reisinger, H. Schaber, and R. E. Doezema, Magnetoconductance studies of accumulation layers on n-InAs, *Phys. Rev. B* **24**, 5960–5969 (1981).
111. R. E. Doezema, M. Nealon, and S. Whitmore, Hybrid quantum oscillations in a surface space-charge layer, *Phys. Rev. Lett.* **45**, 1593–1596 (1980).
112. M. Nealon, S. Whitmore, R. R. Bourassa, and R. E. Doezema, Determination of subband population in tipped magnetic fields, *Surf. Sci.* **113**, 282–286 (1982).
113. Y. Katayama, N. Kotera, and K. F. Komatsubara, Tunable infrared detector using photoconductivity of the quantized surface inversion layer of MOS transistor, *J. Japan. Soc. Appl. Phys. 40, Suppl.*, 214–218 (1971).
114. N. Kotera, Y. Katayama, and K. F. Komatsubara, Magnetoconductance oscillations of n-type inversion layers in InSb surfaces, *Phys. Rev. B* **5**, 3065–3078 (1972).
115. W. Beinvogel and J. F. Koch, Spectroscopy of electron subband levels in an inversion layer on InSb, *Solid State Commun.* **24**, 687–690 (1977).
116. W. Beinvogel and J. F. Koch, Spectroscopy of electron subband levels in an inversion layer on InSb, *Surf. Sci.* **73**, 547–548 (1978).
117. M. Horst, U. Merkt, and J. P. Kotthaus, Cyclotron resonance studies of the electron–phonon interaction in inversion layers of p-InSb, *Surf. Sci.* **113**, 315–317 (1982).
118. K. Arai, M. S. Thesis, University of Tokyo (1977) (in Japanese).
119. Y. Takada, Theory of electronic properties in n-channel inversion layers on narrow-gap semiconductors. II. Inter-subband optical absorption on InSb, *J. Phys. Soc. Japan* **50**, 1998–2005 (1981).
120. K. von Klitzing, Th. Englert, E. Bangert, and D. Fritzche, in Proceedings of the 15th International Conference on the Physics of Semiconductors, Kyoto; *J. Phys. Soc. Japan* **49**, *Suppl. A*, 979–982 (1980).
121. H. C. Cheng and F. Koch, Magnetoconductance studies on InP surfaces, *Solid State Commun.* **37**, 911–913 (1981).
122. H. C. Cheng and F. Koch, Electron subbands on InP, *Surf. Sci.* **113**, 287–289 (1982).
123. H. C. Cheng and F. Koch, Electron subbands on InP, *Phys. Rev. B* **26**, 1989–1998 (1982).
124. H. L. Störmer, R. Dingle, A. C. Gossard, N. Wiegmann, and M. D. Sturge, Two-dimensional electron gas at a semiconductor–semiconductor interface, *Solid State Commun.* **29**, 705–709 (1979).
125. R. Dingle, H. L. Störmer, A. C. Gossard, and W. Wiegmann, 2-D electrical transport in GaAs-Al$_x$Ga$_{1-x}$As multilayers at high magnetic fields, *Surf. Sci.* **98**, 134 (1980).
126. L. L. Chang and L. Esaki, Electronic properties of InAs–GaSb superlattices, *Surf. Sci.* **98**, 70–89 (1980).
127. D. C. Tsui, H. L. Störmer, and A. C. Gossard, Zero-resistance state of two-dimensional electrons in a quantizing magnetic field, *Phys. Rev. B* **25**, 1405–1407 (1982).
128. J. R. Sites and H. H. Wieder, Surface and bulk charge carrier transport in InAs epilayers, *CRC Crit. Rev. Solid-State Sci.* **5**, 385–389 (1975).

129. H. H. Wieder, Charge carrier transport in gate-voltage-controlled heteroepitaxial indium arsenide layers, *Thin Solid Films 41*, 185–195 (1977).
130. H. Terao, T. Ito, and Y. Sakai, Interface properties of InAs-MIS structures and their application to FET, Elec. Eng. Japan, *94*, 127–132 (1974).
131. D. A. Baglee, D. H. Laughlin, B. T. Moore, B. L. Eastep, D. K. Ferry, and C. W. Wilmsen, in: *Gallium Arsenide and Related Compounds* (H. W. Thim, ed.), pp. 259–265, Institute of Physics, Bristol (1980).
132. A. Hartstein, T. H. Ning, and A. B. Fowler, Electron scattering in silicon inversion layers by oxide charge and surface roughness, *Surf. Sci. 58*, 178–181 (1976).
133. E. Yamaguchi and M. Minakata, Magnetoconductance study of inversion layers on InAs metal–insulator–semiconductor field-effect transistors, *Appl. Phys. Lett. 43*, 965–967 (1983).
134. K. F. Komatsubara, H. Kamioka, and Y. Katayama, Electrical conductivity in an n-type surface inversion layer of InSb at low temperature, *J. Appl. Phys. 40*, 2940–2944 (1969).
135. D. L. Lile, D. A. Collins, L. G. Meiners, and L. Messick, n-Channel inversion mode InP MISFET, *Electron. Lett. 14*, 657–659 (1978).
136. L. Messick, D. L. Lile, and A. R. Clawson, A microwave InP/SiO$_2$ MISFET, *Appl. Phys. Lett. 32*, 494–495 (1978).
137. D. Fritzsche, InP-SiO$_2$ MIS structure with reduced interface state density near conduction band, *Electron. Lett. 14*, 51–52 (1978).
138. L. G. Meiners, D. L. Lile, and D. A. Collins, Inversion layers on InP, *J. Vac. Sci. Technol. 16*, 1458–1461 (1979).
139. L. G. Meiners, D. L. Lile, and D. A. Collins, Microwave gain from an n-channel enhancement-mode InP MISFET, *Electron. Lett. 15*, 578 (1979).
140. L. Messick, Power gain and noise of InP and GaAs insulated gate microwave FETs, *Solid-State Electron. 22*, 71–76 (1979).
141. D. Fritzsche, in: *Insulating Films on Semiconductors*, (G. G. Roberts and M. J. Morant, eds.), pp. 258–265, Institute of Physics, Bristol (1980).
142. L. Messick, A high-speed monolithic MISFET integrated logic inverter, *IEEE Trans. Electron. Devices ED-28*, 218–221 (1981).
143. T. Kawakami and M. Okamura, InP/Al$_2$O$_3$ n-channel inversion-mode MISFETs using sulphur-diffused source and drain, *Electron. Lett. 15*, 502–504 (1979).
144. K. Kamimura and Y. Sakai, The properties of GaAs–Al$_2$O$_3$ and InP–Al$_2$O$_3$ interfaces and the fabrication of MIS field-effect transistors, *Thin Solid Films 56*, 215–223 (1979).
145. P. N. Farennec, M. LeContellec, H. L'Haridon, G. P. Pelous, and J. Richard, Al/Al$_2$O$_3$/InP MIS structures, *Appl. Phys. Lett. 34*, 807–808 (1979).
146. M. Okamura and T. Kobayashi, Current drifting behavior in InP MISFET with thermally oxidized InP/InP interface, *Electron. Lett. 17*, 941–942 (1981).
147. M. Okamura and T. Kobayashi, Slow current-drift mechanism in n-channel inversion type InP-MISFET, *Japan. J. Appl. Phys. 19*, 2143–2150 (1980).
148. M. Okamura and T. Kobayashi, Improved interface in inversion-type InP-MISFET by vapor etching technique, *Japan. J. Appl. Phys. 19*, 2151–2156 (1980).
149. Y. Hirota, M. Okamura, E. Yamaguchi, T. Nishioka, Y. Shinoda, and T. Kobayashi, Surface controlled InP-MIS (metal–insulator–semiconductor) triodes, *J. Appl. Phys. 52*, 3498–3503 (1981).
150. T. Kobayashi, M. Okamura, E. Yamaguchi, Y. Shinoda, and Y. Hirota, Effect of pyrolytic Al$_2$O$_3$ deposition temperature on inversion-mode metal–insulator–semiconductor field effect transistor, *J. Appl. Phys. 52*, 6434–6436 (1981).
151. Y. Hirayama, H. M. Park, F. Koshiga, and T. Sugano, Enhancement type InP metal–insulator–semiconductor field-effect transistor with plasma anodic aluminum oxide as the gate insulator, *Appl. Phys. Lett. 40*, 712–713 (1982).

152. K. P. Pande and S. Pourdavoud, Sr., Ge_3N_4–InP MIS structures, *IEEE Trans. Electron. Devices Lett. EDL-2*, 182–184 (1981).
153. J. Woodward, G. T. Brown, B. Cockayne, and D. C. Cameron, Substrate effects on performance of InP MOSFETs, *Electron. Lett. 18*, 415–417 (1982).
154. A. Yamamoto, A. Shibukawa, M. Yamaguchi, and C. Uemura, Low temperature (~77 K) properties of InP MOSFETs using anodic-oxide gate insulator, *Electron. Lett. 18*, 710–711 (1982).
155. T. Sawada and H. Hasagawa, InP high mobility enhancement MISFETs using anodically grown double-layer gate insulator, *Electron. Lett. 18*, 742–743 (1982).
156. Y. Shinoda and T. Kobayashi, InGaAsP n-channel inversion-mode metal–insulator–semiconductor field-effect transistor with low interface state density, *J. Appl. Phys. 52*, 6386–6394 (1981).
157. I. Eisele and G. Dorda, Negative magnetoresistance in n-channel (100) silicon inversion layers, *Phys. Rev. Lett. 32*, 1360–1363 (1974).
158. H. H. Wieder, A. R. Clawson, D. I. Elder, and D. A. Collins, Inversion-mode insulated gate $Ga_{0.54}In_{0.53}As$ field-effect transistors, *IEEE Trans. Electron. Devices Lett. EDL-2*, 73–74 (1981).
159. A. S. H. Liao, R. F. Leheny, R. E. Nahory, and J. C. DeWinter, An $In_{0.53}Ga_{0.47}As/Si_3N_4$ n-channel inversion mode MISFET, *IEEE Trans. Electron. Devices Lett. EDL-2*, 288–290 (1981).
160. A. S. H. Liao, B. Tell, R. F. Leheny, and T. Y. Cheng, $In_{0.53}Ga_{0.47}As$ n-channel native oxide inversion mode field-effect transistor, *Appl. Phys. Lett. 41*, 280–282 (1982).
161. K. Ishi, T. Sawada, H. Ohno, and H. Hasagawa, InGaAs enhancement-mode MISFETs using double-layer gate insulator, *Electron. Lett. 18*, 1034–1036 (1982).
162. H. H. Wieder, in: *Insulating Films on Semiconductors*, (G. G. Roberts and M. J. Morant, eds.), pp. 234–250, Institute of Physics, Bristol (1980).
163. Y. Shinoda, M. Okamura, E. Yamaguchi, and T. Kobayashi, InGaAs n-channel inversion mode MISFET, *Japan. J. Appl. Phys. 19*, 2301–2302 (1980).
164. S. Jodaprawira, W. I. Wang, P. C. Chao, C. E. C. Wood, D. W. Woodward, and L. F. Eastman, Modulation-doped MBE GaAs/n-$Al_xGa_{1-x}As$ MESFETs, *IEEE Trans. Electron. Devices Lett. EDL-2*, 14–15 (1981).
165. K. Muro, S. Narita, S. Hiyamizu, K. Nanbu, and H. Hashimoto, Far-infrared cyclotron resonance of two-dimensional electrons in an $Al_xGa_{1-x}As/GaAs$ heterojunction, *Surf. Sci. 113*, 321–325 (1982).
166. D. C. Tsui, A. C. Gossard, G. Kaminsky, and W. Wiegmann, Transport properties of GaAs-$Al_xGa_{1-x}As$ heterojunction field-effect transistors, *Appl. Phys. Lett. 39*, 712–714 (1981).
167. D. C. Tsui, A. C. Gossard, G. Kaminsky, and W. Wiegmann, Transport properties of GaAs IGFETs, *Surf. Sci. 113*, 464–466 (1982).
168. H. L. Störmer, A. C. Gossard, and W. Wiegmann, Observation of intersubband scattering in a two-dimensional system, *Solid State Commun. 41*, 707–709 (1982).
169. R. D. Thom, F. J. Renda, W. J. Parrish, and T. L. Koch, A Monolithic InSb Charge-Coupled Infrared Imaging Device, International Electron Devices Meeting, Washington, D.C. (December 1978).
170. J. D. Langan and C. R. Visawanthen, Characterization of improved InSb interfaces, *J. Vac. Sci. Technol. 16*, 1474–1477 (1979).

6

Interfacial Constraints on III–V Compound MIS Devices

Derek L. Lile

1. Introduction

The first proposed structure for an active semiconductor device appeared in 1930[1] and consisted of a three-terminal element where current control was exercised by means of a metal–semiconductor barrier. This device and a subsequently modified structure employing an aluminum oxide dielectric spacer[2] were unsuccessful because of an inability at that time to prepare an electrically suitable surface. Interestingly, both these devices were proposed using non-Si semiconductors and although they were notable failures initially, they have since, in somewhat modified form, appeared successfully as, what we now call, the Schottky-gate[3] and MOS[4] field-effect transistor, respectively. These devices were not pursued following their initial proposal because of the emergence in 1948 of silicon and germanium junction bipolar transistors[5-7] which, because of their relative ease of fabrication as well as their primary reliance on charge transport remote from a surface, were both fabricated and shown to operate with little developmental delay. Despite their higher-operating-temperature advantage, we now know that Si devices achieved their overwhelming preeminence primarily because of the excellent bulk and interfacial characteristics of silicon's native thermal oxide, SiO_2.[8] The significance of this one fact cannot be overstated when we consider the vast array of devices, circuits, and systems which rely for their operation on the fortuitous circumstance that Si has a good compatible dielectric.

Derek L. Lile ● Department of Electrical Engineering, Colorado State University, Fort Collins, CO 80523.

In 1980 alone sales of Si discrete devices and ICs amounted to a staggering $13 billion and show every indication of increasing even more in the years ahead.[9] This is not to imply that alternate developmental paths could not have been successfully pursued but such considerations are entirely in the realm of conjecture.

Despite the preeminent position of Si and its native oxide in the historic development of electronics, other materials were not ignored.[10] Most surely the degree of attention attendant these other semiconductors was negligible when compared to the vast concentration of manpower brought to bear on the problem of Si-based systems. However, although this effort was comparatively small, it was not insignificant and has resulted in many devices and prospective devices which in their own way are having as revolutionary an effect as has Si on the solid-state technology. Although other semiconducting combinations of elements, such as HgCdTe, have received and are receiving much attention it is probably true to say that on other than Si, the III–V compounds as a class have attracted most of this effort. In fact, it was interest in AlSb and InSb[11] that generated the initial impetus for a systematic program of research in compound semiconductors in general.

The perceived advantages of the III–V materials are primarily in their transport and optical properties. Table 1 collates the III–V binary compounds with their minimum band gap and electron mobility being shown, together with a number of ternary and quaternary compounds of technological interest. As a class, it can be seen that all of these materials,

Table 1. Band Gap and Representative Room-Temperature Electron Mobility of Some Technologically Significant III–V Compounds

	Band gap (eV)	Mobility (cm^2/V-s)
AlAs	2.14	1,000
GaP	2.25	120
GaAs	1.43	8,500
GaSb	0.7	4,000
InP	1.35	4,500
InAs	0.356	30,000
InSb	0.18	76,000
$In_{1-x}Ga_xAs$	0.75^a	11,000
$Ga_xAl_{1-x}As$	1.96	3,500
$In_{1-x}Ga_xAs_yP_{1-y}$	$0.75 \rightarrow 1.35$	≤11,000

^{a} Because of their composition dependence, band-gap values for the ternaries and quaternary are no more than representative.

when compared to Si, have large low-field electron mobilities and high values of saturated electron velocity and thus are *a priori* promising candidate materials on which to fabricate devices for high-frequency operation. In particular, the high-field transport characteristics of GaAs have been used to advantage in the development of Schottky-gate FETs (MESFET, Metal–Semiconductor Field-Effect Transistors) which, with 1-μm gate length, are capable of operation to frequencies as high as 30 GHz.[12] Although this performance is demonstrably superior to what can be achieved with a Si device of equivalent geometry,[13] the degree to which the speed advantage is due to higher electron mobility has been debated. Confusion arises because of the availability of some III–V materials (eg., GaAs and InP) in a compensated high-resistivity form: Used as substrates for the active devices these semi-insulating materials result in reduced parasitic capacitances which in themselves lead to improved high-frequency response as a result of reduced parasitic charging times. In contrast to Si, many of these compounds also possess subsidiary conduction-band extrema which permit for transferred electron effects prior to avalanche breakdown. Most notable here are GaAs and InP where, in particular, the latter material has been demonstrated to exhibit Gunn-effect oscillations to frequencies in excess of 200 GHz corresponding to wavelengths ~1.5 mm.

These electrical characteristics are certainly in themselves sufficient to warrant attention; however, the III–V materials, in addition, possess optical properties which have led to a large industry in electro-optic devices and subsystems. Specifically, these materials have direct gaps which, in contrast to Si and Ge, permit for their use as efficient light-emitting diodes and lasers. In many cases their band gaps are narrower than Si which means that they can be used for optical detection at wavelengths beyond the reach of the column IV semiconductors. Particular examples of note are GaInAs and GaInAsP for use at wavelengths in the range 1.2–1.7 μm, values compatible with absorptive loss minima in optical fibers. Such extended optical response is also of benefit in solar-cell applications where enhanced conversion efficiencies result from increased total optical absorption. These advantageous characteristics have been widely capitalized as is evidenced by the present commercial availability of a number of devices based on these materials.

The advantages described above result from the distinct characteristics of these materials when compared to the column IV elements, in particular, Si. An impediment to the implementation of these compounds on an even larger scale, however, has resulted from the limited success achieved to date in the development of low defect density surfaces on these materials when prepared in contact with a dielectric. Most certainly none of the III–V's appear to possess a native oxide of the caliber of

SiO$_2$ on Si and all attempts to deposit heteromorphic layers on these materials have met with serious difficulties, as is evidenced by the fact that of all the III–V's only on InSb has a commercially marketable MIS device as yet been developed. The details of this effort to prepare good dielectrics on these materials and their problems and present status are covered elsewhere in this volume. Rather it is the purpose of this chapter to address the question of the current development and future application of dielectrics to devices on this class of semiconductors. To maintain documentation at a manageable level, no attempt has been made to include references to all the extant literature. In fact, in many cases a conscious decision was made to exclude certain work. It is hoped, however, that all principal papers are included and that by reference to these the interested reader will readily be led to the remaining information in the field.

Generally, it can be stated that dielectric layers serve two functions in solid-state electronics. First, they are used as low-loss mechanical spacers as, for example, in isolating interconnection lines where they cross and in maintaining a metal control electrode a fixed distance from the surface of a semiconductor as is the case in the MISFET.* Second, they are used to chemically and/or structurally modify the surface of a semiconductor and then to maintain this modification from further perturbation by acting as a physical barrier to the external environment. The former application, although having been applied with a great deal of ingenuity on many occasions during the history of solid-state electronics, is nevertheless somewhat straightforward because of its mechanistic nature and will in this discourse concern us no further. As a practical matter, the technical requirements on dielectrics for this application are in large measure well met in the materials presently available for use with the III–V's.[14,15] It is the second area of application with which we shall be concerned. It includes: (1) the use of dielectrics for the passivation and stabilization of the electrical properties of surfaces to reduce leakage currents, (2) the development of semiconductor/dielectric interfaces of low recombination rate for electro-optic applications, and (3) the ability to prepare layers of dielectrics on semiconductor surfaces which will allow efficient and controllable transport of charge in the interfacial region of the surface. It is in these more demanding areas that technical deficiencies in the current technology exist.

Although varied, all of these requirements involve a consideration of the characteristics of the semiconductor/dielectric interface. We will

---

*Metal–Insulator–Semiconductor (MIS) rather than Metal–Oxide–Semiconductor (MOS) is used as a more general term which includes MOS as a subclass. It includes the possibility that nonoxide dielectrics may have a part to play in the III–V technology.

begin with a summary of these interface effects pertinent to our discussion and then proceed to a consideration of the consequences of these effects on device performance.

2. Dielectric–Semiconductor Interfacial Phenomena

The termination of a periodic lattice potential as occurs at any crystal surface results in the introduction of additional localized permitted states within the normally forbidden energy gap. These "intrinsic" so-called Tamm or Shockley states, which differ only insofar as the details of the potential termination at the surface are concerned, can exist, in principal, in densities as large as $\sim 10^{15}$ cm^{-2} and can be identified with the dangling or unterminated lattice bonds of the surface atoms. Subsequent contamination of the surface by impurities, oxidation, deposition of a dielectric or metal, or relaxation of the crystal structure in the near-surface region can and, in general, does result in a modification of the distribution or density of these surface states.[16]

In particular, Si, which is believed to have a relatively high density of states on its clean as-cleaved surface, is considerably improved by the formation of a surface thermal oxide.[17] The III–V compounds are also thought to support a large intrinsic surface state density but, because of an energy-minimizing relaxation of the relative positions of the anion and cation components at the surface, these states are moved in energy beyond the conduction- and valence-band edges where their effect as trapping centers is removed.[18] From an applications viewpoint, this would seem to be very satisfactory. However, in practical structures following either the growth or deposition of dielectrics onto the surfaces of these semiconductors the situation is far less encouraging with large surface state densities $>10^{13}$ cm^{-2} eV^{-1}, having been recorded on a number of occasions on a variety of these materials. These results will be discussed for specific cases in more detail later.

In addition to these localized states in the interfacial region, the possibility exists of defects within the dielectric also contributing an effect. Na or K ion drift in SiO_2 is the classic example that can be cited in the case of the Si technology and analogous situations should be expected for dielectrics on the III–V's.

If, for the moment, it is accepted that localized states can be anticipated in general both at the interface and in the bulk of the dielectric layers on these compound semiconductors, then their effects on charge carriers in the near-surface region of the semiconductor will include both trapping and scattering as well as recombination which in almost all cases are phenomena deleterious to device performance. These effects, although present to some extent in all devices, become more pronounced as designs push device geometries to ever-smaller size. Reductions in lateral dimensions result, through scaling requirements, in reduced layer thicknesses with an outcome being that semiconductor components become more and more "a collection of interconnected surfaces."

2.1. Trapping

Figure 1 schematically illustrates the distributions of permitted energy states, in general, to be expected in the interface region of a semiconductor/insulator system. Although such a description may be commonly applied to both elemental as well as compound semiconductors, there are some features and considerations which do set the latter apart. For example, in the case of III–V compounds, because of the widely differing volatilities of the constituent elements, it is generally to be expected that some noncongruent dissociation of the components will occur[19] in the outermost layers of the surface resulting in a hopefully small, yet finite, region of nonstoichiometry.[20] In the case of InP, Farrow[21] has concluded that temperatures in excess of 350°C can lead to dissociation of the surface due to the preferential loss of P which is to be compared to Si which is routinely processed at temperatures as high as 1100°C. Such a "metamorphic layer" has, for example, been proposed to explain some results on GaAs[22] and InAs[23] MIS structures. Con-

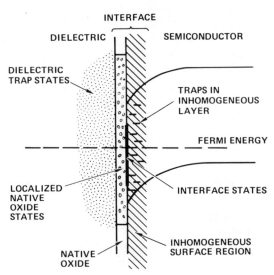

Figure 1. Schematic band diagram illustrating the potential sources of surface and interfacial trapping to be expected in an MIS device.

sequently, in addition to extraneous impurities introduced from outside, it may be expected that some contamination of the oxide grown on these compounds will occur by column III and V atoms dissociated from the semiconductor during dielectric deposition and subsequent processing.

In equilibrium a surface will adopt a configuration in which all states (continuum bulk plus localized surface) below the Fermi level will be filled. This requires an exchange of carriers between these two types of states with the result that the surface will in general acquire a nonzero surface potential V_s, the exact value depending upon the dopant type of the semiconductor and whether the surface states act as donors or acceptors. The time taken for this exchange to be accomplished will vary, with those states close to the band edge in the semiconductor being the first to fill and those deeper in energy and physically more remote from the semiconductor the last.

This trapping of charge at the surface has two consequences. First, it defines and controls the "quiescent" or zero-bias condition of the semiconductor surface. Second, it impedes subsequent changes being induced in the surface potential by externally applied fields. The former effect, for example, can drastically degrade integrated-circuit operation by enhancing or inverting surfaces leading to unwanted and undesirable interdevice surface current leakage paths and it can also result in spurious intradevice conduction. An example of this is seen in Fig. 2, where an enhancement-mode field-effect transistor (FET), designed to exhibit no conduction in the absence of an input signal is, in fact, conducting due to inadvertent surface enhancement. Such shifts in threshold voltage

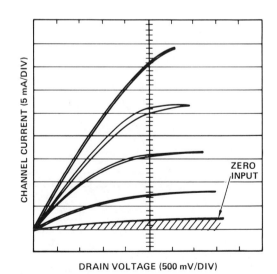

Figure 2. Output characteristics of an InP enhancement mode MIS transistor in the presence of unwanted zero input conduction (shown stippled).

could seriously degrade any circuit based on this device due to non-optimum current bias levels.

Attempts to change the surface potential of a semiconductor by means of an externally applied signal will also be frustrated if within the time of observation the induced charge can be trapped and immobilized at the surface. The qualification "within the time of observation" is important because, even in the presence of large densities of surface levels, fast signals can be propagated unhindered through surface effect devices due simply to an inability of surface traps to respond.[24] This, of course, is of significance for microwave and mm-wave devices where, apart from bias setting, signals of interest are high frequency. An equivalent state of affairs can be generated at lower frequencies by cooling which, because surface states N_{SS} tend to respond in a manner suggesting an associated activation energy, tends to slow the response of the traps.

If the traps can respond and charge is trapped then the effects on device performance can be manifold. In the FET the result is a reduction of current for given voltage settings as well as a degradation in transconductance. This is illustrated in Fig. 3 where the calculated transconduct-

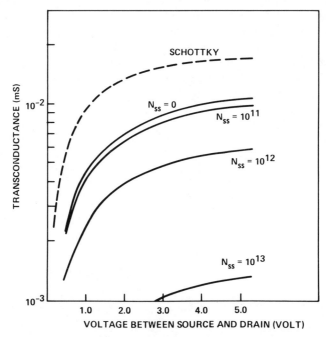

Figure 3. Calculated curves showing the effects of interfacial trapping on the transconductance of a depletion-mode MISFET on GaAs. The calculation is based on a dielectric thickness of 1000 Å, a gate length of 4 μm, and a channel width of 260 μm. Also shown for comparison is the expected performance of an equivalent Schottky-barrier device.

ance of a GaAs MIS-gated depletion device is shown for various values of N_{SS} together with a comparison curve for the Schottky structure. Lile[25] and Shinoda and Kobayashi[26] have shown that for minimal and generally acceptable performance effective values of $N_{SS} \leq 10^{11}$ cm^{-2} eV^{-1} are necessary over the region of surface potential accessed by the device of interest which in the n-channel enhancement FET, for example, means the near-CB-edge region of the surface. In addition to this, the invariably observed distribution in time constants of traps in MIS-gated structures on the III–V's generally results in long-term ($t > 1$ μs) changes or drift in device characteristics. In fact, most interface traps on the III–V's appear to be of long time constant; thus, although this fact is not always clear in the literature, surface state distributions obtained, for example, by quasistatic C–V measurements are often a measure of the same centers responsible for the ubiquitous long-term drift effects observed on these compounds. This effect is not the sole province of the compound semiconductors, however, and for a long time was a severe restraint on the use of Si MOS devices also. In that case, the problem was eventually traced to Na contamination of the SiO_2.[27] The stability problem for the III–V's is severe and will be discussed at length in a later section.

In a charge-coupled device (CCD), charge, rather than being moved in a continuous flow between source and drain as is the case in an FET, is moved incrementally under the control of a number of MIS gates (see Fig. 4). Charge trapping in a CCD results in a removal of charges from

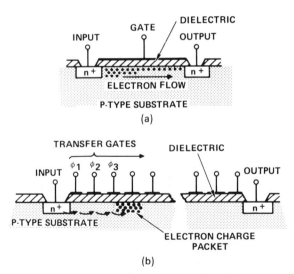

Figure 4. Diagram illustrating schematically the contrasting charge transport mechanisms of (a) an FET and (b) a CCD.

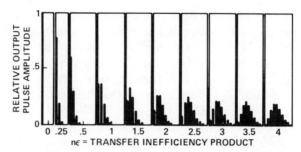

Figure 5. Calculated output pulse sequences from a CCD following injection of a single charge packet for various values of the total transfer inefficiency product as shown. (Taken from Ref. 28, Copyright 1975, Bell Telephone Laboratories, reprinted by permission.)

the leading pulses in a signal train and their subsequent release from the device subsequent to cessation of the signal. The signal pulse "spreading" which results is illustrated in Fig. 5, taken from Ref. 28, for a number of values of the transfer inefficiency product, defined as the product of the number of charge packet transfers n and ε, the fractional loss of charge per transfer. This distortion results in a limit to the number of bits or cells that can be used in the CCD and also appears to represent a fundamental constraint on the ultimate speed of these structures.[29]

From the results to be presented later it will become evident that on no III–V's at room temperature has a method as yet been developed to prepare a dielectric–semiconductor interface of the quality of that which, as judged by its trap density, can be attained on Si where N_{SS} values $\leq 10^{10}$ cm^{-2} eV^{-1} are possible. Although no unambiguous understanding for this has yet emerged, a number of models have been proposed, which include:

1. Vacancies generated by the selective loss of either the anion and/or cation from the surface layers of the semiconductor which could be expected to introduce additional localized levels which could act to trap carriers near the surface.[18]
2. A region of nonstoichiometry resulting from disorder of the constituent components of either the semiconductor or the dielectric in the near-interfacial region of the surface which might result in a heavily trap-laden defect layer.[20]
3. The native oxide itself, which presumably exists on the surface of any semiconductor prior to dielectric deposition, has been indicted as the dominant source of traps in at least the case of GaAs and InP.[30]
4. The selective buildup of the excess metallic column V element at the interface during dielectric growth has been suspect in a number of cases.[31]

No matter which of these models, if any, eventually is substantiated, it is clear that in all cases it is the "compound" nature of the III–V semiconductors which is believed to create the problem, manifesting itself in preferential oxidation or physical loss of one of the components due to their differences in volatility. For this reason low-temperature processing is in general deemed desirable and anneals, when used, must be rigorously limited.

Despite the lack of understanding which has largely hampered efforts to date to comprehend the chemistry of the III–V interface, much empirical and useful data on interfacial trapping have been generated using a number of techniques, including photovoltage,[32] photoemission,[33] gated Van der Pauw,[34] and capacitance measurements under a variety of experimental conditions. The last in particular have received considerable emphasis and have been a fruitful source of knowledge as is reviewed elsewhere in this volume.[35] It has in many cases, however, become clear that only limited projections as to subsequent device performance can be made from such data.[36,37] Ultimate verification or disproof of the suitability of any given surface preparation and dielectric growth technique for device use must await, in the final analysis, the results of measurements made on the devices themselves. Although often overlooked, it is clear that such device measurements can themselves be a powerful source of information in contributing to an understanding of interfacial phenomena[38,39] and, as is clear from the extensive work

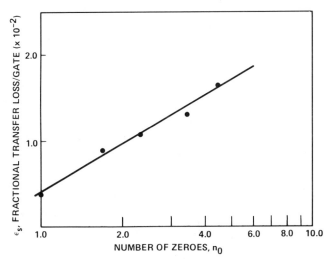

Figure 6. Data obtained on an insulated gate, eight-bit, four-phase surface channel InP CCD showing the dependence of charge loss on the number of preceding zeros. Surface state density follows from $N_{SS} = q\varepsilon_S N_{sig}/kT \ln(4n_0)$.

on the Si/SiO_2 interface, often provide a degree of sensitivity unavailable with more analytically oriented structures.[40] An example for the InP/SiO_2 interface is shown in Fig. 6 where a plot of charge loss in an eight-bit, four-phase CCD due to interface trapping vs. trap discharge time is used to deduce an interface state density $\sim 1.0 \times 10^{11}$ cm^{-2} eV^{-1} using the model of Carnes and Kosonocky[41] with $N_{Sig} = 5 \times 10^{11}$ cm^{-2}; a value in good agreement with results obtained by more conventional techniques.

2.2. Scattering

Many of the devices for which the III–V compounds are potentially applicable involve the transport of charge parallel and close to the external surfaces of the material. Good examples are the enhancement-mode MISFET and the surface channel CCD where charge carriers are being moved in a narrow inversion layer close to the dielectric–semiconductor interface. In such a carrier-confining situation the restraining force on carrier motion balancing the driving force of the electric field is not only the conventional interaction of the carriers with bulk scattering centers, such as defects, impurities, and lattice vibrations but also their "collisions" with the physical surface. The effect of this last interaction is to reduce to some extent the drift mobility of carriers near and within a scattering length of the surface. The magnitude of this degradation in the overall average effective mobility of the carriers depends not only on the extent of the confinement of the carriers near the surface, but also on the nature of the interaction of the carrier with the surface. Factors which potentially affect and determine this interaction include surface topography (that is, surface roughness), surface state density, and the charge occupancy of the surface levels.

Moore and Ferry,[42] for example, have performed calculations and Baglee et al.[43] have conducted experiments on InAs which suggest that the electron mobility in surface inversion layers at 77 K is determined by fixed oxide charges at low surface carrier concentrations, n_s, and by surface roughness for n_s values $\geqslant 1 \times 10^{11}$ cm^{-2}. A similar conclusion was subsequently arrived at for InP,[44] suggesting that the wide variations in values reported for electron inversion-layer mobilities at 300 K on this semiconductor might be ascribed to differences in physical surface structure resulting from differences in sample pretreatment. The conclusions of Von Klitzing et al.[39] from magnetoconductivity measurements of Shubnikov–de Haas oscillations in MISFETs at 4.2 K also support the dominance of the surface structure in limiting mobility at high n_s which in their case occurs for values $>4 \times 10^{12}$ cm^{-2}. Most recently, Shinoda and Kobayashi[45] have presented an analysis of their mobility data on inversion-mode FETs on p-type InP using Al_2O_3 for the dielectric. They propose that the mobility limit in their case is due to a 10–20 mV surface potential fluctuation across the interface which, in combination with a fairly high trap density, can account not only for the magnitude of μ_{FE} but also its observed relative insensitivity to temperature across the range from 360 to 130 K. Whatever mechanism is actually limiting it is clear that both in Si, as well as in the III–V's, surface mobility values rarely exceed two-thirds of the bulk mobility value. In Si, for example,[46] the very best reported value of μ_{FE}, 900 cm^2/V-s at 300 K, is somewhat less than the bulk value for electrons of 1400 cm^2/V-s. In InAs the best reported value of 12,000 cm^2/V-s at 77 K is considerably less than the bulk value, whereas in InP the best report of 2400 compares favorably with a bulk mobility of 3500–4000 cm^2/V-s to be expected for the doping level employed.* Although the reasons for this are not yet clear, for our purposes it is sufficient to note that mobility degradation can occur, that it is presumably exacerbated by the presence of charged surface states and/or surface roughness, and, insofar as carrier mobility is a materials' parameter with direct bearing on device performance, any reduction will tend to degrade the resulting parameters of any devices built around that interface.

2.3. Recombination

FETs, CCDs, memory cells, and so on, are structures whose carrier densities are, in general, calculable on the basis of thermal equilibrium statistics. Most certainly for very small dimensions it may be anticipated that carrier velocities will exceed, in the steady state, those to be expected under equilibrium conditions because of the inability of the carriers to equilibrate with the lattice during their short drift times. These so-called ballistic transport effects, however, are processes which disturb only the energetics of the system and, barring avalanche effects, leave the numbers of carriers involved unaffected. Recombination considerations are thus of no importance. An entirely different situation exists when devices such as light-emitting diodes, CCD sensor arrays, bipolar junction transistors, and avalanche photodetectors (APDs) are considered. These devices operate in the presence of carrier densities in excess of the thermal equilibrium values and in such cases the contribution of surface states to recombination becomes very important. For such devices, steps to reduce the effects of the surfaces are invariably undertaken which often include

---

*Recent results on InP have in fact suggested that, on this compound semiconductor at least, values of surface mobility close to bulk values may be possible. See: M. J. Taylor, D. L. Lile, and A. K. Nedoluha, *J. Vac. Sci. Technol.* **B2**, 522 (1984) and K. P. Pande and D. Gutierrez, *Appl. Phys. Lett.* **46**, 416 (1985).

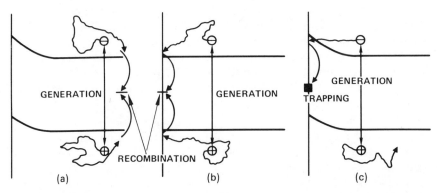

Figure 7. Surface contribution to the recombination of carriers generated remote from the surface, (a), and within a diffusion length of the surface in the absence, (b), and presence, (c) of a surface field.

the use of a dielectric overlayer or "passivation layer" to reduce or inactivate the contribution of the surface.

With reference to Fig. 7 let us consider the sequence of events which follow the injection of a single electron–hole pair into the surface region of a semiconductor. If the charge pair is outside the influence of the surface space-charge field as in Fig. 7(a), it will move randomly until it recombines either directly or, as is in general more likely, by an indirect two-step process involving a bulk recombination site. If the carrier pair is within approximately a diffusion length of the surface, a finite probability exists for the carrier's recombination via a surface level. The simplest case occurs when no surface field is present, as in Fig. 7(b), when by random diffusion both the electron and hole encounter the surface and recombine. This process is additive to the bulk mechanisms and thus acts to reduce the overall recombination lifetime of excess carrier in the material. It can be seen that the surface in this case acts as a sink for electron–hole pairs with a current of positively and negatively charged particles flowing into the surface. For this reason surface recombination is generally described in terms of a velocity s, with units of cm/s. In the more likely case of the presence of a surface field, one carrier will tend to be drawn by drift to the surface whereas the other carrier will be repelled. In the example shown in Fig. 7(c), electrons will be preferentially trapped in the surface. In the absence of the holes required for recombination the electrons will be either reemitted to the CB or, if a hole becomes available, recombination can occur with the trapped electron dropping into the VB. It can be seen that the effect of the field is thus to "screen" the surface from one type of carrier and thus act to reduce its contribution to recombination. Because a surface (or bulk) energy level must interact with both the VB state and CB state to act as a recombination site it is those states near midgap which are most effective in this regard. This is described by ascribing to such "deep" levels a larger capture probability for recombination than those shallow states nearer the band edges.

Surface recombination, which usually is nonradiative, tends to be undesirable in optical devices where it acts to remove carriers from the primary and desired radiative recombination path. Efforts to reduce its effect by reducing N_{SS} or lowering the capture cross section for recombination usually involve the use of a dielectric passivation layer.[47] Such a coating, if chosen properly, can not only reduce s, but can also stabilize the surface by chemically isolating it from the ambient. Heller[48] has discussed the effects of introducing strongly bound species on the surfaces of semiconductors whereby the bond lengths are reduced with a concomitant reduction in the overlap of the surface state wave function with the CB and/or VB leading to a reduction in such a state's contribution to recombination. Ru^{3+} ions in particular appear to be beneficial on the external surface and grain boundaries of $GaAs$[49,50] and Ag^+ on polycrystalline InP has resulted in a 1000-fold improvement in solar-cell efficiency. Such techniques as these have resulted in a reduction in the surface recombination velocity on GaAs from $\sim 10^6$ to 3×10^4 cm/s,[51] a value which should be compared with the 10^3 cm/s and 10^2 cm/s which can be achieved on InP and Ge, respectively. In fact, the observation of luminescence effects in the surface region of a semiconductor can be used as a useful and sensitive indicator of the quality of that surface. Ando *et al.*[52] have reported very large values of photoluminescence in InP from which they have inferred low values of surface recombination, and Hirota *et al.*[53] have demonstrated high-efficiency electroluminescence[54] in MIS triode structures on this same material which they proposed may be useful as a low-voltage high-speed photoemitter.

Traps which act as recombination sites in the presence of excess carriers also contribute to carrier generation when a deficiency of carriers occurs. A surface which has a high density of surface energy levels and thus a large value of s will also act as a ready source of carriers when biased into deep depletion. The effect of this on CCDs, for example, is to reduce storage time and increase dark current by enhancing the generation rate of minority carriers, a situation which in general is deemed to be undesirable insofar as it limits image sensitivity. On InP, for example, surface generation velocities $\sim 10^4$ cm/s have been reported using SiO_2[55] and velocities of 2×10^3 cm/s being obtained using Al_2O_3.[56] It must be appreciated, however, that such data will depend very much on the method of surface treatment employed, in addition to the specific dielectric chosen.

In discussing the dielectric–semiconductor surface to this point we have at least implicitly confined our attention to the *external* surfaces of single-crystal material. In optical applications, particularly those involving solar cells, much consideration from a cost standpoint is being given to the use of relatively inexpensive polycrystals of semiconductor which contain a large "internal" surface area between the grains. These intergrain regions act in many ways like the surfaces we have been discussing, affecting both carrier densities and mobility as well as recombination and current leakage.[57] These internal surfaces, however, are less accessible and are not so amenable to the more usual dielectric passivation techniques we have discussed. They are, however, highly disordered and thus act as regions of high diffusivity permitting for the selective introduction of chemical species by diffusion from the gas or liquid phase. Heller[48] and Chu et al.[47] among others have addressed this question.

Based on considerations such as those outlined above we are now in a position to consider explicitly those devices which are of importance to the III–V compounds and their present status. It is apparent that in many cases the dielectric–semiconductor interface is of considerable importance in the correct operation of these devices and that the relative success or failure of any specific device on any of the III–V's depends critically on the ability to control the electrical properties of the dielectric and interface.

We might mention that such surface effects as have been described in this section are, in large part, generally suppressed in non-MIS structures. In metal–semiconductor interfaces as are, for example, encountered in the GaAs MESFET technology, the presence of an essentially limitless supply of electrons from the metal means that most interface-state electron requirements are met with little noticeable first-order effect on device performance. This is not to say that such effects are entirely absent however; localized interfacial states may and in fact are likely to be present and do exert effects on device results.[58] In junction devices such as the JFET the interface is relegated to the bulk of the crystal lattice and little discontinuity or contamination with resulting surface trapping is in general deemed likely.

3. MIS-Device Characteristics

Devices on the III–V compounds appear to be of interest for one of two reasons, namely, high-frequency operation at frequencies ≥ 2 GHz or optical sensing or emission at wavelengths beyond 1 μm—regions which in general are inaccessible to the established Si technology. In the former case, such devices as FET, CCD, dynamic memories, and so on, are of importance whereas the latter application manifests itself in detectors, CCD imaging arrays, solar cells, phototransistors, and LEDs and lasers. At a superficial level it would appear that not all of these devices are dependent on the insulator–semiconductor interface. All devices do require stable operating characteristics, however, and as such necessitate passivation which in many cases will involve the use of an encapsulating dielectric.

In what follows we will discuss in turn each of the main devices which are being developed for use with the III–V's and which rely for their operation on a dielectric overlayer. Following this, specific results will be presented on a material-by-material basis. Those devices whose only need for such a layer is passivation will be omitted with some general statements concerning this aspect of device development being introduced incidentally. Similarly, we will make only passing comment where deemed appropriate on the many different dielectrics and methods of growth and deposition that have been employed in device studies on these semiconductors. Those readers requiring further information are referred to a number of excellent review articles in the open literature.[15,59] It should be realized that most, if not all of the techniques available permit for the growth of layers with resistivity sufficient for most device applications where a relatively modest value of $\rho \sim 10^{10}$ Ω-cm may well be adequate.[60] It is primarily in the manner of how the deposition techniques perturb and affect the semiconductor surface that differences arise.

3.1. Field-Effect Transistors

Figure 8 illustrates schematically the various configurations possible for an FET. In all cases the device consists of a piece of semiconductor to which two contacts, the source and drain, are attached and between which a control electrode, the gate, is situated to modulate by field-effect control of the carrier density, the "channel" current flowing between the two contacts. As shown in the figure, metal–semiconductor Schottky gates, p–n junctions, heterojunctions, and MIS control electrodes can all be used; the choice of which is to be preferred depending on the application, as well as upon materials constraints.[61,62] For example, the Schottky-gate structure, which has been very successfully applied to the high-speed GaAs FET[63] and which is in principle the simplest of all approaches to implement, is not apparently well suited to InP and GaInAsP devices because of the low metal–semiconductor barrier height which occurs on n-type samples of these compounds.[64] Attempts to increase the effective barrier height of such metal–semiconductor junctions by the use of a thin intervening dielectric layer ≤ 100 Å thick have met with some success,[36,65] but such devices might more properly be considered to be a special form of MIS structure rather than a "pseudo"-Schottky device

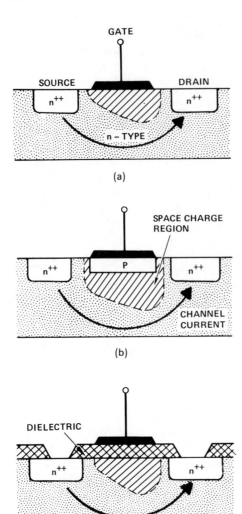

Figure 8. Schematic illustration of the various gating schemes possible in an FET: (a) Schottky, (b) junction (homo or hetero), and (c) MIS.

and for the purposes of this discussion will be treated as such. The p–n junction and heterojunction approach would be expected to be highly radiation tolerant but for high-frequency operation this structure suffers from a somewhat increased gate capacitance due to the contribution of the edges of the gate well, an effect which becomes more pronounced as geometries are reduced. The MIS gate is for many applications a very attractive structure provided that a stable, low interfacial and bulk trap density interface can be made. Although unclear from Fig. 8, the FET can be made in two different electrical configurations—the depletion-

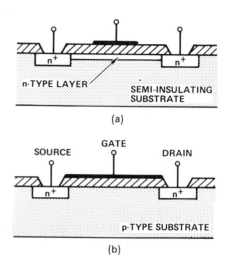

Figure 9. Schematic illustration of the depletion (a) and enhancement (b) FET structure for an insulated gate device.

and the enhancement-mode device. Figure 9 shows these alternatives for the MIS gate.

In Fig. 9(a) current flows between the n^{++} source and drain contacts in the absence of any gate input signal by virtue of electron conduction in an ion-implanted or epitaxially grown channel n layer. It should be recalled that in general it is electron conduction only with which we will be concerned in the III–V's because of the far smaller value of hole mobility in these materials. The channel current in this "normally-ON" device is then reduced by electrostatic depletion of the electron concentration in the channel by means of a negative voltage input to the gate. The enhancement structure differs dramatically in that in the absence of any input there is essentially zero source-to-drain current flow. This can be achieved by either making a depletion-type structure with the channel layer being sufficiently thin to be completely depleted at zero gate voltage, V_G by virtue of the quiescent surface potential of the semiconductor or, as is to be preferred, by making the device from p-type material to which source and drain n^{++}-type regions are then attached. Conduction between the two back-to-back channel contacts is then induced by inversion of the semiconductor surface by means of a positive potential applied to the gate electrode which must, in this case, entirely span the source-to-drain gap. It is clear that an MIS control electrode is mandatory in this latter device to prevent the inversion charge being extracted by current flow to the reverse-biased gate as would occur with a metal or junction barrier. The MIS gate also has the advantage of permitting large (>5 V) input voltage excursions of both positive and negative polarity limited only by the ability of the dielectric to withstand the resulting high fields. Such large dynamic ranges measurably simplify circuit design and prevent

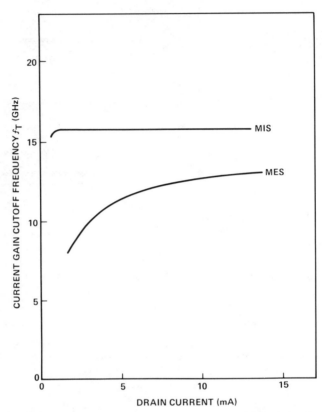

Figure 10. Cutoff frequency dependence on drain current calculated for the case of an insulated-gate and metal-gate FET (taken from Ref. 68 by permission).

premature gate breakdown in the event of inadvertent positive-going voltage transients. Enhancement-mode FETs based on the reduced depletion of a thin channel layer can be made with Schottky[66] or junction[67] control gates but these devices do place stringent demands on threshold voltage control and hence on the thickness and doping uniformity of the very thin (~500 Å) channel layers employed and are limited to positive-going gate potentials ≤ 1 V. As well as restricting circuit design, such low dynamic ranges also preclude device bias into the optimum operating range and such devices are thus confined to operation at reduced g_m, I_{DSS}, and consequently speed.

All these considerations apply to the enhancement device. The depletion FET less clearly benefits from an MIS approach although the physical separation of the potentially reactive metal and semiconductor by an isolating dielectric can only help reliability by minimizing any reaction or interdiffusion degradation of the gate. Although at present somewhat unclear from the available data, it has also been proposed that an MIS approach will have a distinct speed advantage over the equivalent MESFET structure. Yamaguchi and Takahashi[68] in particular have modeled MIS- and MES-gated structures on GaAs and find that although both g_m and C_G are reduced in the MISFET due to the field drop across the dielectric, their ratio, proportional to the current-gain cutoff frequency f_T, is in fact larger. This effect, which results from the decreased fringing capacitance contribution of the MIS gate, is illustrated in Fig. 10 for the case of a 1-μm structure.

3.2. Charge-Coupled Devices

A charge-coupled device is similar to an inversion-layer enhancement-mode FET except that the CCD is equipped with a multiplicity of transfer gates between the input and output electrodes (diodes). These multiple gates are interconnected to form a number of "phases," 2, 3, or 4 being typical. The enhancement FET and CCD differ, however, in that, whereas the signal charge is transported throughout the entire length of the FET channel in a continuous stream, in the CCD the charge is moved incrementally under the controlling influence of clock pulses applied to the gates. Although an inversion-layer charge comprises the signal-carrying medium in both structures, in the CCD it results from selective injection at the input diode to the device in response to an input signal rather than from a continuous flow to maintain a thermal equilibrium carrier density as is the case in the FET.

Because of the necessity to transport inversion charge, albeit in an incremental rather than continuous fashion, the CCD like the FET described previously requires the use of minority-carrier blocking MIS gates. Most certainly this is the case for the surface channel CCD (SCCD) just described where charge flows at the semiconductor–dielectric interface. An alternative approach to CCD design,[69] which was originally proposed to prevent surface effects from degrading device performance by moving the signal-carrying channel carriers away from the interface, relies upon doping variations to position the potential minimum for electrons normal to the surface, some 0.1–1 μm from the interface. This bulk channel approach (BCCD) also permits for the use of alternate gating structures and in fact GaAs BCCD have been designed and operated using Schottky-barrier transfer gates.

To ensure maximum performance in a CCD it is essential that as little as possible of the injected signal packet be lost or even delayed as the charge is propagated along the device. It can be readily appreciated, for example, that if surface states trap some of the signal charge which is then subsequently released, there will result a distortion of the signal on reaching the output (see Fig. 5). The use of a so-called "fat zero" background charge to saturate the surface traps can alleviate this effect at the price of a reduced dynamic range as can the use of a buried channel structure. Neither of these two approaches, however, can prevent degradation of the charge transfer efficiency (CTE), defined as the fraction of the charge packet transferred to the next gate, resulting from interelectrode potential barriers. Ideally, one would choose to have zero spacing between the transfer gates. In practice, of course, this is impossible. Using a single-level gate metalization design as is often used with MIS CCD and as must be used if a non-MIS approach is adopted, a practical limit at the present state of the technology is to have gaps $\sim 1\,\mu$m. Most certainly in Si devices very good values of CTE ~ 0.9995 can be achieved with this approach. In the case of GaAs the situation is not so clear-cut. Using an MIS-gate design the alternative two-level metalization approach can be used which allows for essentially a "zero-gap" separation. This is the approach most generally used in Si CCD structures. The use of such overlapping gates does, however, result in increased interelectrode capacitance which may be expected to degrade to some extent the high-speed performance of the device. Ideally, a "self-aligning" process is to be desired where one level of gate metalization is used to define the second level thereby automatically generating an essentially zero-overlap structure. Such a process technology for FETs will also be necessary if any appreciable degree of integration is to be achieved. Attempts to duplicate the Si process, however, with InP, for example, have led to little success to date[70] because of the inability of the gate dielectrics to maintain good interface properties in the presence of the necessary high-temperature anneals.

In addition to high-speed signal processing, such as, for example, is used in signal delay and filtering, CCDs on the III–V's are of interest for monolithic imaging applications where the long-wavelength response rather than the high mobility of the carriers is used to advantage. In a monolithic imager, signal detection as well as readout occur in the same slab of semiconductor thus avoiding the fabrication and reliability problems inherent in hybrid approaches such as must, for example, be used in long-wavelength imaging focal plane arrays with Si where narrow-gap detectors must be connected by means of many interconnects to the Si readout CCD.[71] Among the III–V's, InSb has attracted much attention in this regard as will be discussed more fully in a later section.

3.3. Integrated Circuits

In giving consideration to the application of dielectrics in the development of ICs on the III–V's, we are getting close to the heart of the significance of these materials. Most certainly this is where most of the commercial interest lies and where most of the advantages of the III–V's will be manifest. High-speed discrete device operation implies that the devices themselves will be small. Any attempt to interconnect devices to perform a circuit function at high speed requires that the interconnect lines be kept short to reduce parasitic charging times. For cost effectiveness the packing density, yield, and reliability must also all be high. These considerations mandate an IC approach and much effort in recent years has been devoted to developing such a technology for the III–V's.[72–74]

Dielectrics are required to serve many functions on an IC including their use to isolate metal crossovers, to act as gate isolation in active devices such as FETs, and to passivate the circuit by encapsulating it with a material which will both remove surface leakage as well as maintain a constant operating characteristic for the circuit. Such long-term stability is paramount if useful performance is to be achieved. The importance of the dielectric can be appreciated when it is realized that in the absence of a compatible dielectric for Si all ICs as we now know them would cease to exist. Even bipolar circuits which do not require a dielectric in their basic device structure employ insulating layers in critical aspects of their design.

It should be appreciated, of course, that the demands placed on the dielectric in these various functions vary greatly. The most stringent requirements are placed on dielectrics for use in active devices where not only must the insulator permit high performance by supporting a low interfacial trap density with the semiconductor, but it must also maintain that performance with high stability in the often severe environment of elevated temperatures and high electric fields. Dielectrics for passivation need only prevent and maintain an absence of surface leakage and physically isolate sensitive device regions, such as surface $p-n$ junctions, from the influence of external factors such as water vapor. The least demanding requirements are placed on those insulating layers used to isolate metal interconnects, for example, in conductor crossovers. The layer here simply has to act as a reliable pinhole free insulator as can in fact be at present readily achieved on the III–V's by a variety of deposition procedures.

If we consider the most stringent test, namely that of dielectrics for IC use in active devices, the question needs to be asked as to whether such an application is mandatory: Are MIS-based active devices, such as FETs and CCDs, really necessary or are their benefits, if any, outweighed by the problems associated with their implementation? GaAs is a good case in point as, as will be discussed in detail in the next chapter, the surface of this semiconductor appears to support a large density of surface traps and thus, as far as the present technology is concerned, MIS-based active devices appear to be eliminated from consideration for all but a limited range of dynamic IC designs. As a result almost all of the IC development in GaAs has been based around the MESFET. If we confine our attention to digital designs, that is, logic circuitry, it becomes immediately apparent that the absence of MISFETs has serious implications. This is illustrated in Fig. 11, where three alternative approaches for implementing a basic logic operation are shown. As was discussed earlier, the MESFET is inherently a depletion device and using such a device the output from a logic gate is displaced in voltage compared to the input. This is shown in Fig. 11(a) where a "level-shifting" circuit must be interposed between logic stages to make the output compatible with the input of the next device. This level shifting typically requires the use of diodes and an FET as well as an additional power supply, and must be performed between every gate in the logic design. This vast increase in component count consumes circuit real estate, reduces yield and reliability, and adds significantly to the power consumed and hence heat generated by the circuit.

Some improvement in this situation using depletion MESFETs has been achieved using the Schottky-diode FET logic approach[72] shown in Fig. 11(b), where the logic functions are performed by the Schottky diodes and the FET acts solely to provide signal inversion and gain. It is clear, however, that the circuit is still complex when compared to the direct-coupled approach shown in Fig. 11(c) which is the simplest implementation possible. The DCL (direct-coupled logic) approach, however, does require the use of an enhancement-mode driver. Attempts in this direction are being made using the enhancement MESFET;[73] however, with its limited dynamic range, values of V_{DD} are restricted to less than 0.6 V. Such low drive levels mandate strict control on threshold voltage variations with standard deviations typically having to be held to less than 30 meV. This imposes such extreme demands on the technology that it

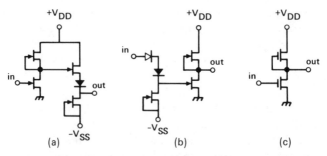

Figure 11. Schematic circuit diagrams for a basic inverter using (a) depletion MESFET transistor–transistor logic, (b) depletion MESFET Schottky-diode–transistor logic, and (c) enhancement/depletion direct-coupled MISFET logic.

may never be cost effective. In contrast, an MISFET enhancement design based on inversion-layer conduction is not restricted to intolerably small values of V_{DD}, which makes feasible direct-coupled logic with all its attendant advantages of simplicity and low power. This is the basis of the NMOS approach in Si which could be implemented in the III-V's if compatible dielectrics were available. In the absence of such layers, circuit design can be used to circumvent the device restriction but, as can be seen from the example cited above and shown in Fig. 11, a price in terms of complexity and power must be paid. Despite these advantages, serious questions have been raised, and justifiably so, concerning the possibilities of implementing digital MIS circuit designs on the III-V's. A survey of the literature reveals widely differing opinions on this point and resolution must await the outcome of future experiments. It is the opinion of the present author that success or failure in this area will depend in large part on the application and the semiconductor chosen, with InP, for example, already showing every indication of being a prospective contender for an integrated MIS technology.

3.4. Optical Devices

Essentially all optical devices, including light-emitting diodes, solid-state lasers, $p-n$ junctions, and pin-diode detectors, as well as phototransistors, solar cells, and optical photoconductive switches, can and do benefit from the use of dielectric overlayers for passivation, encapsulation, and antireflection coating. Examples include the work of Capasso and Williams[75] on the passivation of a variety of surfaces, including those of InGaAs detectors, using nitridization of a surface previously reduced in a hydrogen plasma and the results shown in Fig. 12 where the reverse $I-V$ characteristics of a p^+-n InP avalanche photodiode (APD) are displayed in the presence and absence of a surface overlayer. At the high reverse biases at which these structures are required to operate, the reverse leakage in the absence of passivation is large enough to cause heating and stability problems. Diadiuk et al.[76] have reported that in the case of both InP and GaInAsP the reverse leakage of these APDs can be reduced by as much as four orders of magnitude by the application of selected surface coatings. In particular, plasma-deposited Si_3N_4, polyimide, and SF_6 propelled photoresist all appeared to be effective. In addition to the above, dielectrics are of interest for more direct incorporation in a variety of optical structures, examples of particular interest for the III-V's being solar cells and optical imagers.

In solar-energy conversion direct improvements in solar-cell performance have resulted in both Si and GaAs from the use of MIS-type structures rather than Schottky barriers.[77] Increased values of open

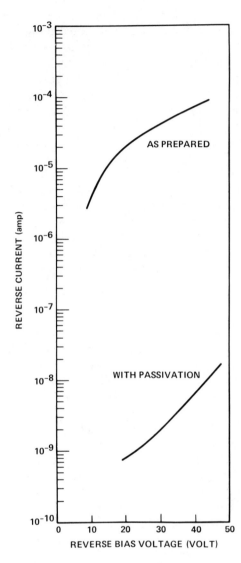

Figure 12. Reverse current–voltage characteristics for an InP avalanche photodiode in the presence and absence of passivation (taken from Ref. 76 by permission).

circuit voltage V_{oc} and efficiency η of conversion of sunlight to electricity of 615 mV and 13.1%, respectively, have been reported from the use of such thin dielectric interface layers in metal–semiconductor junction cells on Si. GaAs MIS cells have, in fact, attained efficiencies of 18% and values as high as 26% have been proposed for such structures on InP under AM2 conditions.[78] Because of their simplicity, Schottky barriers are attractive candidate structures for low-cost terrestrial solar converters where the present system goal is to achieve ~10% efficiency at less than 30 cents/peak watt. Schottky diodes, however, often suffer from large values of leakage dark current which result in reduced values of V_{oc} and η. Introduction of a thin intervening dielectric can reduce this leakage current significantly resulting in an appreciable improvement in cell performance. Calculations indicate that efficiencies as high as 25% should be possible on Si under AM1 conditions compared to values rarely >13% which are obtained in practice. Larger efficiencies should be possible with some of the III–V's because of their narrower gaps and hence extended wavelength response. In fact InP with an absorption peak near 1.0 μm is nearly optimum in this regard.[79] Also of some benefit are the larger thermal conductivities and band gaps of many of these compounds compared to Si which permits for higher-temperature, higher-energy density operation and thus for the use of solar concentrators rather than the presumably more expensive extended solar panels. Values of $\eta \sim 19\%$ have in fact already been reported for GaAs cells operating at concentration ratios as high as 1700.[80]

MIS enhancement of solar-cell performance has been studied since 1975 and during this time Si, GaAs, and InP have all been shown to benefit from this approach. Although some details of the device remain unclear,[81] it is evident that this is an attractive method to increase cost effectiveness in solar conversion. Presumably these devices act like minority-carrier nonequilibrium MIS tunnel diodes which rely upon a sufficient

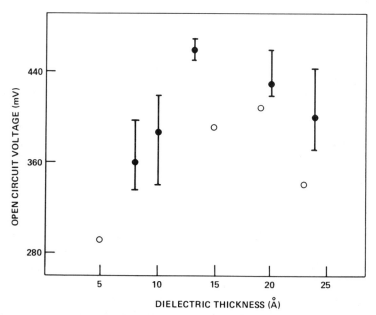

Figure 13. Open circuit voltage in a Si MOS solar cell as a function of intervening dielectric thickness: (●) p-type and (○) n-type (taken from Ref. 77 by permission).

tunnel current of minority carriers through the dielectric layer to extract the dc power with minimum series loss.[82] Insulator thickness is thus likely to be very important. The results shown in Fig. 13 suggest that for Si a layer ~15 Å thick appears to be optimal, in contrast to GaAs where a thickness somewhat greater, ~30 Å, may be preferred.[77]

Surface trapping effects in these structures are similarly not well documented. For very thin insulators it seems that their effect is in general to be one of reducing η by virtue of a decreased barrier height. Surprisingly perhaps, in thicker dielectric structures where series resistance is limiting, the effect of surface traps can be beneficial as they act to increase η by enhancing tunneling.

A number of examples also exist for the advantages of MIS-type devices in optical detection and sensing. In particular, these occur where more simple approaches such as Schottky barriers cannot be applied. InAs and InSb, for example, do not sustain good blocking metal–semiconductor barriers in general and, in such cases, an MIS approach becomes attractive. Lile and Wieder[83] have shown, that MIS surface photodiodes on polycrystalline InSb may be made with performance equal to or exceeding that of bulk single-crystal devices and that detectivity values as high as 3×10^{10} cm/W-s$^{1/2}$ at 5.0 μm may be achieved at 80 K.

For long wavelength imaging, MIS arrays on InSb are of considerable interest and Thom et al.[84] and Kim[85] have both reported on detector arrays of such devices for integration with CCD structures for serial readout. Monolithic intrinsic imaging arrays, combining sensing elements with CCDs, attain high sensitivity by operation in an integration and hold mode, where optically generated electrons (or holes) are accumulated over an extended "acquisition time" in the depletion wells of the CCD. Such operation of course requires a very low rate of intrinsic carrier generation so as to keep dark current low and storage times high. Measurements of this parameter on p-type InSb,[86] using the pulsed C-t response and Zerbst plots of MIS diodes, have yielded storage times as large as 0.5 s at 77 K. Such a value implies that the dark current is dominated by bulk generation and that surface recombination on these devices was less than 20 cm/s. Although shorter than state-of-the-art Si/SiO$_2$, these values of storage time are likely to be more than adequate for imaging applications at the moderate clocking frequencies usually employed and in fact Kim[87] has presented evidence that storage times are sufficiently long to permit for near-background limited performance (BLIP) in MIS charge injection detectors in InSb operating at 77 K.

For very-high-sensitivity imaging, where extended spectral response is not of overwhelming importance, the larger gap materials, such as InP and GaAs, become of interest. For bulk limited performance, which would be expected in InP with its low surface state density, generation rate varies exponentially with E_g/kT and thus the approximately 0.2-eV difference between InP and Si in this parameter can result in orders-of-magnitude difference in the dark current.

3.5. Memory Cells

The availability of high-speed memories is equally critical to the implementation of a fast mainframe computer as is the development of high-speed CPUs. Since the first application of semiconductor memory to data storage for computers was initiated 15 years ago,[88,89] much effort has been devoted to increasing speed and reducing power requirements of such elements, as well as to developing nonvolatile memory structures[90] which retain the advantages of more conventional approaches without the risk of loss of data during power off. At present, MOS-type memories in Si are as fast as bipolar but are far more compact, and thus the future of Si MOS memories seems assured. Evaluations of the use of MIS-type transistors in static memory elements, such as flip-flops, reduce to a consideration of the relative merits of MIS *vs.* MES for FETs as has already been discussed at length earlier in this chapter. Of more specific interest for our purposes is the availability of two distinct methods of memory implementation unique to MIS-type structures: dynamic memory and nonvolatile memory. Dynamic memory exists when, due to signal decay, a refresh of the stored data must occur on some regular basis. One example of a dynamic memory cell in Si consists of an MOS transistor and an MOS capacitor where information is stored in the inversion layer of the diode and must be rewritten when it has decreased to some predetermined percentage of its initial level by leakage through the dielectric. For such an application, a very-high-quality low-leakage insulator is thus required. Nonvolatile memory refers to a method of storage which is retained even in the absence of power applied to the unit. This is very useful to prevent loss of stored programs during inadvertent power down of computers and has been achieved in Si using devices such as the MAOS and MNOS (Metal–Aluminum Oxide– and Metal–Nitride Oxide–Semiconductor), where digital data are stored via the presence or absence of injected electrons trapped in interfacial states between two different levels of dielectric in a MIS transistor. Little work has been done on any type of nonstatic memory using III–V materials. Exceptions include the results of Bayraktaroglu *et al.*,[91] where interfacial trapping at the GaAs/dielectric interface was proposed as a possible mechanism for a nonvolatile memory element and the results of Hasegawa and Sakai[92] on GaP. In neither case was it clear how reproducible or stable were these characteristics. However, in principle these approaches would appear to be attractive and might warrant further attention.

4. Device Results

4.1. Gallium Arsenide

Apart from Si, GaAs is the most widely studied and best understood of all semiconductors. Most certainly among the III–V's this material has received the most attention, both with regard to materials growth as well as device development. In the latter category the outstanding performance characteristics and simplicity of the GaAs MESFET are to be noted.[63] Because of this prior attention to this semiconductor it is perhaps understandable that a considerable effort has also been addressed to the development of MIS-based devices on this material.

The first report of an MIS-gated FET on GaAs appeared in 1965 when Becke *et al.*[93] reported on n-channel depletion-mode devices using CVD SiO_2 for the dielectric. These devices, with 7-μm-long diffused channels, exhibited g_m values ~16 ms/mm of gate width and effective (field-effect) channel mobilities ~1780 cm^2/V-s. An inability to invert the surfaces of the GaAs, which was ascribed to a high density of surface states $N_{SS} > 10^{12}$ cm^{-2} eV^{-1}, prevented operation of enhancement devices and also led to an increasing value of g_m with frequency in the depletion structures. These first results were quickly followed by work on an insulated-gate FET prepared on polycrystalline GaAs,[94] but further efforts to develop MIS devices were somewhat eclipsed by the spectacular success of the MESFET which was proposed in the same year.

Charlson and Weng[95] were the next to report a thin-film MIS transistor in 1968 and then in 1974 Ito and Sakai[96] published a paper describing the development of an inversion-mode FET exhibiting a channel mobility ~2240 cm^2/V-s. Since that time numerous results have appeared on both the normally-on and normally-off structures using various dielectric preparation techniques for gate insulation,[97] including wet chemical[98–104] and plasma anodization,[105] plasma oxidation, oxidation,[106,107] pyrolysis of SiO_2,[108] Al_2O_3,[109] $Si_xO_yN_z$,[110] and Ge_3N_4,[111] and thermal oxidation.[112] Without exception the results have shown that depletion MISFET may be made on GaAs which exhibit attractive output characteristics as might, for example, be displayed on a curve tracer. By way of illustration, the photomicrograph and resulting output characteristics shown in Fig. 14 were obtained on such a GaAs device made using an anodic dielectric. By means of such measurements, g_m values as high as 55 ms/mm of gate width with corresponding values of $\mu_{FE} \sim 3200$ cm^2/V-sec have been reported.[97] Results such as these, however, are somewhat misleading. Surface state studies using a wide variety of insulator preparation techniques on the GaAs dielectric interface without exception have revealed large densities of surface traps which would be expected to severely degrade devices such as MISFETs on this material. Promising results such as those in Fig. 14 are only possible because the frequency of measurement (100 or 120 Hz) is beyond that at which the majority of the defects can respond. In fact, in some cases it has apparently been found necessary to use pulsed measurements of the output characteristics to overcome the debilitating effects of the slow trap centers.[103] This is clearly evidenced in the dispersion of both C–V characteristics,[25] as well as transconductance reported by a number of

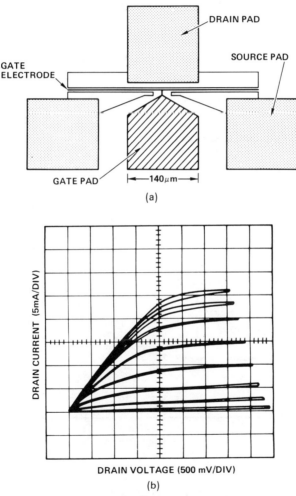

Figure 14. (a) Schematic drawing of a typical high-frequency MISFET geometry. (b) Representative 120-Hz output characteristics for a 4-μm-gate-length GaAs MISFET.

authors on reducing signal frequency below ~100 Hz.[25,93,112,113] Figure 15 shows the results of such a measurement where a fit to a theoretical model based on a trap density of $3.2 \times 10^{12}\,\text{cm}^{-2}\,\text{eV}^{-1}$ is demonstrated.[114] The consequences of such a low-frequency characteristic for these devices is quite severe. Particularly it implies a lack of dc and hence bias control and would seem to preclude the use of these structures in anything other than a dynamic mode where dc signal levels do not have to be held for more than perhaps 10 ms. One such application has in fact been demonstrated where a ring oscillator, consisting of a continuously

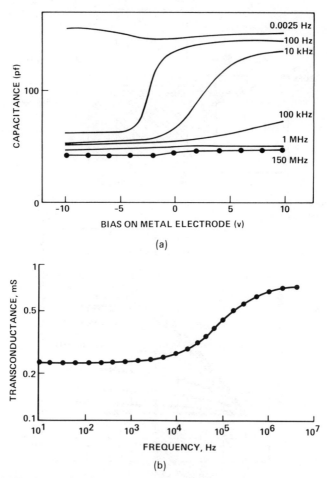

Figure 15. (a) Representative C–V data for a GaAs MIS diode obtained at various measurement frequencies. (b) Theoretical curve and experimental data points showing the effects of $3.2 \times 10^{12}\,\text{cm}^{-2}\,\text{eV}^{-1}$ surface states on the low-frequency dispersion of MISFET response (taken from Ref. 114 by permission).

switching series connection of 13 inverter stages, has been demonstrated to operate successfully with a propagation delay of 110 ps at a power level of 2 pJ for a 2-μm gate length.[115] Although dynamic memories and logic approaches where continuous signal refresh is employed are conceptually possible, experience with such logic designs on Si has demonstrated some severe disadvantages, including high power consumption and more complex circuitry.

Such debilitating densities of long-time-constant surface traps would not be expected, however, to impede or degrade the high-frequency performance of such devices and that in fact has been shown to be the case.[24,105,116–119] Figure 16 demonstrates such microwave response where maximum stable gain, maximum available gain, and deduced unilateral

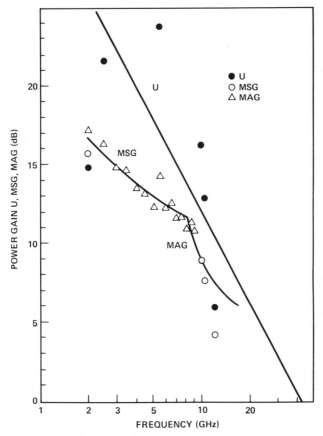

Figure 16. Calculated curves and measured values of the maximum stable gain (MSG), maximum available gain (MAG), and unilateral gain (U) for a 1.0-μm-gate-length depletion MISFET on GaAs (taken from Ref. 105, by permission).

gain for a 1.0-μm-gate-length depletion GaAs MISFET are shown. Mimura et al.[119] have also reported on a 2.0-μm-gate-length enhancement structure exhibiting a maximum frequency of oscillation f_u of 13 GHz. A 1.8-μm depletion device of comparable geometry had a f_u of 22 GHz, which is somewhat larger than that to be expected from a MESFET of the same gate length. This improvement was ascribed to the reduced gate parasitic capacitances of the MOSFET structure and appears to be consistent with the subsequent analysis and conclusions of Yamaguchi and Takahashi.[68] Such performance is certainly comparable to what would be expected on a GaAs MESFET of similar geometry although without stable low-frequency response the practical merit of such devices must be seriously called into question.

An even more restrictive impediment to the development of an MIS technology on GaAs is that, with the one exception of the early work of Ito and Sakai[96] and Miyazaki et al.,[120] despite numerous more recent efforts, no convincing demonstration of other than minimal inversion-layer conduction on GaAs has been forthcoming.[98] This is consistent with the surface studies reported in detail elsewhere in this volume, namely that the n-type GaAs surface is possessed of a quiescent surface potential ~ 0.85 eV, and a very high density of surface states $N_{SS} > 10^{12}$ cm^{-2} eV^{-1}. By operating at sufficiently high frequencies, > 100 Hz, adequate surface modulation is possible to permit for good depletion-mode operation. Any attempt to drive the surface into inversion, however, is thwarted not only by the large value of N_{SS} but also because the surface must be initially moved through ~ 0.8 eV to even achieve the onset of inversion. The use of high frequencies to incapacitate the surface traps is precluded in the inversion range of operation because of the decreased response time of traps close to the conduction-band edge. Despite the distinctive lack of success in recent attempts to produce good inversion-layer performance on GaAs, the early results remain. Although definitive explanations are not possible, it is tempting to conjecture that the results observed were due to some phenomenon other than inversion. A primary candidate for GaAs is the inadvertent formation of an n-type conduction layer on the surface due to thermal degradation during processing. Such a layer might be interpreted from FET performance as inversion conduction and is consistent with the historically well-documented observation of surface thermal-type conversion in this semiconductor.

Enhancement-mode devices, relying upon reduction of the depletion region in a thin epitaxial n-type conduction layer on GaAs, have been demonstrated[117,119] with microwave gain being reported to 8 GHz for a 2.0-μm-gate-length device. These transistors, however, would presumably be expected to suffer from the same low-frequency response restrictions as the depletion structure. The above results suggest that, unless a way can be found to either reduce or otherwise incapacitate the effects of what now is generally conceded to be a high trap density at the GaAs interface, then the prospects for practical implementation of such devices seem remote. Attempts to improve interfacial properties by the use of different dielectrics or by variations in deposition conditions or the use of thermal annealing has generally resulted in only limited success. More recently, attempts to reduce the state density have primarily been based on the postulate that oxidation of the GaAs surface is responsible for the high state density, perhaps via the mechanism of vacancy or vacancy complex generation, and that this problem is intrinsic to the GaAs/oxide interface.[121] The theoretical arguments and results of Spicer et al.[122] are not inconsistent with such a model although the adsorption of materials other than oxygen has also elicited similar effects. Similarly, the experimental results of Suzuki and Ogawa[123] and others,[124,125] who have reported dramatic reductions in photoluminescence intensities from cleaved single-crystal GaAs surfaces on their exposure to an oxidizing ambient, are pertinent. If oxidation is inherently undesirable, then the prospects for overcoming this degradation in practical devices seems remote. Attempts have been made to remove the native oxide and replace it with a nonoxidizing layer such as Ge_3N_4 or Si_3N_4[126] but to date the results must be considered as somewhat tentative. The data of Ahrenkiel et al.,[127] based on the suggestion of Lucovsky that an oxyfluoride dielectric may form a more complete bond to the GaAs surface, are similarly tentative and require further examination. Alternative approaches which circumvent the problem using novel circuit modes[128] or "leaky" dielectrics[129] where surface charge is "drawn off" through the insulator have led to some success but at the expense of other parameters such as noise or complexity of approach.

To overcome this restriction of poor surface properties, a number of attempts have been made to engineer a buried channel-type MISFET device where the dielectric is either a high-resistivity region of proton-bombarded GaAs[130,131] or an O_2-doped layer of GaAlAs.[132] Although limited at present, further effort in this direction, including perhaps the use of an epitaxially grown high-resistivity layer of GaAs or other wide-gap semiconductor such as ZnSe,[133] may prove successful. Of much recent interest has been the use of modulation-doped superlattice structures between GaAs and GaAlAs, first proposed by Dingle et al.[134] and implemented into a transistor structure by Mimura et al.[135] These high-electron-mobility devices (HEMT), which rely for their good performance on the diffusion of electrons from the highly doped large-band-gap GaAlAs into the high-purity, high-mobility channel layer of GaAs, have conventionally been implemented using Schottky gates. Hotta et al.,[136] however, have recently reported on such devices fabricated with MIS control electrodes. Their results indicate effective mobilities in these structures at 77 K of 27,000 cm^2/V-s using MBE techniques for the semiconductor and Al$_2$O$_3$ for the insulator. These MISS (metal–insulator–semiconductor–semiconductor) FETs are perhaps the most promising of all the insulator-based devices to appear to date on GaAs.

In addition to the FET, the application of insulating layers to solar-cell efficiency enhancement on GaAs has received some attention. In fact, because of the relaxed interfacial and dielectric requirements placed on solar-cell applications, successful improvements in conversion efficiency and open circuit voltage V_{oc} have been achieved in this area using MIS configurations on this semiconductor. Increases in V_{oc} of 62% using thin (~20 Å) evaporated layers of SnO$_2$ and short circuit current improvements by as much as 60% have been reported by Stirn and Yeh.[137]

Polycrystalline material would have the same advantage in GaAs as in Si,[138] namely lower cost, and attempts have been made to implement such devices. Pande et al.[57] in particular have reported an anodic surface passivation technique for such GaAs devices which eliminates shunting effects of the grain boundaries resulting in a five to six orders-of-magnitude improvement in reverse leakage.

4.2. Indium Phosphide

The first indications that InP might possess favorable surface properties for device applications appeared in 1975 when first Wilmsen[139] and then Lile and Collins[140] reported low values of N_{SS} from $C-V$ data on MIS diodes prepared by anodic oxidation. The subsequent observations of Casey and Buehler[141] and others[124,125,142] of enhanced photoluminescence intensities from InP surfaces when compared to GaAs tended to confirm these results and implied that surface recombination velocities for this semiconductor were no more than 10^3 cm/s. This work was quickly followed by the demonstration of depletion-mode FET performance by Messick et al.[143] who, using a pyrolytically deposited layer of SiO$_2$, prepared devices exhibiting similar output characteristics to those previously reported on GaAs MISFETs except that low-frequency response, as exemplified in Fig. 17, did not decrease due to the degrading effects of surface traps at input signal frequencies <100 Hz.[144] As was discussed earlier, this fact is of extreme importance in considering the application of these devices to circuits where it is in general essential to be able to set dc levels.

The superior performance of InP appears to result from a lower density of long-time-constant surface traps in the dielectric/semiconductor interface region with minimum values in the range 10^{11}–10^{12} typically being reported.[145] Because of this, as well as the fact that in the absence

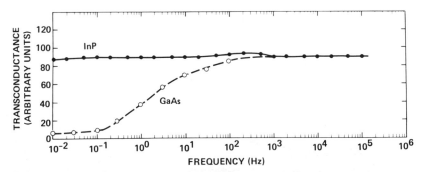

Figure 17. Low-frequency dispersion in transconductance for representative InP and GaAs MISFETs.

of externally applied perturbations the Fermi level passes at the surface within 0.1–0.2 eV of the CB edge on both n- and p-type material,[145] it has also proven possible to induce strong electron inversion-layer conduction on this semiconductor.[39,146] This was first demonstrated by Lile et al. in 1978[147] when they reported on an enhancement-mode–inversion-layer transistor on InP in the form of a MISFET on a p-type substrate using pyrolytic SiO_2 for the gate dielectric. Subsequent work at a number of laboratories has resulted in considerably improved device characteristics being obtained using a variety of device geometries and dielectrics, including pyrolytic SiO_2,[148] Si_3N_4,[149] and Al_2O_3[150] as well as P_3N_5[151] and the plasma[152] and wet chemical anodic native oxide.[153] Figure 18(a) shows the resulting output characteristics obtained on an InP MIS normally off FET whose source and drain were made by ion implantation of Si, an n-type dopant, into (100)-oriented p-type bulk single-crystal material. The gate dielectric was SiO_2 deposited by thermal pyrolysis. Noteworthy is the relative lack of hysteresis as well as the good saturation and near-ideal quadratic behavior of the output current I_D with gate bias V_G obtained in these devices. This is illustrated in Fig. 18(b) where $I_D^{1/2}$ is plotted vs. V_G. Similarly, InP enhancement-mode devices have been shown to obey the classical inverse dependence of g_m on channel length l, which gives encouragement to the possibility of even better performance on smaller geometry designs.[154]

Of interest for these devices is their performance characteristics when considered for high-frequency operation. A figure of merit here is the current-gain cutoff frequency or gain bandwidth product (GBW) given, for a device width of w, by[155,156]

$$f_T = \frac{w}{l}\mu_{\text{eff}}$$

where the effective or field-effect mobility is related to the transconduct-

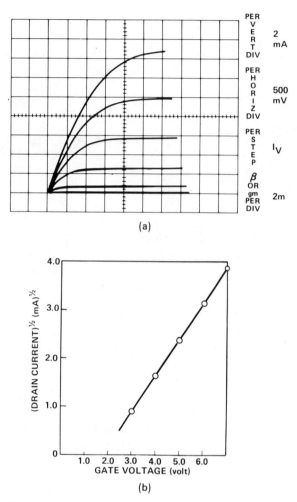

Figure 18. (a) Typical output characteristics for an InP MIS enhancement-mode FET. (b) The same data plotted at fixed drain voltage ($V_D = 3$ V).

ance in the linear low-field region of operation by

$$\mu_{\text{eff}} = \frac{l}{w} \frac{g_m}{CV_D}$$

Early enhancement devices exhibited values of $\mu_{\text{eff}} \sim 400 \text{ cm}^2/\text{V-s}$, which is considerably below the bulk value of electron mobility to be expected in InP of this doping level ($\sim 3000 \text{ cm}^2/\text{V-s}$ at 10^{17} cm^{-3}).[157] Admittedly some degradation might be anticipated from the effects of carrier scattering at the surface; however, additional degrading mechanisms are suspect. Surface trapping, for example, will reduce μ_{eff} as will the effect of surface damage resulting from possible thermal degradation during postimplant anneal (typically ~700°C for 15 min). An additional effect, poor channel contacts, will also lead to reduced performance. The source contact in particular is required to supply very high values of current density into the perhaps 100-Å-thick surface inversion layer and this requires careful contact design.[60] Early InP devices of this type fabricated using alloyed metal contacts frequently suffered from this problem which often can reveal its presence in the form of nonlinear or "swayback" low-field current–voltage characteristics.[60,158] With these considerations in mind more recent device structures have included the use of epitaxial layers,[60,159] diffusion,[150] and ion implantation to form the contact areas either by selective deposition of the ions via a photoresist, oxide, or metal pattern[60] or by subsequent selective removal of unwanted regions following uniform implantation.[160] In addition a variety of surface preparation techniques and dielectrics have been employed for the gate isolation including the use of an HCl pre-etch which has been proposed as a way of removing the perhaps problematic native oxide interface layer.[161] The result of these efforts has been a steady improvement in published values of electron mobility as is illustrated in Fig. 19, which shows the best mobility values reported as a function of time. Despite the wide spread in results, it is apparent that values of $\mu_{\text{eff}} \sim 2000$ cm^2/V-s are presently achievable in enhancement-mode devices on InP (ENFET) with associated values of $g_m \sim 60$ ms/mm of gate width being attained. These results should be compared with what is achievable in Si,[46] where values of $\mu_{\text{FE}} \sim 850$ cm^2/V-s are generally considered to be the maximum obtainable in inversion n-channels.*

In comparison, depletion devices based on a full ion-implant technology have been reported using a variety of insulating layers, including SiO$_2$,[60,162] Al$_2$O$_3$[160] and organic polymers[163] with values of $\mu_{\text{eff}} \sim 2200$ cm^2/V-s and g_m as large as 120 ms/mm for gate lengths ~2 μm. This should be compared with 1-μm-gate-length GaAs MESFET devices, where g_m values of 100 ms/mm are considered typical. The p-channel devices based upon inversion of an n-type InP substrate wafer oriented (100) and doped with $n \sim 1 \times 10^{16}$ cm^{-3} have also been reported by Okamura and Kobayashi.[164] These devices, which employed AuZn/Au sintered source and drain contacts and an HCl surface etch prior to the deposition of the CVD Al$_2$O$_3$, exhibited field-effect hole mobilities ~ 16 cm^2/V-s and g_m values of 0.64 ms. These values would seem to be too low to encourage

---

*Recent results on InP have in fact suggested that, on this compound semiconductor at least, values of surface mobility close to bulk values may be possible. See: M. J. Taylor, D. L. Lile, and A. K. Nedoluha, *J. Vac. Sci. Technol.* **B2**, 522 (1984) and K. P. Pande and D. Gutierrez, *Appl. Phys. Lett.* **46**, 416 (1985).

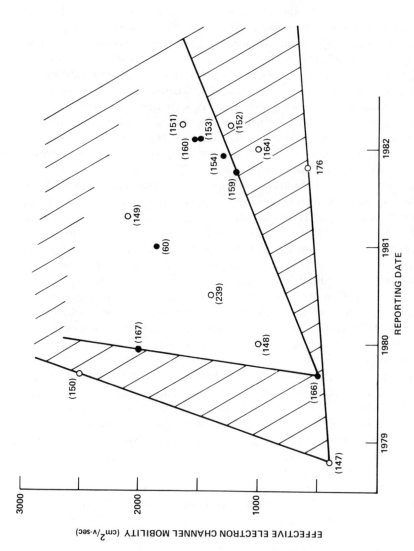

Figure 19. Best mobility values for electron enhancement channel conduction on p-type (○) and semi-insulating (●) InP plotted vs. reporting date in the open literature. Stippled area encompasses spread of data on SI material.

any immediate rationale for a CMOS approach on this semiconductor; however, they do substantiate the low surface state claims for this interface by evidencing directly the possibility of moving the surface Fermi level across the entire band gap into the VB.

Despite the apparently good performance of the ENFETs described above they should neither be expected to, nor in-practice do they, show any microwave gain because of the signal shorting effects of the parasitic capacitances associated with the gate pads on the high carrier concentration p-type substrate. Fortunately, it has proven possible on InP to duplicate the ENFET performance on devices built on Fe-doped semi-insulating material using a variety of dielectrics[165] somewhat analogous to the use of silicon on sapphire.[34,166,167] Because semi-insulating material is very slightly n-type as typically grown, it is more accurate to refer to the operation of these structures as being due to the gate-induced induction of an accumulation layer between the n^{++} source and drain contacts on the semi-insulating substrate surface rather than of an inversion layer as is the case on the p-type material.[34] Other than for the conductivity of the substrate, however, these accumulation-mode devices are identical in construction to those on the p-type material. In fact, some of the data in Fig. 19 were obtained on such semi-insulating substrate structures. Moreover, low-frequency transconductance as well as channel current are similar on both devices.[167] Some differences in performance, however, are to be expected theoretically and in fact are observed in practice: Sze[168] has shown that under a variety of simplifying assumptions, including the neglect of all surface trapping and the assumption of constant mobility, the channel current I_D in an enhancement MISFET may be written

$$I_D = \frac{w}{l} \mu_{FE} C_{ox} \left[\left(V_G - 2\psi_B - \frac{V_D}{2} \right) V_D - \tfrac{2}{3} K 2^{1/2} [(V_D + 2\psi_B)^{3/2} - (2\psi_B)^{3/2}] \right]$$

with $K = (\varepsilon_s q N_A)^{1/2} C_{ox}^{-1}$, where C_{ox} is the dielectric gate capacitance/unit area, V_g and V_D are the gate and drain bias, respectively, ε_s is the dielectric constant of the semiconductor, N_A is the acceptor concentration, w and l are the channel width and length, respectively, and

$$\psi_B = \frac{kT}{q} \ln \left(\frac{N_A}{n_i} \right)$$

The above equation is plotted in Fig. 20 for typical device parameters and for the case of p-type and high-resistivity substrate material. It can be seen that use of p-type material results in an increased threshold voltage, V_{th}, and reduced saturation voltage, V_{sat}, due to the requirement

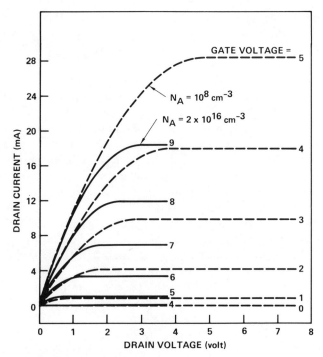

Figure 20. Calculated output characteristics for enhancement-mode FETs fabricated on p-type (solid) and semi-insulating (dashed) substrates.

for the gate field to initially deplete the p-material prior to the induction of the n-channel. Although the differences in V_{th} are somewhat masked in practice due to the variable effects of trapped dielectric charge, the differences in V_{sat} are clearly seen as is exemplified in Fig. 21, where typical output characteristics for p-based and semi-insulating-material-based devices of comparable geometry are shown.

Both the enhancement FET on semi-insulating material, as well as the depletion-mode devices,[118,169] have shown microwave power gain as is illustrated in Fig. 22. Early results[166] indicated a more rapid than expected falloff in gain for the normally off device, believed due to the effect of parasitic gate-to-source and drain capacitances resulting from electrode overlap. More recent results of Ohata et al.,[159] however, do not show this effect, as is evident in the figure, which were obtained on devices with 1.5-μm gate length and 280-μm width using pyrolytic SiO_2 for the dielectric and vapor-phase epitaxial growth of n^+ InP for the channel contacts. At 4 GHz, these devices exhibited a 7.2-dB maximum power gain and a 4.2 dB minimum noise figure with 4.0 dB associated gain. Their current-gain cutoff frequency of 15 GHz was also superior

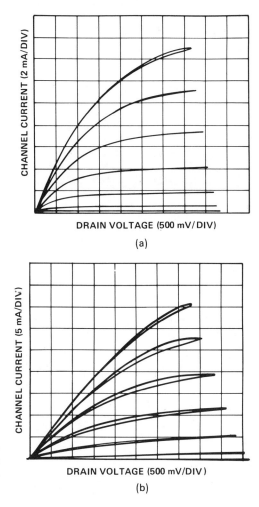

Figure 21. Output characteristics measured on representative transistors on (a) p-type and (b) semi-insulating substrates.

to what was achieved on comparable Si MOS and GaAs MES devices,[170] although it was not as high as the 15-dB gain at 8 GHz reported by Morkoc et al.[36] on 1.0-μm-gate-length depletion FETs on InP made using thin SiO_2 layers for the gate dielectric.

These results suffice to establish the potential of these devices for high-speed/low-power direct-coupled logic applications and have resulted in a number of attempts to implement and initiate an MIS-based IC technology in InP. The first such circuit, reported by Messick in 1980,[171] consisted of a 4-μm-gate-length enhancement drive and depletion load inverter. This circuit, operating at 12.4 V exhibited a delay time of 350 ps with an associated average power dissipation typically ~25 mW. Consider-

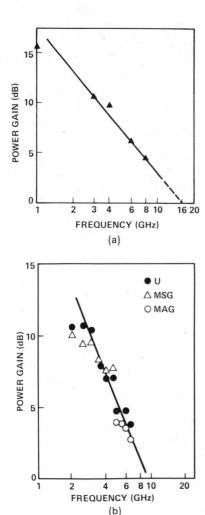

Figure 22. Frequency dependence of the microwave gain of MISFETs on InP. (a) Depletion device with 1.8-μm gate. (b) Enhancement device with 1.5-μm gate. [Fig. 22(b)] was taken from Ref. 159 by permission.

ing the far-from-optimum design of this circuit, these results are not inconsistent with the data in Fig. 22 and predict low-power, high-speed operation for devices of reduced geometry.

Ring oscillator data have been reported by Kinell[172] on a 15-stage circuit giving a delay time of 170 ps. This result, for the 2-μm design rules chosen, is consistent with the 4-μm device performance reported by Messick. More recent results by Clark and Jullens[173] have extended these results to delay times as low as 120 ps using 1.5-μm-design-rule 15-stage circuits.

Despite these successes InP MISFETs, as well as those on other III–V's, still suffer from a number of deficiencies: Lack of uniformity and reproducibility of device data has been a continuing problem, as exemplified by seemingly random differences in g_m and I_{DSS} between purportedly identical devices across a wafer as well as between wafers. This has generally been believed to be due to uncontrolled variations in both dielectric and interface properties as well as to nonuniformity in the starting bulk InP which is invariably oriented (100) for device fabrication to give more easily handled square cleaved chips. These (100)-oriented slices have, however, often been cut from (111)-pulled boules which means that the wafer is sampling along a significant length of the ingot and thus is likely to exhibit nonuniformities due to the segregation of intentional and unintentionally introduced impurities during the growth process. (100)-pulled InP is now becoming available and developments in surface preparation and dielectric deposition procedures are resulting in significant improvements in this area. It seems that with advances in process control this problem will be alleviated. Less encouraging perhaps

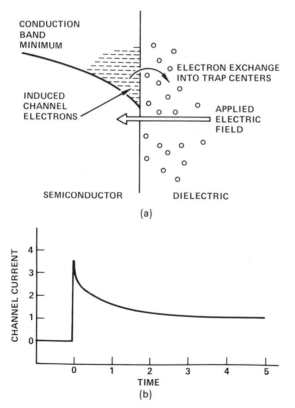

Figure 23. (a) Schematic band diagram and (b) typical time-dependent channel current response of a MISFET following a step change in input.

has been the progress made in understanding and overcoming drift effects which now remain as the dominant limitation in the performance of MIS devices, not only on InP but on all the III–V's.[174] This phenomenon, which is believed to result from injection of electrons from the InP channel into localized interface states either in the near-surface region of the semiconductor or the dielectric, most noticeably manifests itself in the form of hysteresis in C–V measurements and in looping and drift of FET characteristics. The latter is exemplified by Fig. 23, where a schematic band-structure diagram and the manifestation of this effect in a single-stage common-source InP MISFET amplifier is shown following the application of a step voltage to the input gate. This effect, which is particularly troublesome in analog circuit applications, is far more pronounced in GaAs MIS structures[144] where, because of the much larger surface trap density, essentially all low-frequency input fields are terminated on trapped charge. Although far less problematic on InP the effect is nevertheless still sufficient to warrant attention and efforts in this direction are continuing to be pursued at a number of laboratories.

Fritzsche was the first to discuss this problem for InP.[148] His data, obtained on inversion-mode transistors with P-doped SiO_2 gate insulation to retard surface degradation, exhibited a power law dependence of the channel conductivity on time. Despite a decreasing magnitude of drift with reduced temperature, these results were interpreted in terms of tunnel injection of carriers into traps distributed throughout the native oxide surface layer which must inevitably underlie the deposited SiO_2. This conclusion is seemingly consistent with the earlier report of Fritzsche concerning the beneficial effect on transconductance of a reducing pre-etch with HCl which presumably would tend to remove, at least in part, any surface oxides.[161] More recently, Okamura and Kobayashi[175] have discussed at some length drift effects in their inversion-type FETs, where they propose as beneficial a severe gaseous HCl surface treatment prior to the deposition of their Al_2O_3 gate insulator. This vapor-etching technique (VET) consisted of etching the InP surface in a 50% HCl–50% H_2 gas stream at a temperature of 200°C for 10 min, resulting in the removal of approximately 1500 Å of semiconductor. Dielectric growth at temperatures $>330°C$[176] was then initiated in the same reaction chamber hopefully minimizing the possibility of any intervening reoxidation. Significant improvements in drift were reported by means of this process as can be seen in Fig. 24 where representative data for unetched and VET inversion-mode MISFETs on p-type InP are shown. Although apparently beneficial for device stability it might be expected that such a severe surface treatment would be deterimental to channel mobility and in turn to g_m and I_{DSS}. Whether that was the case in the present work was, however, unclear from the data. It is also of interest that severe

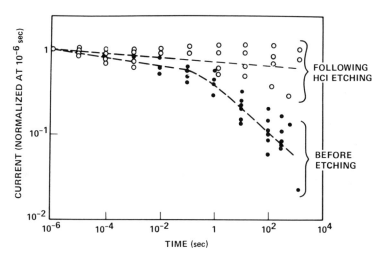

Figure 24. Normalized drain current *vs.* time characteristics for an enhancement-mode FET on InP following application of a +2-V change in input gate voltage (taken from Refs. 30 and 175 by permission).

drift effects were observed even subsequent to the HCl pre-etch if the Al_2O_3 was deposited at temperatures below 330°C. Although perhaps incidental it is not without interest to note that this is also the temperature threshold reported by Wager and Wilmsen[177] for the onset of significant thermal oxidation for InP.

Okamura and Kobayashi[30] have analyzed their results in terms of a two-layer dielectric model whereby a single trapping level TR-1 of density $N_{tr}^{(1)}$ resides in the deposited dielectric, Al_2O_3 in their case, and a second level TR-2 of density, $N_{tr}^{(2)}$ is located higher in energy in any intervening native oxide. A band diagram for their model and the resulting current–time behavior calculated as a function of temperature are shown in Fig. 25.

The expression they derive for the current at time t on which Fig. 25 is based is

$$\frac{I_D(t)}{I_D(t_0)} \cong 1 - \frac{1}{N_0 a}\left[N_{tr}^{(1)} + N_{tr}^{(2)} \exp\left(\frac{-2E_s^{(2)}}{kT}\right)\right]\ln\left(\frac{t}{t_0}\right)$$

where N_0 is the area density of electrons in the inversion layer and a, with a value of 3×10^7 cm^{-1} for Al_2O_3, is a tunneling coefficient.

The model presupposed that tunneling is the primary transport mechanism and thus that charge exchange with TR-1, lying below E_F, is relatively temperature insensitive. It is this trap that contributes the long-time-constant component to the drift in both devices in Fig. 24 and remains in effect at low temperature in Fig. 25. TR-2 on the other hand

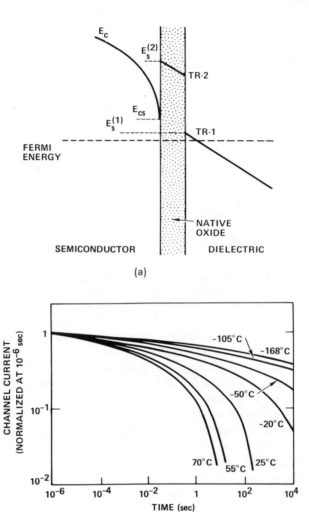

Figure 25. (a) Electron trap model for the InP–dielectric interface assuming a single monoenergetic trap in the native oxide interlayer and a single level in the deposited dielectric. (b) Calculated current–time and temperature characteristics resulting from this model (taken from Ref. 30 by permission).

is above E_F and relies heavily, for a ready supply of carriers, on thermal excitation of electrons from the CB. Lowering the temperature, as well as reducing $N_{tr}^{(2)}$ by, for example, removal of the native oxide, will thus lead to a reduced contribution of the second term in the equation. In contrast $N_{tr}^{(1)}$ is a function of the properties of the deposited dielectric and should to first order be unaffected by surface treatment prior to growth. These conclusions appear to be substantiated by the data shown in Fig. 24 as well as by the results reported by these same authors on a comparison of InP and Si MISFETs prepared with the same Al_2O_3 dielectric.[30] Fitting their data to the model they arrive at a trap density of 1.5×10^{17} cm^{-3} for the native oxide on InP and 8×10^{15} cm^{-3} for the deposited alumina.

Despite the apparent correlation of this model with experiment, however, some severe discrepancies and questions remain: Why, for example, it should be necessary to remove as much as 1500 Å of the semiconductor surface if all that is required for stable performance is the removal of the native oxide, is unclear. Although results on one semiconductor should not be taken as necessarily normative for another it is not without significance that optimized surface properties are believed to be obtained on InSb[178] and HgCdTe if attempts are made to preserve, rather than remove, the native oxide by the use of minimally disruptive dielectric deposition techniques. Even more difficult to reconcile are results subsequently published by Okamura and Kobayashi[179] in which they report essentially identical improvements in drift to those obtained by HCl etching by thermally preoxidizing the p-type InP surface prior to deposition of the Al_2O_3 gate dielectric!

These results,[30,175,179] together with those of Fritzsche[148] and Lile and Taylor,[37] are summarized in Fig. 26. Lile and Taylor have discussed the results of their drift studies on enhancement-mode FETs on InP using a variety of surface preparation treatments, including preoxidation and HCl etching, and have concluded that although some of these treatments are certainly beneficial in improving electron surface transport as measured by μ_{FE}, g_m, and I_{DSS} values, they are not necessary, or in some cases even beneficial, in improving the stability of their devices. This case is exemplified by the highly stable transistor performance shown in Fig. 26, which was obtained on a FET prepared with a plasma CVD SiO_2 layer deposited without benefit of any surface pretreatment of the InP other than detergent and solvent washing.

These results would more nearly seem to support the model of Sawada and Hasegawa,[20] wherein a disordered or contaminated metamorphic layer located within the near-interfacial region of the semiconductor–dielectric interface acts to trap carriers and that minimum device drift will accompany any technique which attempts to minimize surface disorder, including the use of etchants to remove surface damage and low-temperature processing including dielectric growth to minimize loss of volatile components. The preoxidation results reported previously would not be inconsistent with such a model wherein the effect of the anneal would be perhaps to oxidize to completion any excess P and/or In in the native oxide.

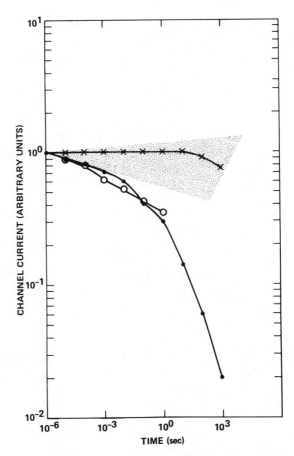

Figure 26. Normalized plots of drain current *vs.* time following the application of an input voltage step to an InP MISFET. The stippled area encompasses the spread of data following the etching shown in Fig. 24: (●) data from Ref. 175 with no etching, (○) data from Ref. 148, and (×) data from Ref. 37.

Most certainly the chemistry[180] and role of Cl and HCl in improving the properties of the InP/dielectric interface as well as that of other III–V's[181] at present remains unclear. Ho and Sugano,[182] for example, have noted the beneficial effects of a Cl anneal in reducing the interface state density on plasma-anodized Si to a value as low as 10^{10} states/cm^2-eV and Gaind and Kasprzak[183,184] have reported that HCl doping leads to reduced trap densities in CVD layers of SiO_2 on Si, as compared to the thermal process, due to the effect of HCl in preventing the incorporation of free Si in the oxide. Such a mechanism might be extrapolated to the III–V's wherein the HCl "ties up" any excess anion or cation species incorporated in the near-interface region of the dielectric due to dissociation of the semiconductor during oxide growth. At present, however, such reasoning is no more than conjecture.

In addition to use for FETs, InP has also been used in the fabrication of surface channel MIS-gated CCDs. Figure 27 shows a photograph and schematic cross section of an eight-bit, four-phase structure which has been used to demonstrate high-efficiency charge transfer in the surface inversion layer in this semiconductor.[185] The device employs a two-level overlapping gate design using CVD plasma-enhanced pyrolysis of SiO_2 for insulation. The input and output diodes and channel stop were prepared by selective ion implantation into the (100) face of bulk single-crystal p-type InP. An MIS field electrode has also been used to define the channel boundaries in some cases. The devices were exercised in a conventional manner with four-phase clocking pulses being applied to

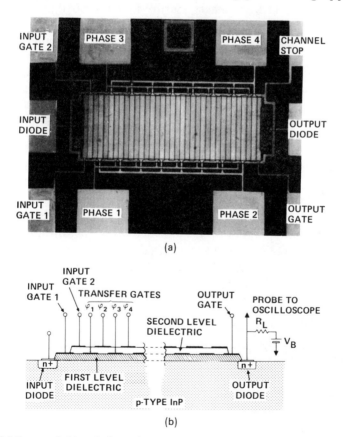

Figure 27. Micrograph (a) and schematic cross section (b) of an eight-bit, four-phase surface channel CCD on InP fabricated using an insulated gate design. For clarity the channel stop has been omitted in Fig. 27(b).

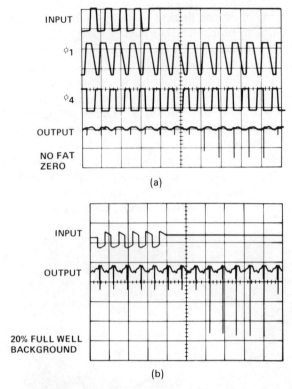

Figure 28. Typical response of a surface channel CCD on InP to a series of five input pulses. The upper diagram (a) shows the type of performance obtained with no background charge. The lower diagram (b) is for a 20% full well fat zero.

the transfer gates and the input signal being applied to either the input diode or input gate. Figure 28 shows the type of response obtained, with and without the application of any "fat zero" background, to a series of five input pulses. From the charge losses from the first response pulse, a transfer efficiency ~0.99–0.995 was obtained with no fat zero, whereas the response rises to ~0.998 with a 20% full well background charge. No high-speed data have yet appeared on these devices although a large 40-μm-gate-length demonstration CCD on InP did operate to 50 MHz[186] indicating the potential of these structures for very-high-speed performance.* In contrast with the FET results these devices are apparently very stable, presumably because of the relatively low current densities at which they operate. It is known, for example, that MIS-gated FETs on both GaAs and InP exhibit increasing levels of hysteresis as their drive levels

---

*Recent results have shown operation of these devices to ~16 Hz. See: D. L. Lile and D. A. Collins, *IEEE Electron Device Lett.* EDL-5, 335 (1984).

are increased. At large values of drain voltage especially, such devices can suffer both permanent as well as reversible changes in their output characteristics due perhaps to field-enhanced charge injection across the surface into the dielectric as has been reported for MOS structures on Si.

4.3. Other Binary Compounds

GaAs and InP are by far the most extensively studied and developed of the III-V binary compounds both with regard to materials development as well as device fabrication. Despite this, other binaries are of considerable interest for a number of applications.

The narrower-gap semiconductors InSb and InAs, for example, have attracted attention both because of their small carrier electron effective masses, and hence very high electron mobilities, as well as because of their narrow optical gaps—a parameter of interest for detector applications at long wavelength. Photodetectors operating at 77 K have, in fact, been available for many years with responses to 5.5 and 3 μm, respectively, using these two materials with commercial MIS detectors being available on InSb. Both of these semiconductors appear to possess surface state distributions when in contact with suitably prepared dielectrics which at least in magnitude resemble those on InP, namely N_{SS} values $\leqslant 10^{12}$ cm^{-2} eV^{-1}.[187,188] Because of their smaller values of E_g, however, which lead to higher intrinsic carrier densities and faster surface trap response times, it is often found necessary in device applications to operate these materials at reduced temperature which for convenience is usually chosen to be 77 K. At this temperature it has been reported[188] that N_{SS} on InAs reduces to $\sim 2.8 \times 10^{11}$ cm^{-2} eV^{-1}, whereas InSb in the presence of a properly prepared oxide can support a surface state distribution as low as 10^{10} cm^{-2} eV^{-1}.[178] From the extant data, however, it is unclear that these reportedly reduced values at lower temperatures are not simply due to a decreased surface state response time and thus an inability of the traps to respond at the frequency of characterization. A physical reduction in the actual number of surface states present with decreasing temperature would appear to be difficult to reconcile in light of our present understanding of the origin of surface traps.

Like InP, it has, in general, been found that n-type samples of both InSb and InAs are normally accumulated, whereas p-type samples are inverted.[23,189] In either material the application of a surface field at 77 K allows the Fermi level at the surface to be moved across the entire band gap.[187,189] These observations imply that n-channel inversion as well as depletion transistors on both these materials should be feasible, and in fact inversion-layer transport on both InSb and InAs was observed as early as 1965.[189,190]

Because of their small band gaps, Schottky-gate and p–n junction structures on InSb and InAs exhibit considerable reverse leakage and thus MIS-type devices have historically been implicitly assumed to be the approach of first choice. Frantz[191] was the first to report an FET device on this class of materials. This transistor was made using an approximately 200-Å thick polycrystalline layer of InSb evaporated onto a glass substrate held at room temperature. Evaporated SiO was used as the gate dielectric with Al and Au or Al being employed for the gate and channel contacts, respectively. Hall measurements gave effective electron mobilities of 560 cm^2/V-s and carrier concentrations $\sim 3.7 \times 10^{17}$ cm^{-3} in these layers. Despite the fact that these transport parameters are far less than what might be expected in bulk material, the devices themselves exhibited appreciable field-effect modulation with values of $g_m \sim 1$ ms/mm of gate width being obtained for channel lengths $\sim 25\ \mu$m. This work was soon followed by the first report by Brody and Kunig[192] of a depletion FET in InAs. Their devices were fabricated on layers of semiconductor prepared by the three-temperature method, which had Hall mobilities μ_H at 300 K as high as 8000 cm^2/V-s and carrier densities in the range 10^{17}–2×10^{18} cm^{-3}. Coplanar MISFET structures were fabricated using reactively evaporated SiO as the gate insulator with a source–drain length of 100 μm and a width of 1400 μm. Transconductance values as high as 7 ms/mm of gate width were observed in these devices with effective mobilities of 1800 cm^2/V-s being obtained with the highest mobility layers. In fact, a remarkably good correlation was reported between the values of μ_{FE} and μ_H obtained on these films prior to device fabrication.

Since these early reports a number of other groups have published results on MIS transistors on these binary compounds. Spinulescu-Carnaru,[193] Lile and Anderson,[194] and Luo and Epstein[195] have all reported on depletion-type structures on InSb and Kunig[196] presented further results on their InAs FETs in 1968. In all cases these devices were thin-film structures prepared on evaporated layers of the semiconductor using SiO$_x$ for the gate insulation, except in the work of Lile and Anderson where anodically prepared Al$_2$O$_3$ was used. The best results indicated values of μ_{FE} at 300 K as high as 3000 cm^2/V-s in InAs.[196] In general, all devices of this type reported to date have exhibited some instabilities presumably associated with either charge trapping in the dielectric or at the intergrain boundaries present in the polycrystalline starting layers used. Sewell and Anderson[174] in fact proposed a tunneling model very similar to that currently being used for InP to explain their observed drift effects in InSb MISFETs and Baudrand et al.[197] have recently presented an analysis of the effects of the polycrystalline layer on device performance. The effect of incomplete surface oxidation on frequency dispersion has been discussed by Heime and Pagnia.[198]

The transistors described above all rely for their operation on the modulation of the majority-carrier density in a thin layer of n-type semiconductor deposited on an insulating substrate. Little FET device data have yet been reported for the binary III–V's where current control is by means of an inversion layer at the surface of the semiconductor. As has been previously stated, electron inversion layers have been reported from capacitance and transport measurements on both InAs and InSb with increasing conduction during depletion operation of MISFETs on InSb having also been attributed to inversion-layer formation at the drain end of the channel.[194] Thom *et al.*[86] have observed good p-channel enhancement-mode FET performance on InSb with inversion-layer mobility values ~ 500 cm^2/V-s being obtained using an Al$_2$O$_3$ gate insulator at 77 K; however, in no case to our knowledge has a report appeared on the implementation of an n-channel inversion-channel FET device in this material. In contrast, inversion-mode InAs MISFETs have been studied by Terao *et al.*[23] using Al$_2$O$_3$ grown by the CVD decomposition of aluminum isopropoxide at 125°C. The deposition temperature of 320°C was chosen as a compromise between high-temperature growth which resulted in higher density, more stable layers, and the need to preserve the surface of the InAs against thermal decomposition. Using p-type substrates, of carrier density $\sim 1.5 \times 10^{17}$ cm^{-3}, 40-μm-gate-length FET were fabricated which exhibited effective electron inversion-layer mobilities as large as 5000 cm^2/V-s and g_m values ~ 20 ms/mm of gate width at 77 K.

Despite these successes these mobilities remain far below the 35,000 cm^2/V-s and 78,000 cm^2/V-s potentially feasible in InAs and InSb, respectively, at room temperature. Moreover, n-channel inversion-mode devices rely upon the availability of low reverse leakage source and drain contacts to block the flow of the majority carrier which at room temperature are difficult to obtain on these low-gap materials.[23] If these problems can be solved, by perhaps operation at reduced temperatures, or alternatively, if operation at low drive voltages (~ 1–2 V) is accepted then high-performance enhancement- and depletion-mode FET structures should be feasible. Data already exist to suggest that electron inversion-layer mobilities of at least 5000 cm^2/V-s in InSb at 4.2 K should be possible,[199] with values as high as 12,000 cm^2/V-s having been reported for InAs.[43]

Interest in FETs on the narrow-gap intermetallics has been primarily motivated by the possibility of high-speed operation due to the extremely large values of electron mobility possible in these compounds. Most certainly, high-speed CCDs might also in principal be based on these materials but to date most interest in this device has been driven by imaging applications based on the extended spectral response of these compounds. Phototransistors might equally be conceived for these materials but for imaging systems applications the CCD offers some distinct advantages.

A number of groups have made extensive contributions in this area with operational CCD structures having recently been demonstrated working at high efficiency on InSb at 77 K.[200] The earliest references to an InSb CCD appeared in 1975 when Thom et al.[86] reported on a p-channel four-phase, four-bit overlapping gate device fabricated on a Te-doped n-type single-crystal material of carrier concentration 0.4–1 × 10^{15} cm^{-3}. The dielectrics used were electron-beam-deposited Al_2O_3 and thermally evaporated SiO, Cd diffusion was used to define input and output p–n junctions and because of an inability to ion implant or diffuse n-type regions an MIS-field-electrode channel stop was employed. This 50-μm-gate-length surface channel device demonstrated charge transfer action at 77 K in a variety of drive modes with the best transfer efficiency of 0.92 per transfer being obtained at 5 kHz with a 50% full well fat zero input. This transfer efficiency, while not as large as might be desired, is consistent with the geometry of these devices and the 6×10^{11} surface states/cm^2-eV deduced from C–V measurements on MIS diodes which were prepared adjacent the CCDs. More recent results have indicated that an improvement in surface properties with a concomitant increase in CCD performance can be achieved on this semiconductor by going to alternate dielectric preparation techniques. Langen, in particular, has reported a low-temperature chemical vapor deposition of SiO_2 which supposedly preserves the interfacial native oxide and results in a value of $N_{SS} \sim 10^{10}$ cm^{-2} eV^{-1} at 77 K as deduced from C–V and G/ω data.[178]

Based on such an improved interface–dielectric combination and a refined fabrication technology employing Be ion implantation for input/output diodes, Thom et al.[84] reported in 1978 a 20-bit, 12.5-μm-gate-length surface channel CCD operating in the surface channel mode with a transfer efficiency value of 0.995. This value, lower than the 0.9995 to be expected for these low N_{SS} surfaces, was in fact believed not to be limited by surface trapping but rather to result from lateral surface potential variations arising from oxide granularity. Such a phenomenon would produce potential wells resulting in "pockets" of trapped charge leading to a degradation of operation. Whatever the mechanism involved, the high efficiency reported for this device encourages the hope that such devices can indeed be implemented as readout elements in imagers. In fact, the CCD under discussion was reported as part of a monolithic optical array employing 20 InSb MIS diodes for detection.[200] Operating in such an array imaging mode, this circuit exhibited a detectivity D^* of 8×10^{12} cm Hz$^{1/2}$-W^{-1} clocking at a frequency of 500 Hz.

Gallium antimonide[201] and gallium phosphide[202] have received relatively scant attention when compared to the other binary compounds despite the fact that GaP, in particular, because of its large band gap (~2.3 eV) might be of interest for high-temperature applications. What information is available, however, tends to suggest that the surfaces of these compounds are similar to GaAs, being possessed of large densities of interface traps. Weimann,[203] for example, has reported $C-V$ data on GaAs, GaSb, and GaP using anodic oxides prepared by wet chemical processing, which show in all cases large values of trapping and hysteresis. To this author's knowledge, no data exist on any MIS-device application of these binary compounds, and unless improvements in interface preparation yield a reduction in the trap densities of their surfaces it seems unlikely that these materials will prove particularly significant in the near-term development of MIS systems.

4.4. Ternary and Quaternary Alloys

The ternary and quaternary alloys when considered for device applications are possessed of two attractive characteristics. First, it is possible to engineer their lattice parameters by a judicious choice of the relative concentrations of the constituent elements to effect a lattice match to a suitable substrate matrix,[204] and second, in many cases, they possess large values of electron mobility and peak velocity which suggest their use in high-speed devices.[205] The quaternary alloy $In_{1-x}Ga_xAs_yP_{1-y}$ is noteworthy in that over a range of x and y it can be grown epitaxially on both InP (for $y = 2.2x$) and GaAs (for $y = 2.2x - 1.2$).[204] In particular, the InP lattice-matched ternary $In_{0.53}Ga_{0.47}As$ end-point composition ($x = 0.47$, $y = 1$) has attracted much attention because, as is shown in Fig. 29, it has both the highest mobility[206] as well as a spectral response which extends to 1.7 μm, which is compatible with minimum-attenuation low-loss optical fibers. The possibility thus exists of combining long-wavelength optical detection, in devices fabricated on epitaxial layers of the ternary, with on-chip data reduction and signal processing at high speed using MIS circuitry built in the underlying InP. Data transmission off chip could even be envisaged using optical emitters where the typical problems of limited bandwidth encountered in conventional electrical readout are avoided. Such monolithic electro-optic circuitry is of considerable interest and some highly promising results have already been reported for an integrated GaInAs laser and InP MISFET driver.[207] In addition to a higher low-field mobility with values of 12,000 cm^2/V-s being obtained at present at room temperature,[208] GaInAs also exhibits larger high-field electron velocities than either GaAs or InP[209] as well as an experimentally

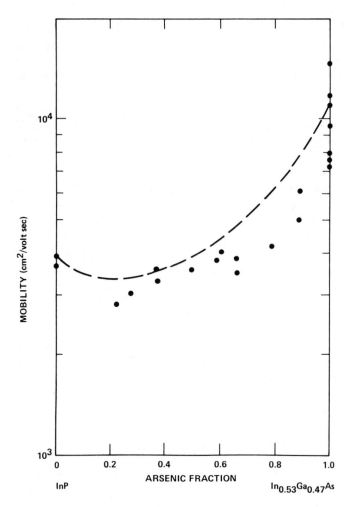

Figure 29. Variation of measured mobility with composition for $In_{1-x}Ga_xAs_yP_{1-y}$ grown lattice matched to InP (taken from Ref. 206 by permission).

demonstrated advantage in FET applications of a factor of 2 over GaAs in saturation velocity with $v_s = 2.95 \times 10^7$ cm/s.[210]

Application of these compounds, however, is hampered by the low values of barrier height ~0.2 eV obtained in metal–semiconductor junctions from which follows an inherent inability to fabricate good MESFET devices.[211,212] Alternative gating mechanisms are thus considered attractive, including the possibility of junction[206] and MIS structures.

The first report on the possibility of employing MIS techniques to this class of materials appeared in 1980 when Shinoda et al.[213] presented results in agreement with the earlier projections of Wieder[214] on an n-channel inversion-mode MISFET on $In_{0.87}Ga_{0.13}As_{0.29}P_{0.71}$ grown epitaxially on InP which exhibited an apparent channel mobility ~ 2600 cm^2/V-s. The device was fabricated using ion-implanted Si to form n^+ contact regions in the quaternary active layer and approximately 1000 Å of Al_2O_3 for the gate dielectric grown by means of the pyrolytic decomposition of aluminum triisopropoxide at 350°C.

These results were quickly followed by the demonstration of inversion-layer transport in $In_{0.53}Ga_{0.47}As$ by Wieder et al.,[64] who reported field-effect electron mobilities of 30 cm^2/V-s and g_m values ~ 0.3 ms/mm of gate width in MISFET structures prepared with SiO_2 gate dielectrics on p-type heteroepitaxial layers grown on semi-insulating InP. Improved performance values were subsequently reported by Liao et al.[215,216] in n-channel InGaAs devices using Si_3N_4 for the gate isolation, where channel mobilities as high as 325 cm^2/V-s and g_m values ~ 4 ms/mm were observed. Using pulsed gate voltage, Liao et al.[217] subsequently observed an order-of-magnitude improvement in g_m with values of 33 ms/mm being reported. More recent devices fabricated with plasma anodic dielectrics[218] and with SiO_2 deposited by thermal CVD, sputtering, and electron-beam evaporation[219] have been reported, where values of g_m as high as 40 ms/mm for 2.0-μm gate lengths and channel mobilities of 1000 cm^2/V-s have been observed.[219,220] A related pseudo-MESFET device made with a thin interfacial dielectric layer of plasma CVD Si_3N_4 between the 1.2-μm-length gate metal and the ternary-alloy channel gave dc transconductances as high as 130 ms/mm,[221] implying saturated velocities for the channel electrons in this semiconductor of $2.0 \pm 0.5 \times 10^7$ cm/s—a value 60–70% higher than that obtained in GaAs and in close agreement with the value reported by Bandy et al.[210] Morkoc et al.[36] have also reported such thin MISFET structures on InP and GaInAsP using ~ 50 Å of SiO_2 under the gate metal. Their results indicated an improvement in device rf characteristics with decreasing dielectric thickness with maximum values of $g_m \sim 100$ ms/mm being obtained in these 1-μm-gate-length structures on InP.

It is significant that both the reported observation of p-channel conduction on n-type material,[216,222] with an effective mobility as high as one-half that of the bulk hole mobility, as well as the inversion-type characteristics of $C-V$ data on p- and n-type epilayers,[223] suggest that the surface of this compound can be modulated across the entire band-gap region.

Most recently, Shinoda and Kobayashi[26,224] have reported a series of measurements on inversion-mode MISFETs on lattice-matched ($y = 2.2x$) LPE layers of p-type $In_{1-x}Ga_xAs_yP_{1-y}$ grown with various compositions in the range $0 \le y \le 0.55$ on (100)-oriented InP. These devices, of a similar geometry to that reported previously,[213] had 10-μm channel lengths, an Al_2O_3 gate dielectric, and were prepared concurrently with MIS diodes for $C-V$ evaluation of the interface. Pulsed dc transconductance measurements on FETs prepared on $0.7-5 \times 10^{17}$ cm^{-3} p-doped material indicated a monotonically increasing value of field-effect mobility with increasing Ga and As mole fraction from a minimum of ~ 500 cm^2/V-s for InP to a maximum of 1000 cm^2/V-s for an InGaAsP layer of composition $x = 0.24$, $y = 0.55$. The largest value of μ_{FE} observed was 2300 cm^2/V-s obtained on a more lightly doped quaternary p-layer with $N_A = 3 \times 10^{15}$ cm^{-3}. These data, together with the related surface state distributions deduced from a Terman analysis of 1-MHz $C-V$ results, are shown in Fig. 30. It was concluded that the differences in μ_{FE} observed as a function of layer composition result primarily from variations in N_{SS} with increasing mobilities accompanying decreasing surface state concentrations for increasing x and y. Smaller temperature dependences of μ_{FE} and larger variations in threshold voltage with temperature in InP compared to the quaternary were similarly ascribed to the effect of larger surface trap densities at least over the surface potential range of interest for these devices near the CB edge. Extrapolation of this data suggests that N_{SS} should be minimized and the mobility maximized for the ternary-alloy composition InGaAs. In contrast Fritzsche et al.[225] have reported results on $In_{0.53}Ga_{0.47}As$ obtained using 1-MHz $C-V$ and $G-\omega$ measurements and DLTS between 20 and 300 K which suggests that the surface of this material is somewhat similar to InP.

Depletion MISFETs as well as n-channel inversion devices have been fabricated and evaluated by Gardner et al.[226] using VPE growth for the GaInAs and CVD SiO_2 for the gate dielectric. Enhancement device results were similar to those already reported by other groups with values of $\mu_{eff} \sim 300$ cm^2/V-s being typical. The depletion structures, however, had mobilities as high as 5200 cm^2/V-s, which is the largest figure reported to date for any room-temperature FET. Microwave results on these 3-μm-gate-length structures gave power gains of 4 dB at 6 GHz with power outputs of 57 mW at power-added efficiencies of 19.7%. These values should be compared with those reported by Morkoc et al.[36] on thin-oxide MISFETs on GaInAsP, where values of power gain of 9 dB at 8 GHz were obtained using SiO_2 for the gate dielectric. These first depletion FET high-speed results are particularly encouraging and, when taken in combination with the enhancement FET data reported by Shinoda and Kobayashi, would seem to give cause for optimism in judging the potential feasibility of applying these alloys to MIS-device applications.

In addition to GaInAs, a number of other ternaries have attracted some attention, including $GaAs_xP_{1-x}$[227] and $Ga_xAl_{1-x}As$.[228] In particular, $Ga_xAl_{1-x}As$ is of considerable interest and importance. Grown

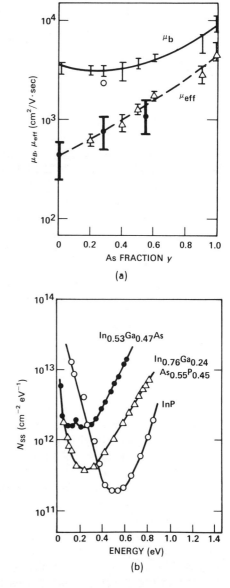

Figure 30. (a) Compositional dependence of $In_{1-x}Ga_xAs_yP_{1-y}$ MISFET electrical properties with bulk mobility μ_B and effective mobility μ_{eff} being plotted as a function of As mole fraction. (b) Energy dependence of the interface state density for three compositions of the $In_{1-x}Ga_xAs_yP_{1-y}$ system. (Taken from Refs. 26 and 224 by permission.)

epitaxially on GaAs over a range of lattice-matching values of x, this material has a band gap of ~ 2.0 eV and a low-field μ of ~ 3000 cm^2/V-s. Although of shorter-wavelength response and lower mobility than the quaternary, this semiconductor is of much interest, and is perhaps the most widely studied of all the multicomponent III–V's because of its wide applicability to optical emitters and detectors and cascade solar cells, as well as its more recently identified potential in superlattice and high-electron-mobility transistor structures and as the detector material for monolithic integration with a GaAs CCD for imaging. All of these applications and devices, however, are only peripherally related to the dielectric interface and thus little work to date has addressed this question.

5. Epilogue

In the preceding sections we have attempted to give an overview of the status and applications of dielectrics to device development on the III–V's and to identify the problems and successes which have accompanied this rapidly growing field. Of necessity, some areas have received more emphasis than others, and although this inevitably will reflect to some extent the bias of the author it primarily results from the uneven attention given, and consequently data generated on, the many materials comprising this family of semiconductors. GaAs, InP, and InSb, for example, have received by far the lion's share of the work and thus must occupy a dominant position in a review of this type. From what has been said it is clear that many advantages would accrue from the availability of compatible dielectrics for use with these semiconductors. This observation is not new of course, having been eloquently stated in one of the very earliest papers on MOS devices on GaAs.[229] What is new is that now, in contrast with the mid-1960s, it is possible to justify by means of the extant data, enthusiasm for success in this endeavor. InP is a case in point where many of the results generated during the last few years give hope that a working MIS technology on this III–V at least is potentially feasible. This is not to gloss over or ignore the remaining problems; most certainly reproducibility and uniformity of device performance must be improved as must the level of device stability typically achieved. This latter problem, in particular, is one that has plagued all the III–V's and will need improvement before practical systems based on these semiconductors become a reality.

Central to this entire subject of course is the development of surface preparation and dielectric growth procedures fitted to the restraints imposed by these compounds. Low-temperature processing, for example, has led to the development of a variety of new procedures, including two unique dielectric growth methods, namely photoassisted pyrolysis[230] and indirect plasma-enhanced CVD (IPEC),[231] which may well have considerable impact on the Si technology also, a fact perhaps surprising to those who see the symbiotic relationship between Si processing and all else as being a one-way street. Despite these successes and the increased understanding of the physics and chemistry of these interfaces, there still remain fundamental and critical aspects of MIS systems on the III–V's which remain obscure: What specific surface processing to employ prior to dielectric deposition is an example of an area that has eluded understanding despite its identification as a topic crucial to future success of these materials. For example, is an overpressure of the column V element really beneficial as some recent results would suggest?[232] Even which dielectric or dielectrics would be most appropriate is still unclear. Most certainly, pyrolytic SiO_2 and Al_2O_3 seem at present to be the materials of choice but whether eventually they will be superseded by alternatives, such as Ge_3N_4,[75,111,233,234] a thermal nitride,[235,236] or P_3N_5[151] and whether novel techniques such as MBE for dielectric growth,[237] for example, will be of importance remain unknown. It is worth repeating, however, that no matter which material is chosen, a method of deposition which minimizes the energy imparted to the semiconductor surface is to be desired. Sputtering, for example, should be employed only with extreme caution[238] as also should other high-energy or chemically reactive techniques, such as direct plasma-enhanced CVD which potentially could cause problems through dissociation of the surface.[158,239]

These statements are not meant to cast gloom on the promise presented by the III–V's but rather are stated to illustrate that, despite the remarkable successes recorded to date, particularly in light of the relatively small fiscal and manpower investments made so far, much remains to be done before these materials reach maturity in an MIS technology.

It is clear that in a review of this type much will be omitted both by oversight as well as by intent. In the latter category, we have consciously chosen to exclude such topics as the use of dielectrics for capping of samples during high-temperature processing as, for example, is encountered during post ion-implantation activation. We have also said nothing concerning the influence of crystal orientation on surface-related device effects. It is well known in the Si technology that the (111) surface supports in general a larger density of traps and thus for the most demanding applications as, for example, are encountered in low-light-level imaging CCDs, (100)-oriented material is to be preferred.[184] To date such effects on the III–V's would seem to be of second order and thus have been largely ignored. Exceptions exist of course as is exemplified in the paper by Fritzsche et al.,[240] where higher channel mobility was reported for devices fabricated on the (111)-oriented surface of InP as compared to the (100). Scatter in the extant data between various groups, however, makes it difficult to conclude much in this area and it seems likely that resolution of this issue must await improvements in other aspects of device technology.

Although it is not without attendant risks, it is always tempting to try and forecast what future developments may be in store. Most certainly the ever-present push to higher frequencies of operation is unlikely to abate and as devices shrink in size to the submicron range the notion of discrete devices interconnected into "integrated" circuits is likely to suffer some changes. As device dimensions shrink and concomitant packing densities increase, parasitic effects due to field fringing will tend to introduce unwanted device interactions which in the limit are likely to render conventional circuit designs inoperative. In consequence, for the very highest frequencies of operation, it is likely that distributed designs will evolve where circuit operation is defined by the positional control and direction of mobile charge regions by means of overlying metal patterns. Such circuits, somewhat analogous to conventional CCDs in their mode of operation, will likely of necessity be MIS in nature because of their need to overlap metalizations over large areas of circuit. Such truly integrated circuits, with no possible discrete component analogues are as yet but a gleam in the eye of those looking to ultrahigh-frequency signal processing, but nevertheless represent, over and above all that has already been said, another further practical inducement to pursue the development of MIS systems for these materials.

Although the final responsibility for the opinions and conclusions expressed in this chapter must rest with the author, it is clear that a review of this type must draw extensively on the thoughts and ideas as well as the data presented by many other people in the scientific literature as well as in personal conversation. To all of these "co-contributors," far too numerous to name individually, I acknowledge my indebtedness and express my sincere thanks. Particularly helpful and encouraging have been my colleagues at NOSC, including Art Clawson, Dave Collins, Larry Meiners, Louis Messick, Fred Nedoluha, Marylin Taylor, Harry Wieder, and Carl Zeisse.

Finally, I wish to record a sincere debt of gratitude to Mrs. Diana Griffin who labored with much skill and consummate patience in the typing and retyping of the many iterations of this manuscript.

References

1. J. E. Lilienfeld, Method and Apparatus for Controlling Electrical Currents, U.S. Patent No. 1,745,175 (filed 8 October 1926, issued 28 January 1930).
2. J. E. Lilienfeld, Device for Controlling Electric Current, U.S. Patent No. 1,900,018 (filed 28 March 1928, issued 7 March 1933).
3. C. A. Mead, Schottky barrier gate field-effect transistor, *Proc. IEEE 54*, 307–308 (1966).
4. S. R. Hofstein and F. P. Heiman, The silicon insulated-gate field-effect transistor, *Proc. IEEE 51*, 1190–1202 (1963).
5. J. Bardeen and W. H. Brattain, The transistor, a semi-conductor triode, *Phys. Rev. 74*, 230–231 (1948).
6. J. Bardeen and W. H. Brattain, Physical principles involved in transistor action, *Phys. Rev. 75*, 1208–1225 (1949).

7. W. Shockley, The theory of p-n junctions in semiconductors and p-n junction transistors, *BSTJ* **28**, 435–489 (1949).
8. M. M. Atalla, E. Tannenbaum, and E. J. Scheibner, Stabilization of Silicon Surfaces by Thermally Grown Oxides, *BSTL* **38**, 749–783 (1959).
9. J. T. Mendel, GaAs—A technological catch-22, *Microwave J. March 1981*, 24–32.
10. D. A. Jenny, The status of transistor research in compound semiconductors. *Proc. IRE* **46**, 959–968 (1958).
11. H. Welker, Ueber Neue Halbleitende Verbindungen, *Z. Naturforsch A* **7**, 744–749 (1952).
12. R. Zuleeg and K. Lehovec, High frequency and temperature characteristics of GaAs junction field-effect transistors in the hot electron range, *Inst. Phys. Conf. Ser.* **9**, 240–250 (1970).
13. W. Baechtold, K. Daetwyler, T. Forster, T. O. Mohr, W. Walter, and P. Wolf, Si and GaAs 0.5 μm—gate Schottky-barrier field-effect transistors, *Electron. Lett.* **9**, 232–234 (1973).
14. C. W. Wilmsen and S. Szpak, MOS processing for III–V compound semiconductors: Overview and bibliography, *Thin Solid Films* **46**, 17–45 (1977).
15. B. L. Sharma, Inorganic dielectric films for III–V compounds, *Solid-State Technol.* **1978**, 48–53; **1978**, 122–126.
16. For a review of surface effects see A. Many, Y. Goldstein and N. B. Grover, *Semiconductor Surfaces*, North-Holland, Amsterdam (1971).
17. P. E. Gregory, W. E. Spicer, S. Ciraci, and W. A. Harrison, Surface state band on GaAs (110) face, *Appl. Phys. Lett.* **25**, 511–514 (1974).
18. W. E. Spicer, I. Lindau, P. Pianetta, P. W. Chye, and C. M. Garner, Fundamental studies of III–V surfaces and the (III–V)-oxide interface, *Thin Solid Films* **56**, 1–18 (1979).
19. A. R. Clawson, W. Y. Lum, and G. E. McWilliams, Control of substrate degradation in InP LPE growth with PH_3 partial pressure, *J. Cryst. Growth* **46**, 300–303 (1979).
20. T. Sawada and H. Hasegawa, Interface state band between GaAs and its anodic native oxide, *Thin Solid Films* **56**, 183–200 (1979).
21. R. F. C. Farrow, The evaporation of InP under Knudsen (equilibrium) and Langmuir (free) evaporation conditions, *J. Phys. D.* **7**, 2436–2448 (1974).
22. H. Hasegawa and T. Sawada, Electronic properties of interface between GaAs and its anodic native oxide, Proceedings of the Seventh International Vacuum Congress, Vienna, 1977, pp. 549–552.
23. H. Terao, T. Ito, and Y. Sakai, Interface properties of InAs MIS structures and their application to FET, *Elec. Eng. Japan* **94**, 127–132 (1974).
24. D. L. Lile, D. A. Collins, L. Messick, and A. R. Clawson, A microwave GaAs insulated gate FET, *Appl. Phys. Lett.* **32**, 247–248 (1978).
25. D. L. Lile, The effect of surface states on the characteristics of MIS field effect transistors, *Solid-State Electron.* **21**, 1199–1207 (1978).
26. Y. Shinoda and T. Kobayashi, InGaAsP n-channel inversion-mode metal–insulator–semiconductor field-effect transistor with low interface state density, *J. Appl. Phys.* **52**, 6386–6394 (1981).
27. E. H. Snow, A. S. Grove, B. E. Deal, and C. T. Sah, Ion transport phenomena in insulating films, *J. Appl. Phys.* **36**, 1664–1673 (1965).
28. C. H. Sequin and M. F. Tompsett, *Charge Transfer Devices*, Academic Press, New York (1975).
29. For a review of CCDs see J. D. E. Beynon and D. R. Lamb, *Charge-Coupled Devices and Their Applications*, McGraw-Hill, New York (1980).
30. M. Okamura and T. Kobayashi, Slow current-drift mechanism in n-channel inversion type InP-MISFET, *Japan. J. Appl. Phys.* **19**, 2143–2150 (1980).

31. R. P. H. Chang, T. T. Sheng, C. C. Chang, and J. J. Coleman, The effect of interface arsenic domains on the electrical properties of GaAs MOS Structures, *Appl. Phys. Lett.* **33**, 341–342 (1978).
32. L. G. Meiners, Capacitance–voltage and surface photovoltage measurements of pyrolytically deposited SiO_2 on InP, *Thin Solid Films* **56**, 201–207 (1979).
33. D. L. Lile, Surface photovoltage and internal photoemission at the anodized InSb surface, *Surf. Sci.* **34**, 337–367 (1973).
34. L. G. Meiners and H. H. Wieder, Charge-carrier transport in semi-insulating InP surface layers, *Semi-Insulating III–V Materials*, Shiva Press, Orpington, England (1980), pp. 198–205.
35. L. G. Meiners, Chapter 4 of present volume.
36. H. Morkoc, T. J. Drummond, and C. M. Stanchak, Schottky barriers and ohmic contacts on n-type InP based compound semiconductors for microwave FETs, *IEEE Trans. Electron. Devices* **ED-28**, 1–5 (1981).
37. D. L. Lile and M. J. Taylor, The effect of interfacial traps on the stability of MIS devices on InP, in *J. Appl. Phys.* **54**, 260–267 (1983).
38. J. R. Sites and H. H. Wieder, Magnetoresistance mobility profiling of MESFET channels, *IEEE Trans. Electron. Devices* **ED-27**, 2277–2281 (1980).
39. K. von Klitzing, Th. Englert, and D. Fritzsche, Transport measurements on InP inversion metal-oxide semiconductor transistors, *J. Appl. Phys.* **51**, 5893–5897 (1980).
40. R. J. Kriegler, T. F. Devenyi, K. D. Chik, and J. Shappir, Determination of surface-state parameters from transfer-loss measurements in CCDs, *J. Appl. Phys.* **50**, 398–401 (1979).
41. J. E. Carnes and W. F. Kosonocky, Fast interface-state losses in charge coupled devices, *Appl. Phys. Lett.* **20**, 261–263 (1972).
42. B. T. Moore and D. K. Ferry, Scattering of inversion layer electrons by oxide polar mode generated interface phonons, *J. Vac. Sci. Technol.* **17**, 1037–1040 (1980).
43. D. A. Baglee, D. K. Ferry, C. W. Wilmsen, and H. H. Wieder, Inversion layer transport and properties of oxides on InAs, *J. Vac. Sci. Technol.* **17**, 1032–1036 (1980).
44. D. A. Baglee, D. H. Laughlin, B. T. Moore, B. L. Eastep, D. K. Ferry, and C. W. Wilmsen, Inversion layer transport and insulator properties of the indium-based III–Vs, *Inst. Phys. Conf. Ser.* **56**, 259–265 (1980).
45. Y. Shinoda and T. Kobayashi, Effective electron mobility in inversion-mode Al_2O_3–InP MISFETS, *Solid-State Electron.* **25**, 1119–1124 (1982).
46. Y. Omura, Bulk doping effect on field-effect mobility of MOSFETs, *Japan. J. Appl. Phys.* **20**, 1985–1986 (1981).
47. T. L. Chu, S. S. Chu, C. L. Lin, Y. C. Tzeng, L. L. Kazmerski and P. J. Ireland, Reduction of grain boundary effects in indium phosphide films by nitridation, *J. Electrochem. Soc.* **128**, 855–859 (1981).
48. A. Heller, Chemical control of recombination at grain boundaries and liquid interfaces: Electrical power and hydrogen generating photoelectrochemical cells, *J. Vac. Sci. Technol.* **21**, 559–561 (1982).
49. W. D. Johnston, Jr., A. J. Leamy, B. A. Parkinson, A. Heller, and B. Miller, Effect of ruthenium ions on grain boundaries in gallium arsenide thin film photovoltaic devices, *J. Electrochem. Soc.* **127**, 90–95 (1980).
50. B. A. Parkinson, A. Heller, and B. Miller, Enhanced photoelectrochemical solar energy conversion by gallium arsenide surface modification, *Appl. Phys. Lett.* **33**, 521–523 (1978).
51. R. J. Nelson, J. S. Williams, H. J. Leamy, B. Miller, H. C. Casey, Jr., B. A. Parkinson, and A. Heller, Reduction of GaAs surface recombination velocity by chemical treatment, *Appl. Phys. Lett.* **36**, 76–79 (1980).

52. K. Ando, A. Yamamoto, and M. Yamaguchi, Surface band bending effects and photoluminescence intensity in n-InP Schottky and MIS diodes, *Japan. J. Appl. Phys.* **20**, 1107–1112 (1981).
53. Y. Hirota, M. Okamura, E. Yamaguchi, T. Nishioka, Y. Shinoda, and T. Kobayashi, Surface controlled InP-MIS (metal–insulator–semiconductor) triodes, *J. Appl. Phys.* **52**, 3498–3503 (1981).
54. C. N. Berglund, Electroluminescence using GaAs MIS structures, *Appl. Phys. Lett.* **9**, 441–444 (1966).
55. J. Stannard and R. L. Henry, Minority-carrier generation in n-InP/SiO$_2$ capacitors, *Appl. Phys. Lett.* **35**, 86–88 (1979).
56. P. N. Favennec, M. LeContellec, H. L'Haridon, G. P. Pelous, and J. Richard, Al/Al$_2$O$_3$/InP MIS structures, *Appl. Phys. Lett.* **34**, 807–808 (1979).
57. K. P. Pande, Y-S. Hsu, J. M. Barrego, and S. K. Ghandi, Grain boundary edge passivation of GaAs films by selective anodization, *Appl. Phys. Lett.* **33**, 717–719 (1978).
58. A. Zylbersztejn, G. Bert, and G. Nuzillat, Hole traps and their effects in GaAs MESFETs, *Inst. Phys. Conf. Ser.* **45**, 315–325 (1979).
59. C. W. Wilmsen, Chemical composition and formation of thermal and anodic oxide/III–V compound semiconductor interfaces, *J. Vac. Sci. Technol.* **19**, 279–289 (September/October 1981).
60. D. L. Lile, D. A. Collins, L. G. Meiners, and M. J. Taylor, A microwave MIS FET technology on InP, *Inst. Phys. Conf. Ser.* **56**, 493–502 (1980).
61. H. H. Wieder, Problems and prospects of compound semiconductor field-effect transistors, *J. Vac. Sci. Technol.* **17**, 1009–1018 (1980).
62. H. H. Wieder, Materials options for field-effect transistors, *J. Vac. Sci. Technol.* **18**, 827–837 (1981).
63. C. A. Liechti, Microwave field-effect transistors—1976, *IEEE Trans. Microwave Theory Technol.* **MTT-24**, 279–300 (1976).
64. H. H. Wieder, A. R. Clawson, D. I. Elder, and D. A. Collins, Inversion mode insulated gate Ga$_{0.47}$In$_{0.53}$As field-effect transistors, *IEEE Trans. Electron. Devices Lett.* **EDL-2**, 73–74 (1981).
65. K. P. Pande and C. C. Shen, The electrical and photovoltaic properties of tunnel metal–oxide–semiconductor devices built on n-InP substrates, *J. Appl. Phys.* **53**, 749–753 (1982).
66. K. R. Gleason, H. B. Dietrich, M. L. Bark, and R. L. Henry, Enhancement-mode ion implanted InP FETs, *Electron. Lett.* **14**, 643–644 (1978).
67. K. Lehovec and R. Zuleeg, I–V characteristics of enhancement-mode GaAs JFETs, *Inst. Phys. Conf. Ser.* **33**, 263–274 (1977).
68. K. Yamaguchi and S. Takahashi, Theoretical characterization and high speed performance evaluation of GaAs IGFETs, *IEEE Trans. Electron. Devices* **ED-28**, 581–587 (1981).
69. F. L. Schuermeyer, R. A. Belt, C. R. Young, and J. M. Blasingame, New structures for charge coupled devices, *Proc. IEEE* **60**, 1444–1445 (1972).
70. D. K. Kinell and M. A. Kiesle, Indium Phosphide MOS Transistor Self-Aligning Gate Process Technology, Presented at the Workshop on Dielectric Systems for the III–V Compounds, San Diego, Calif. (1980).
71. R. M. Hoendervoogt, K. A. Kormos, J. P. Rosbeck, J. R. Toman, and C. B. Burgett, Hybrid InSb focal plane array fabrication, *Proc. IEDM 1978*, 510–512.
72. R. C. Eden, B. M. Welch, R. Zucca, and S. I. Long, The prospects for ultrahigh-speed VLSI GaAs digital logic, *IEEE Trans. Electron. Devices* **ED-26**, 299–317 (1979).
73. G. Nuzillat, G. Bert, T. P. Ngu, and M. Gloanec, Quasi-normally-off MESFET logic for high-performance GaAs ICs, *IEEE Trans. Electron. Devices* **ED-27**, 1102–1109 (1980).

74. N. Yokoyama, T. Mimura, and M. Fukuta, Planar GaAs MOSFET Integrated Logic, *IEEE Trans. Electron. Devices ED-27*, 1124–1128 (1980).
75. F. Capasso and G. F. Williams, A proposed hydrogenation/nitridization passivation technique for III–V semiconductor devices, including InGaAs long-wavelength photodetectors, *J. Electrochem. Soc. 129*, 821–824 (1982).
76. V. Diadiuk, C. A. Armiento, S. H. Groves, and C. E. Hurwitz, Surface passivation techniques for InP and InGaAsP p-n junction structures, *IEEE Trans. Electron. Devices Lett. EDL-1*, 177–178 (1980).
77. D. L. Pulfrey, MIS solar cells: A review, *IEEE Trans. Electron. Devices ED-25*, 1308–1316 (1978).
78. R. Singh and J. Shewchun, A possible explanation for the photovoltaic effect in indium tin oxide on InP solar cells, *J. Appl. Phys. 49*(8), 4588–4591 (1978).
79. M. Wolf, Limitations and possibilities for improvement of photovoltaic solar energy converters, *Proc. IRE 48*, 1246–1263 (1960).
80. L. W. James and R. L. Moon, GaAs Concentrator Solar Cells, Proceedings of the 11th IEEE Photovoltaic Specialist Conference, Scottsdale, Ariz. (1975), pp. 402–408.
81. S. J. Fonash and S. Ashok, On the pinhole model for MIS diodes, *Solid-State Electron. 24*, 1075–1076 (1981).
82. J. Shewchun, R. Singh, and M. A. Green, Theory of metal–insulator–semiconductor solar cells, *J. Appl. Phys. 48*, 765–770 (1977).
83. D. L. Lile and H. H. Wieder, The thin film MIS surface photodiode, *Thin Solid Films 13*, 15–20 (1972).
84. R. D. Thom, F. J. Renda, W. J. Parrish, and T. L. Koch, A monolithic InSb charge-coupled infrared imaging device, *Proc. IEDM 1978*, 501–504.
85. J. C. Kim, InSb charge-injection device imaging array, *IEEE Trans. Electron. Devices ED-25*, 232–240 (1978).
86. R. D. Thom, R. E. Eck, J. D. Phillips, and J. B. Scorso, InSb CCDs and other MIS devices for infrared applications, *Proc. IEDM 1975*, 31–41.
87. J. C. Kim, InSb MIS technology and CID devices, *Proc. IEDM 1975*, 1–17.
88. B. Agusta, A 64-bit planar double diffused monolithic memory chip, *Int. Solid State Circuits Conf. 1969*, 38–39.
89. J. K. Ayling, R. D. Moore, and G. K. Tu, A high-performance monolithic store, *Int. Solid State Circuits Conf. 1969*, 36–37.
90. For a review see the special issue *IEEE Trans. Electron. Devices ED-25* (1978).
91. B. Bayraktaroglu, S. J. Hannah, and H. L. Hartnagel, Stable charge storage of MAOS diodes on GaAs by new anodic oxidation, *Electron. Lett. 13*, 45–46 (1977).
92. H. Hasegawa and T. Sakai, Anodic oxides on gallium phosphide for optoelectronic device and processing applications, *J. Appl. Phys. 49*, 4459–4464 (1978).
93. H. Becke, R. Hall, and J. White, Gallium arsenide MOS transistors, *Solid-State Electron. 8*, 813–823 (1965).
94. D. Darmagna and J. Reynaud, A GaAs thin-film transistor, *Proc. IEEE 54*, 2020 (1966).
95. E. J. Charlson and T. H. Weng, Gallium Arsenide Thin Film Transistors, Proceedings of the 20th Annual Southwestern IEEE Conference and Exhibition, 6A1–6A5 (April 1968).
96. T. Ito and Y. Sakai, The GaAs inversion-type MIS transistors, *Solid-State Electron. 17*, 751–759 (1974).
97. T. Mimura and M. Fukuta, Status of the GaAs metal–oxide–semiconductor technology, *IEEE Trans. Electron. Devices ED-27*, 1147–1155 (1980).
98. B. Bayraktaroglu, E. Kohn, and H. L. Hartnagel, First anodic-oxide GaAs MOSFETs based on easy technological processes, *Electron. Lett. 12*, 53–54 (1976).

99. D. L. Lile, A. R. Clawson, and D. A. Collins, Depletion-mode GaAs MOSFET, *Appl. Phys. Lett.* 29, 207–208 (1976).
100. E. Kohn and A. Colquhoun, Enhancement-mode GaAs MOSFET on semi-insulating substrate using a self-aligned gate technique, *Electron. Lett.* 13, 73–74 (1977).
101. B. Weiss, E. Kohn, B. Bayraktaroglu, and H. L. Hartnagel, Native oxides on GaAs for MOSFETs: Annealing effects and inversion-layer mobilities, *Inst. Phys. Conf. Ser.* 33, 168–176 (1977).
102. E. Kohn, A. Colquhoun, and H. L. Hartnagel, GaAs enhancement/depletion n-channel MOSFET, *Solid-State Electron.* 21, 877–886 (1978).
103. A. Colquhoun, E. Kohn, and H. L. Hartnagel, Improved enhancement/depletion GaAs MOSFET using anodic oxide as the gate insulator, *IEEE Trans. Electron. Devices* ED-25, 375–376 (1978).
104. H. L. Hartnagel, MOS-gate technology on GaAs and other III–V compounds, *J. Vac. Sci. Technol.* 13, 860–866 (1976).
105. T. Sugano, F. Koshiga, K. Yamasaki, and S. Takahashi, *IEEE Trans. Electron. Devices* ED-25, 449–455 (1980).
106. T. Mimura, N. Yokoyama, Y. Nakayama, and M. Fukuta, Plasma-grown oxide gate GaAs deep depletion MOSFET, *Japan. J. Appl. Phys.* 17 (Suppl. 17-1), 153–157 (1978).
107. N. Yokoyama, T. Mimura, K. Odani, and M. Fukuta, Low-temperature plasma oxidation of GaAs, *Appl. Phys. Lett.* 32, 58–60 (1978).
108. T. Miyazaki, N. Nakamura, A. Doi, and T. Takuyama, n-Channel gallium arsenide MISFET (unpublished).
109. K. Kakimura and Y. Sakai, The properties of GaAs–Al$_2$O$_3$ and InP–Al$_2$O$_3$ interfaces and the fabrication of MIS field-effect transistors, *Thin Solid Films* 56, 215–223 (1979).
110. L. Messick, A GaAs/Si$_x$O$_y$N$_z$ MISFET, *J. Appl. Phys.* 47, 5474–5475 (1976).
111. G. D. Bagratishvili, R. B. Dzhanelidze, N. I. Kurdiani, Yu. I. Pashintsev, O. V. Saksaganski, and V. A. Skarikov, GaAs/Ge$_3$N$_4$/Al structures and MIS field-effect transistors based on them, *Thin Solid Films* 56, 209–213 (1979).
112. H. Takagi, G. Kono, and I. Teramoto, Thermal oxide gate GaAs MOSFETs, *IEEE Trans. Electron. Devices* ED-25, 551–552 (1978).
113. N. Yokoyama, T. Mimura, and M. Fukuta, Surface states in an n-GaAs/plasma grown native oxide—A modified deep level transient spectroscopy measurement, *Surf. Sci.* 86, 826–834 (1979).
114. L. Schrader, The influence of the interface states on the dynamic transconductance of MIS-FETs, *Solid-State Electron.* 20, 671–674 (1977).
115. N. Yokoyama, T. Mimura, H. Kusakawa, K. Suyama, and M. Fukuta, GaAs MOSFET high-speed logic, *IEEE Trans. Microwave Theory Technol.* MTT-28, 483–486 (1980).
116. H. Tokuda, Y. Adachi, and T. Ikoma, Microwave capability of 1.5 μm-gate GaAs MOSFET, *Electron. Lett.* 13, 761–762 (1977).
117. T. Mimura, K. Odani, N. Yokoyama, and M. Fukuta, New structure of enhancement-mode GaAs microwave MOSFET, *Electron. Lett.* 14, 500–502 (1978).
118. L. Messick, Power gain and noise of InP and GaAs insulated gate microwave FETs, *Solid-State Electron.* 22, 71–76 (1979).
119. T. Mimura, K. Odani, N. Yokoyama, Y. Nakayama, and M. Fukuta, GaAs microwave MOSFETs, *IEEE Trans. Electron. Devices* ED-25, 573–579 (1978).
120. T. Miyazaki, N. Nakamura, A. Doi, and T. Tokuyama, Electrical properties of gallium arsenide–insulator interface, *Japan. J. Appl. Phys.*, Suppl. 2, 441–443 (1974).
121. G. Lucovsky and R. S. Bauer, Local atomic order in native III–V oxides, *J. Vac. Sci. Technol.* 17, 946–951 (1980).
122. W. E. Spicer, I. Lindau, P. Skeath, and C. Y. Su, Unified defect model and beyond, *J. Vac. Sci. Technol.* 17, 1019–1027 (1980).

123. T. Suzuki and M. Ogawa, Degradation of photoluminescence intensity caused by excitation-enhanced oxidation of GaAs surfaces, *Appl. Phys. Lett. 31*, 473–475 (1977).
124. H. Nagai and Y. Noguchi, Ambient gas influence on photoluminescence intensity from InP and GaAs cleaved surfaces, *Appl. Phys. Lett. 33*, 312–314 (1978).
125. H. Nagai, S. Tohno, and Y. Mizushima, Properties of ambient-enhanced photoluminescence from InP and GaAs surfaces, *J. Appl. Phys. 50*, 5446–5448 (1979).
126. M. D. Clark and C. L. Anderson, Improvements in GaAs plasma-deposited silicon nitride interface quality by pre-deposition GaAs surface treatment and post deposition annealing, *J. Vac. Sci. Technol. 21*, 453–456 (1982).
127. R. K. Ahrenkiel, R. S. Wagner, S. Pattillo, D. Dunlavy, T. Jervis, L. L. Kazmerski, and P. J. Ireland, Reduction of fast surface states on p-type GaAs, *Appl. Phys. Lett. 40*, 700–703 (1982).
128. F. L. Schuermeyer, GaAs IGFET digital integrated circuits, *IEEE Trans. Electron. Devices ED-28*, 541–545 (1981).
129. T. L. Andrade and N. Braslau, GaAs Lossy Gate Dielectric FET, Presented at the Device Research Conference, Santa Barbara, Calif. (June 1981).
130. B. R. Pruniaux, J. C. North, and A. V. Payer, A semi-insulated gate gallium-arsenide field-effect transistor, *IEEE Trans. Electron. Device ED-19*, 672–674 (1972).
131. H. M. Macksey, D. W. Shaw, and W. R. Wisseman, GaAs power FETs with semi-insulated gates, *Electron. Lett. 12*, 192–193 (1976).
132. H. C. Casey, Jr., A. Y. Cho, D. V. Lang, E. H. Nicollian, and P. W. Foy, Investigation of heterojunctions for MIS devices with oxygen-doped $Al_xGa_{1-x}As$ on n-type GaAs, *J. Appl. Phys. 50*, 3484–3491 (1979).
133. E. J. Bawolek and B. W. Wessels, ZnSe/GaAs Heterojunctions for MIS Devices, Presented at the Workshop on Dielectric Systems for the III–V Compounds, San Diego, Calif. (June 1982).
134. R. Dingle, H. L. Stormer, A. C. Gossard, and W. Wiegmann, Electron mobilities in modulation-doped semiconductor heterojunction superlattices, *Appl. Phys. Lett. 33*, 665–667 (1978).
135. T. Mimura, K. Joshin, S. Hiyamizu, K. Hikosaka, and M. Abe, High electron mobility transistor logic, *Japan. J. Appl. Phys. 20*, L598–L600 (1981).
136. T. Hotta, H. Sakaki, and H. Ohno, A new AlGaAs/GaAs heterojunction FET with insulated gate structure (MISSFET), *Japan. J. Appl. Phys. 21*, L122–L124 (1982).
137. R. J. Stirn and Y. C. M. Yeh, Technology of GaAs metal–oxide–semiconductor solar cells, *IEEE Trans. Electron. Devices ED-24*, 476–483 (1977).
138. W. A. Anderson, G. Rajeswaran, V. J. Rao, and M. Thayer, Cr-MIS solar cells using thin epitaxial silicon grown on poly-silicon substrates, *IEEE Trans. Electron. Devices Lett. EDL-2*, 271–274 (1981).
139. C. W. Wilmsen, The MOS/InP interface, *Crit. Rev. Solid-State Sci. 5*, 313–317 (1975).
140. D. L. Lile and D. A. Collins, An InP MIS diode, *Appl. Phys. Lett. 28*, 554–556 (1976).
141. H. C. Casey, Jr. and E. Buehler, Evidence for low surface recombination velocity on n-type InP, *Appl. Phys. Lett. 30*, 247–249 (1977).
142. T. Suzuki and M. Ogawa, *In Situ* measurements of photoluminescence intensities from cleaved (110) surfaces of n-type InP in a vacuum and gas ambient, *Appl. Phys. Lett. 34*, 447–449 (1979).
143. L. Messick, D. L. Lile, and A. R. Clawson, A microwave InP/SiO_2 MISFET, *Appl. Phys. Lett. 32*, 494–495 (1978).
144. D. L. Lile and D. A. Collins, The dielectric and interfacial characteristics of MIS structures on InP and GaAs, *Thin Solid Films 56*, 225–234 (1979).
145. L. G. Meiners, Electrical properties of SiO_2 and Si_3N_4 dielectric layers on InP, *J. Vac. Sci. Technol. 19*, 373–379 (1981).

146. L. G. Meiners, D. L. Lile, and D. A. Collins, Inversion layers on InP, *J. Vac. Sci. Technol. 16*, 1458–1461 (1979).
147. D. L. Lile, D. A. Collins, L. G. Meiners, and L. Messick, n-Channel inversion-mode InP MISFET, *Electron. Lett. 14*, 657–659 (1978).
148. D. Fritzsche, Interface studies on InP MIS inversion FETs with SiO_2 gate insulation, *Inst. Phys. Conf. Ser. 50*, 258–265 (1980).
149. D. C. Cameron, L. D. Irving, G. R. Jones, and J. Woodward, MISFET and MIS Diode Behavior of Some Insulator–InP Systems, Presented at INFOS'81 held at Erlangen (April 1981).
150. T. Kawakami and M. Okamura, InP/Al_2O_3 n-channel inversion-mode MISFETs using sulphur-diffused source and drain, *Electron. Lett. 15*, 502–504 (1979).
151. T. Kobayashi and Y. Hirota, Inversion-mode InP MISFET employing phosphorus–nitride gate insulator, *Electron. Lett. 18*, 180–181 (1982).
152. Y. Hirayama, H. M. Park, F. Koshiga, and T. Sugano, Enhancement type InP metal–insulator–semiconductor field-effect transistor with plasma anodic aluminium oxide as the gate insulator, *Appl. Phys. Lett. 40*, 712–713 (1982).
153. A. Yamamoto and C. Uemura, Anodic oxide film as gate insulator for InP MOSFETs, *Electron. Lett. 18*, 63–64 (1982).
154. W. F. Tseng, M. L. Bark, H. B. Dietrich, A. Christou, R. L. Henry, W. A. Schmidt, and N. S. Saks, A virtual self-aligned process for n-channel InP IGFETs (or MISFETs), *IEEE Trans. Electron. Devices Lett. EDL-2*, 299–301 (1981).
155. P. Wolf, Microwave properties of Schottky-barrier field-effect transistors, *IBM J. Res. Dev. 14*, 125–141 (1970).
156. K. E. Drangeid and R. Sommerhalder, Dynamic performance of Schottky-barrier field-effect transistors, *IBM J. Res. Dev.* 82–94, March (1970).
157. W. Walukiewicz, J. Lagowski, L. Jastrzebski, P. Rava, M. Lichtensteiger, C. H. Gatos, and H. C. Gatos, Electron mobility and free-carrier absorption in InP; determination of the compensation ratio, *J. Appl. Phys. 51*, 2659–2668 (1980).
158. A. J. Grant, D. C. Cameron, L. D. Irving, C. E. Greenshalgh, and P. R. Norton, A study of deposited dielectrics and the observation of n-channel MOSFET action in InP, *Inst. Phys. Conf. Ser. 50*, 266–270 (1980).
159. K. Ohata, T. Itoh, H. Watanabe, T. Mizutani, and Y. Takayama, Investigation on SiO_2/InP MIS systems and enhancement-mode MISFETs, Presented at the International Symposium on Gallium Arsenide and Related Compounds, Oiso (1981); *Inst. Phys. Conf. Ser. 63*, 353–358 (1982).
160. L. Henry, D. Lecrosnier, H. L'Haridan, J. Paugam, G. Pelous, F. Richau, and M. Salvi, n-Channel MISFETs on semi-insulating InP for logic applications, *Electron. Lett. 18*, 102–103 (1982).
161. D. Fritzsche, InP–SiO_2 MIS structure with reduced interface state density near conduction band, *Electron. Lett. 14*, 51–52 (1978).
162. Y. Ohmachi and T. Nishioka, Ion implanted n-channel InP IGFET and its low frequency characteristics, *Japan. J. Appl. Phys. 19*, 1425–1426 (1980).
163. G. G. Roberts, K. P. Pande, and W. A. Barlow, InP/Langmuir-film MISFET, *Solid-State Electron. Devices 2*, 169–175 (1978).
164. M. Okamura and T. Kobayashi, Reduction of interface states and fabrication of p-channel inversion-type InP-MISFET, *Japan. J. Appl. Phys. 19*, 599–602 (1980).
165. H. Hasegawa and T. Sawada, Photoionization cross section and threshold of interface states in GaAs and InP MOS structures, Presented at the International Symposium on Gallium Arsenide and Related Compounds, Oiso (1981); *Inst. Phys. Conf. Ser. 63*, 335–340 (1982).
166. L. G. Meiners, D. L. Lile, and D. A. Collins, Microwave gain from an n-channel enhancement-mode InP MISFET, *Electron. Lett. 15*, 578 (1979).

167. T. Kawakami and M. Okamura, n-Channel formation on semi-insulating InP surface by MISFET, *Electron. Lett. 15*, 743 (1979).
168. S. M. Sze, *Physics of Semiconductor Devices*, Wiley, New York (1969), pp. 515–524.
169. L. Messick, A D.C. to 16 GHz indium phosphide MISFET, *Solid-State Electron. 23*, 551–555 (1980).
170. K. Ohata, T. Itoh, H. Watanabe, T. Mizutani, and Y. Takayama, Enhancement-mode InP MISFETs with a CVD-SiO_2 gate insulator, *Electron. Commun. Tech. Rep.* (in Japanese), *81*, 59–66 (1982).
171. L. J. Messick, A high-speed monolithic InP MISFET integrated logic inverter, *IEEE Trans. Electron. Devices ED-28*, 218–221 (1981).
172. D. K. Kinell, An Indium Phosphide MISFET Integrated Circuit Technology, Presented at the 39th Device Research Conference, Santa Barbara, Calif. (June 1981).
173. M. D. Clark and R. A. Jullens, Indium Phosphide MISFET Integrated Circuits, Presented at the Workshop on Dielectric Systems for the III–V Compounds, San Diego, Calif. (June 1982).
174. H. Sewell and J. C. Anderson, Slow states in $InSb/SiO_x$ thin film transistors, *Solid-State Electron. 18*, 641–649 (1975).
175. M. Okamura and T. Kobayashi, Improved interface in inversion-type InP MISFET by vapor etching technique, *Japan. J. Appl. Phys. 19*, 2151–2156 (1980).
176. T. Kobayashi, M. Okamura, E. Yamaguchi, Y. Shinoda, and Y. Hirota, Effect of pyrolytic Al_2O_3 deposition temperature on inversion-mode InP MISFET, *J. Appl. Phys. 52*, 6434–6436 (1981).
177. J. F. Wager and C. W. Wilmsen, Thermal oxidation of InP, *J. Appl. Phys. 51*, 812–814 (1980).
178. J. D. Langan and C. R. Viswanathan, Characterization of improved InSb interfaces, *J. Vac. Sci. Technol. 16*, 1474–1477 (1979).
179. M. Okamura and T. Kobayashi, Current drifting behavior in InP MISFET with thermally oxidized InP/InP interface, *Electron. Lett. 17*, 941–942 (1981).
180. V. Montgomery, R. H. Williams, and R. R. Varma, The interaction of chlorine with indium phosphide surfaces, *J. Phys. C 11*, 1989–2000 (1978).
181. H. Huff, S. Kawaji, and H. C. Gatos, Field effect measurements on the A and B 111 surfaces of indium antimonide, *Surf. Sci. 5*, 399–409 (1966).
182. Vu Quoc Ho and Takuo Sugano, An improvement of the interface properties of plasma anodized SiO_2/Si system for the fabrication of MOSFETs, *IEEE Trans. Electron. Devices ED-28*, 1060–1065 (1981).
183. A. K. Gaind and L. A. Kasprzak, Determination of distributed fixed charge in CVD-oxide and its virtual elimination by use of HCl, *Solid-State Electron. 22*, 303–309 (1979).
184. L. A. Kasprzak and A. K. Gaind, Near-ideal $Si-SiO_2$ interfaces, *IBM J. Res. Dev. 24*, 348–352 (1980).
185. D. L. Lile and D. A. Collins, An 8-bit, 4-phase surface channel charge-coupled device on InP, *IEEE Trans. Electron. Devices ED-29*, 842–845 (1982).
186. D. L. Lile and D. A. Collins, An insulated-gate charge transfer device on InP, *Appl. Phys. Lett. 37*, 552–553 (1980).
187. J. L. Davis, Surface states on the (111) surface of indium antimonide, *Surf. Sci. 2*, 33–39 (1964).
188. R. J. Schwartz, R. C. Dockerty, and H. W. Thompson, Jr., Capacitance voltage measurements on n-type InAs MOS diodes, *Solid-State Electron. 14*, 115–124 (1971).
189. L. L. Chang and W. E. Howard, Surface inversion and accumulation of anodized InSb, *Appl. Phys. Lett. 7*, 210–212 (1965).
190. S. Kawaji and Y. Kawaguchi, Galvanomagnetic properties of surface layers in indium arsenide, *J. Phys. Soc. Japan 21, Supp. 1966*, 336–339.

191. V. L. Frantz, Indium antimonide thin-film transistor, *Proc. IEEE* 53, 760 (1965).
192. T. P. Brody and H. E. Kunig, A high-gain InAs thin-film transistor, *Appl. Phys. Lett.* 9, 259–260 (1966).
193. I. Spinulescu-Carnaru, ZnTe and InSb thin-film transistors, *Electron. Lett.* 3, 268–269 (1967).
194. D. L. Lile and J. C. Anderson, The application of polycrystalline layers of InSb and PbTe to a field-effect transistor, *Solid-State Electron.* 12, 735–741 (1969).
195. F. C. Luo and M. Epstein, Coplanar-electrode thin film InSb transistor, *Proc. IEEE* 60, 997–999 (1972).
196. H. E. Kunig, Analysis of an InAs thin film transistor, *Solid-State Electron*, 11, 335–342 (1968).
197. H. Baudrand, E. Hamadto, and J. L. Amalric, An experimental and theoretical study of polycrystalline thin film transistor, *Solid-State Electron.* 24, 1093–1098 (1981).
198. A. Heime and H. Pagnia, Influence of the semiconductor–oxide interlayer on the AC-behavior of InSb MOS-capacitors, *J. Appl. Phys.* 15, 79–84 (1978).
199. K. F. Komatsubara, Y. Katayama, N. Kotera, and T. Kobayashi, Transport properties of electrons in inverted InSb surface, *J. Vac. Sci. Technol.* 6, 572–575 (1969).
200. J. D. Thom, W. J. Parrish, and T. L. Koch, Monolithic InSb CCD Array Technology, Presented at the IRIS Detector Specialty Group Meeting, Minneapolis, Minn. (June 1979).
201. C. W. Fischer, N. Leslie, and A. Etchells, Properties of the native oxide on GaSb, *J. Vac. Sci. Technol.* 13, 59–63 (1976).
202. G. Sixt, K. H. Ziegler, and W. R. Fahrner, Properties of anodic oxide films on *n*-type GaAs, $GaAs_{0.6}P_{0.4}$ and GaP, *Thin Solid Films*, 56, 107–116 (1979).
203. G. Weimann, Oxide and interface properties of anodic oxide MOS structures on III–V compound semiconductors, *Thin Solid Films* 56, 173–182 (1979).
204. H. Kressel, Materials for heterojunction devices, *Ann. Rev. Mater. Sci.* 10, 287 (1980) (edited by R. A. Huggins, R. H. Bube, and D. A. Vermilyea).
205. J. H. Marsh, P. A. Houston, and P. N. Robson, Compositional dependence of the mobility, peak velocity and threshold field in $In_{1-x}Ga_xAs_yP_{1-y}$, *Inst. Phys. Conf. Ser.* 56, 621–630 (1980).
206. R. F. Leheny, R. E. Nahory, M. A. Pollack, A. A. Ballman, E. D. Beebe, J. C. DeWinter, and R. J. Martin, An $In_{0.53}Ga_{0.47}As$ junction field-effect transistor, *IEEE Trans. Electron. Devices Lett.* EDL-1, 110–111 (1980).
207. U. Koren, K. L. Yu, T. R. Chen, N. Bar-Chaim, S. Margalit, and A. Yariv, Monolithic integration of a very low threshold GaInAsP laser and metal–insulator–semiconductor field-effect transistor on semi-insulating InP, *Appl. Phys. Lett.* 40, 643–645 (1982).
208. T. P. Pearsall, G. Beuchet, J. P. Hirtz, N. Visentin, M. Bonnet, and A. Raizes, Electron and hole mobilities in $Ga_{0.47}In_{0.53}As$, *Inst. Phys. Conf. Ser.* 56, 639–649 (1980).
209. M. A. Littlejohn, J. R. Hauser, and T. H. Glisson, Velocity-field characteristics of $Ga_{1-x}In_xP_{1-y}As_y$ quaternary alloys, *Appl. Phys. Lett.* 30, 242–244 (1977).
210. S. Bandy, C. Nishimoto, S. Hyder, and C. Hooper, Saturation velocity determination for $In_{0.53}Ga_{0.47}As$ field-effect transistors, *Appl. Phys. Lett.* 38, 817–819 (1981).
211. K. Kajiyama, Y. Mizushima, and S. Sakata, Schottky barrier height of n-$In_xGa_{1-x}As$ diodes, *Appl. Phys. Lett.* 23, 458–459 (1973).
212. H. H. Wieder, Fermi level and surface barriers of $Ga_xIn_{1-x}As$ alloys, *Appl. Phys. Lett.* 38, 170–171 (1981).
213. Y. Shinoda, M. Okamura, E. Yamaguchi, and T. Kobayashi, InGaAsP *n*-channel inversion mode MISFET, *Japan. J. Appl. Phys.* 19, 2301–2302 (1980).
214. H. H. Wieder, Surfaces and dielectric–semiconductor interfaces of some binary and quaternary alloy III–V compounds, *Inst. Phys. Conf. Ser.* 50, 234–250 (1980).

215. A. S. H. Liao, R. F. Leheny, R. E. Nahory, and J. C. DeWinter, An $In_{0.53}Ga_{0.47}As/Si_3N_4$ N-channel inversion mode MISFET, *IEEE Trans. Electron. Devices Lett. EDL-2*, 288–290 (1981).
216. A. S. H. Liao, R. F. Leheny, R. E. Nahory, J. C. DeWinter, and R. J. Martin, $In_{0.53}Ga_{0.47}As/Si_3N_4$ n-Channel and p-Channel Inversion Mode MISFETs, Proceedings of the International Electron Devices Meeting, Washington, D.C. (1981), pp. 637–639.
217. A. S. H. Liao, B. Tell, R. F. Leheny, R. E. Nahory, and T. Y. Chang, A Plasma Oxide Insulated Gate $In_{0.53}Ga_{0.47}As$ FET, Presented at the Workshop on Dielectric Systems for the III–V Compounds, San Diego, Calif. (June 1982).
218. A. S. H. Liao, B. Tell, R. F. Leheny, and T. Y. Chang, $In_{0.53}Ga_{0.47}As$ n-channel native oxide inversion mode field-effect transistor, *Appl. Phys. Lett. 41*, 280–282 (1982).
219. J. Selders and H. Beneking, The $Ga_{0.47}In_{0.53}As–SiO_2$ System and Its Application to n-Channel Inversion Mode MISFETs, Presented at the Workshop on Dielectric Systems for the III–V Compounds, San Diego, Calif. (June 1982).
220. R. Kaumanns, J. Selders, and H. Beneking, Surface states and field effect on $Ga_{0.47}In_{0.53}As$ Layers, Presented at the International Symposium on Gallium Arsenide and Related Compounds, Oiso (1981); *Inst. Phys. Conf. Ser. 63*, 329–334 (1982).
221. P. O'Connor, T. P. Pearsall, K. Y. Cheng, A. Y. Cho, J. C. M. Hwang, and K. Alavi, $In_{0.53}Ga_{0.47}As$ FETs with insulator-assisted Schottky gates, *IEEE Trans. Electron. Devices Lett. EDL-3*, 64–66 (1982).
222. A. S. H. Liao, B. Tell, R. F. Leheny, R. E. Nahory, J. C. DeWinter, and R. J. Martin, An $In_{0.53}Ga_{0.47}As$ p-channel MOSFET with plasma-grown native oxide insulated gate, *IEEE Trans. Electron. Devices Lett. EDL-3*, 158–160 (1982).
223. B. Tell, R. E. Nahory, R. F. Leheny, and J. C. DeWinter, Native grown plasma oxides and inversion layers on InGaAs, *Appl. Phys. Lett. 39*, 744–746 (1981).
224. Y. Shinoda and T. Kobayashi, High Mobility $In_{1-x}Ga_xAs_yP_{1-y}$ Inversion-Mode MISFETS, Proceedings of the Ninth International Symposium on GaAs and Related Compounds, Japan (1981).
225. D. Fritzsche, E. Kuphal, G. Weimann, and H. Burkhard, CVD and Sputtered Dielectric films on LPE $In_{0.53}Ga_{0.47}As$ interface properties with respect to MIS applications, Presented at the Electronic Materials Conference, Santa Barbara, Calif. (June 1981).
226. P. D. Gardner, S. Y. Narayan, S. Colvin, and Yong-Hoon Yun, $Ga_{0.47}In_{0.53}As$ metal insulator field effect transistors (MISFETs) for microwave frequency applications, *RCA Rev. 42*, 542–556 (1981).
227. R. K. Ahrenkiel, F. Moser, S. L. Lyu, and T. J. Coburn, Electronic properties of anodic oxides grown on $GaAs_{0.6}P_{0.4}$, *Thin Solid Films 56*, 117–128 (1979).
228. H. Kressel and J. K. Butler, *Semiconductor Lasers and Heterojunction LEDs*, Academic Press, New York (1977).
229. H. W. Becke and J. P. White, Gallium arsenide FETs outperform conventional silicon MOS devices, *Electronics*, 82–90, June 12 (1967).
230. J. W. Peters, Low Temperature Photo-CVD Oxide Processing for Semiconductor Device Applications, Proceedings of the International Electron Devices Meeting, pp. 240–243, Washington, D.C. (1981).
231. L. G. Meiners, Indirect plasma deposition of silicon dioxide, *J. Vac. Sci. Technol. 21*, 655–658 (1982).
232. K. P. Pande and D. Gutierrez, Channel mobility enhancement in InP metal-insulator-semiconductor field-effect transistors, *Appl. Phys. Lett. 46*, 416–418 (1985).
233. B. Bayraktaroglu, R. L. Johnson, D. W. Langer, and M. G. Mier, Germanium (oxy)nitride based surface passivation technique as applied to GaAs and InP, *The*

Physics of MOS Insulators (G. Lucovsky, S. T. Pantelides, and F. L. Galeener, eds.), pp. 207–211, Pergamon Press, New York (1980).
234. K. P. Pande and S. Pourdavour, Sr., Ge_3N_4–InP MIS structures, *IEEE Trans. Electron. Devices Lett. EDL-2*, 182–184 (1982).
235. M. Yamaguchi, Thermal nitridation of InP, *Japan. J. Appl. Phys. 19*, L401–L404 (1980).
236. Y. Hirota, M. Okamura, and T. Kobayashi, The effects of annealing metal–insulator–semiconductor diodes employing a thermal nitride–InP interface, *J. Appl. Phys. 53*, 536–540 (1980).
237. K. Ploog, A. Fischer, and R. Trommer, MBE-grown insulating oxide films on GaAs, *J. Vac. Sci. Technol. 16*, 290–294 (1979).
238. K. Tsubaki, S. Ando, K. Oe, and K. Sugiyama, Surface damage in InP induced during SiO_2 deposition by RF sputtering, *Japan. J. Appl. Phys. 18*, 1191–1192 (1979).
239. D. T. Clark and T. Fok, Surface modification of InP by plasma techniques using hydrogen and oxygen, *Thin Solid Films 78*, 271–278 (1981).
240. D. Fritzsche, E. Kuphal, and G. Weimann, InP/SiO_2 Inversion n-Channel MISFETs: Device Performance Related to Interface Properties, Presented at the Sixth European Specialist Workshop on Active Microwave Semiconductor Devices, Darmstadt (1980).

7

Oxide/III–V Compound Semiconductor Interfaces

C. W. Wilmsen

1. Introduction

Native oxides readily grow on the III–V compounds and can form very thick layers. They also often appear at deposited insulator/III–V interfaces where they can strongly affect the electrical properties. The native oxides can be used to seal or passivate the III–V surface or serve as an insulating layer. This chapter provides insight into the mechanisms of oxide growth and how the oxide/III–V interface is formed. Unfortunately, the growth of oxides on the III–Vs is far more complex than on Si. This is caused by a "competition" between the two or more elements of the III–V substrate, for example, one element may diffuse, evaporate, or dissolve faster than the other. Thus, the chemical composition of bulk oxide layer or its interface may be highly nonuniform.

This chapter discusses oxides formed by the following techniques:

Air oxidation at room temperature
Chemical oxidation
Thermal oxidation
Anodic oxidation
Plasma oxidation

In order to aid in the understanding of these oxidation processes, some general information on oxidation is given in this introductory section.

C. W. Wilmsen ● Department of Electrical Engineering, Colorado State University, Fort Collins, CO 80523.

1.1. Initial Oxidation

The oxidation of a solid surface usually begins with the adsorption of the oxidizing species, for example, O_2, H_2O, OH^-, and CO. The adsorbed molecules form bonds with one or more surface atoms with the strength of these bonds varying from ~0.1 to 3 eV. Weak-bonding adsorption is called physical adsorption since the molecules more or less physically rest on the surface. The stronger bonding is called chemisorption since there is a transfer of charge between the surface and the adsorbing molecule. However, true oxidation does not take place until bonds of the solid are broken and a rearrangement of the atoms occurs. Often the initial oxidation occurs only with the aid of heat, photon, or plasma excitation, as has been observed[1] in the Ga and As 3d X-ray

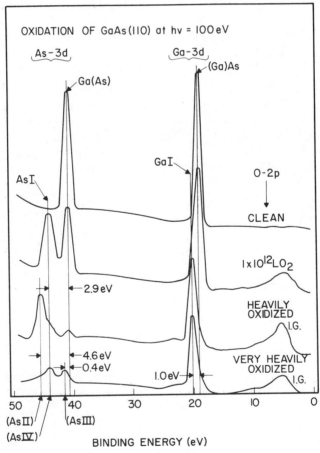

Figure 1. Initial oxidation of a cleaved GaAs surface (Pianetta et al.[1]).

photoemission spectroscopy (XPS) core shifts after exposure of cleaved GaAs to O_2 shown in Fig. 1. The As 3d line is shifted as a result of O_2 adsorption but the Ga3d is not. This indicates that the oxygen molecules have transferred charge with the As atoms but not with the Ga atoms. At this point oxidation has not taken place. Upon exposing the surface to excited oxygen, the Ga 3d level shifts as does the As 3d and the oxidation process has been initiated. While catalytic action may be required for atomically smooth surfaces, oxidation has been found to readily occur at defects where the surface atoms have more broken bonds and there is a lower barrier to oxidation. On GaAs the initial oxidation at defects proceeds without the need of an excitation.

After the initial layer of oxide is formed, further oxidation cannot take place without diffusion of the reactant species, for example, for GaAs, some combination of Ga, As, and/or O must diffuse. This is clear because in order to react, the oxygen and the substrate atoms must be in contact. If the oxide layer is only one monolayer thick, then the diffusion is relatively easy and a second (or possibly a third) oxide layer may form at room temperature. However, without some form of excitation the oxide growth rate is greatly reduced after one to three monolayers of growth and the thickness will appear to self-limit after 10–25 Å.

The initial growth of oxides on the III–V's appears to be uniform across the surface except during anodic oxidation where the oxidation begins with an island stage[2-4] which can become relatively thick before the entire surface is covered with oxide. For thermal-, air-, and plasma-grown oxides, islands have not been observed.

The composition of the oxide layers is determined by kinetic factors, such as diffusion, evaporation, dissolution, and reaction rates, and by thermodynamics which predict the equilibrium oxidation products. Since these parameters vary widely, it is not possible to construct a generalized model which holds for the oxidation of all III–V compounds under all growth conditions. Table 1 provides some of the properties of the various oxides and III–V elements. The information given in this table plus that given in the following sections provides background for the understanding of oxide growth.

1.2. Thermodynamics

Equilibrium thermodynamics predicts the end products of a reaction which has overcome all kinetic barriers. Overcoming these barriers often requires some form of excitation such as heat. A simple example of this is the reaction between As_2O_3 and GaAs. Using the Gibbs free energy of these compounds, thermodynamics predicts that the GaAs will reduce

Table 1. Properties of III-V Oxides

Material	Structure[5]	Melting point (°C)	Density (g/cm²)	Index of refraction N_α	N_β	N_γ	E_g (eV)	$\Delta G_F^{(8)}$ (kcal/mol)	$\delta H_F^{(8)}$ (cal/mol)	$C_p^{(8)}$
vp β-Ga_2O_3	Monoclinic	1715	5.95[5]	1.93[11]			4.7[12]	−238.6	−260.3	22.0
$GaPO_4$	Trigonal	1670	3.567[5]	1.593[5]	1.589			−310.1		
$GaAsO_4$	Tetragonal		2.98							
$GaAsO_4$	Hexagonal		4.20[5]					−212.8	−240.6[18]	
$GaSbO_4$	Tetragonal		6.57[5]					−218.4	−246.2[18]	
vp In_2O_3	Cubic	~2000[6]	7.117[10]		1.9		3.55[12,16]	−198.55	−221.27	22.0
$InPO_4$	Orthorhombic	>900[7]	4.828[10]	1.608[10]	1.618	1.623	4.5[13]	−287 ± 25[7]		
$InAsO_4$								−201.4	−236.1	
$InSbO_4$								−198.4	−226.7	
vp P_2O_3	Monoclinic	23.8[9]	2.135[9]					−176 ± 10[7]	−196.0	
vp P_2O_5	Tetragonal	580[5]	2.30[5]	1.545	1.599[5]	1.624	6–10[14]	−322.4 ± 8[7]	−356.6	25.30
	Orthorhombic	563	2.72		1.578	1.589				
vp As_2O_3	Monoclinic	313	4.15[11]	2.01	1.92[6]	1.87	4.0[15]	−137.91	−156.5	22.86
	Cubic	275	3.865		1.755					
As_2O_5	Hexagonal	>827	4.09[5]					−187.0		
vp Sb_2O_3	Cubic	655	5.2[6]		2.08[11]		6.72[6]	−151.5	−221.05	27.85
	Orthorhombic		5.67(	2.35	2.35[11]	2.18		−149.7	−169.3	24.23

the As_2O_3 to form elemental As and Ga_2O_3 via the reaction

$$As_2O_3 + 2GaAs \rightarrow Ga_2O_3 + 4As.$$

However, mixing powders of As_2O_3 and GaAs does not result in a reaction unless the mixture is heated. The elevated temperature provides the necessary energy to overcome the chemical and diffusion barriers to the reaction.

While thermodynamics may not always predict the composition of an oxide layer grown on the III–V's, it does provide a guide to possible layer products and their stability during thermal annealing. The phase diagram is the most convenient method of displaying the possible reactions between the elements and compounds. For the III–V compounds, these phase diagrams have been worked out by Schwartz and co-workers[17–19,22] for GaAs, GaSb, GaP, InP, InAs, and $In_xGa_{x-1}As$ and for InSb by Smirnov et al.[20]

The ternary phase diagram for a III–V compound and oxygen is constructed by placing one of the elements at each of the apexes of an equilateral triangle and then placing the known stable compounds on lines connecting the proper elements or compounds. Thus, Ga_2O_3 is placed on the line connecting Ga and O, as shown in Fig. 2. $GaAsO_4$ is placed on a line connecting Ga_2O_3 and As_2O_5. If elements or compounds are connected by a line (called a tie line), then no reaction is possible between these two, for example, Ga and GaAs do not react to form other compounds. Conversely, if two elements or compounds are not connected by a tie line, then a reaction can occur. The products of such a reaction are those at each end of the tie lines cut by a line connecting the two reactants.

The phase diagrams for the Ga- and In-based compounds are given in Figs. 2 and 3. From these the equilibrium composition of oxide/III–V

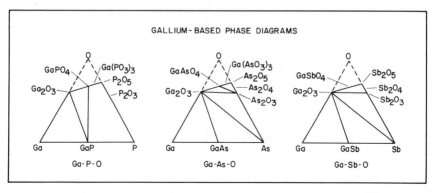

Figure 2. Ternary phase diagrams of the Ga compounds (Schwartz[19]).

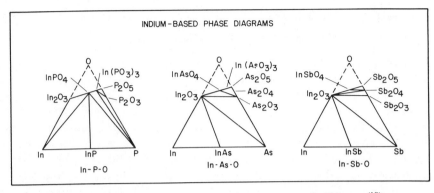

Figure 3. Ternary phase diagrams of the In compounds (Schwartz[19]).

interfaces can be predicted by drawing a line between the III–V compound point and the oxygen apex. If tie lines are cut, then the compounds at the ends of the line will appear at the interface unless they are reduced by the III–V substrate. Table 2 summarizes the equilibrium interface composition and the composition observed after different types of oxide growth. Table 2 reflects the fact that oxide growth is often not near equilibrium.

1.3. Vapor Pressure

The vapor pressure of the III–V elements and their oxides is an important parameter in the thermal oxidation, annealing, and deposition of insulators at elevated temperatures. Figures 4 and 5 summarize these vapor pressures. It is known that the column V elements, As, P, Sb, evaporate to form polyatomic species with As_4 and P_4 dominant for these elements.[27] The column III elements probably evaporate as single atoms. The column V elements sublime, while those of column III do not. The properties of the trioxides of the III–V's are also different. As_2O_3, Sb_2O_3, and P_2O_3 all have low melting points[28] and do not dissociate upon evaporation but rather form the dimer M_4O_6. Al_2O_3, Ga_2O_3, and In_2O_3 have high melting points,[29] low vapor pressures, and dissociate into M, O, MO, and M_2O vapor species with the M_2O the most abundant for In_2O_3 and Ga_2O_3. Al_sO_3 dissociates primarily into Al and O.

2. The Chemically Cleaned Surface

2.1. Polishing and Exposure to Air

To prepare the surface, the III–V compounds are usually soaked in HCl for a few minutes in order to remove the surface oxides and then

Table 2. Oxide/III-V Interface Composition

III-V Compound	Equilibrium[19]	Thermal oxide[21]	Anodic oxide[21]	Anodic oxide (annealed)	Plasma oxide	Plasma oxide (annealed)	Air/Chemical oxide
GaAs	$Ga_2O_3 + As$	$Ga_2O_3 + As$	$Ga_2O_3 + As_2O_3$ [a]	$Ga_2O_3 + As$ [17]	$Ga_2O_3 + As_2O_3$ [25]	$Ga_2O_3 + As$ [25]	Ga_2O_3 [b] or $Ga_2O_3 + As$ or $Ga_2O_3 + As_2O_3$
GaSb	$Ga_2O_3 + Sb$	$Ga_2O_3 + Sb$	$Ga_2O_3 + Sb_2O_3$	$Ga_2O_3 + Sb$ [18]	Unknown	Unknown	
GaP	$GaPO_4$	$GaPO_4$	$Ga_2O_3 + P_2O_5$	Unknown	Unknown	Unknown	
InAs	$In_2O_3 + As$	$In_2O_3 + As$ $In_2O_3 + As_2O_3$	$In_2O_3 + As_2O_3$	$In_2O_3 + As$ [22]	Unknown	Unknown	
InSb	$In_2O_3 + Sb$	$In_2O_3 + Sb$	$In_2O_3 + Sb_2O_3$	$In_2O_3 + Sb$ [23]	Unknown	Unknown	$In_2O_3 + Sb_2O_3$
InP	$InPO_4$	$InPO_4 + P$ or $InPO_4$	$In_2O_3 + P_2O_5$	$InPO_4$ [24]	$InPO_4$ [26]	Unknown	$InPO_4$

[a] Mixed experimental results have been reported.
[b] The composition depends on growth conditions.

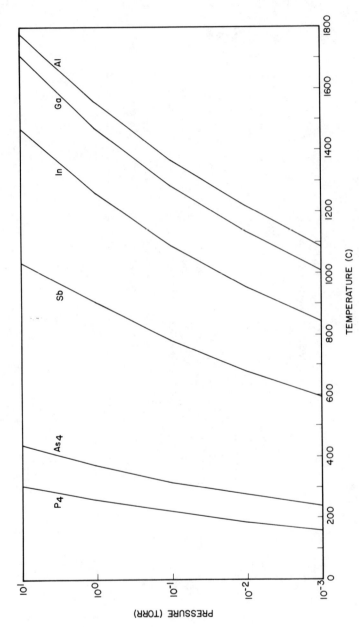

Figure 4. Vapor pressure of the III–V elements as a function of temperature.

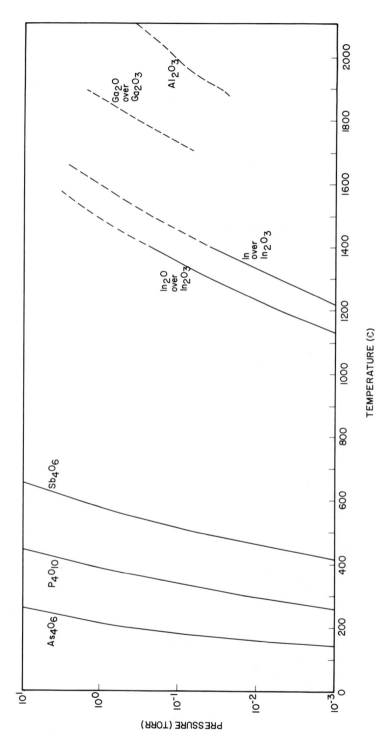

Figure 5. Vapor pressure of the III–V oxides as a function of temperature.

chemomechanically polished on a soft pad using a solution of bromine–methanol (Br–MeOH). For polishing, a 0.5–2.0% Br–MeOH solution is commonly used with a weaker solution used for the final polish. The bromine is removed by rinsing with methanol and sometimes this is followed by a water rinse. Ellipsometric measurements in air immediately following the polish and rinse indicate an oxide-layer thickness of 5–20 Å. Water rinsing does not appear to strongly affect the oxide thickness measured in this way.[31]

The Br–MeOH etch itself, if done properly, does not leave an oxide residue. This has been demonstrated by Aspnes and Studna[31] on GaAs and InP by etching the polished surface with 0.05% Br–MeOH under flowing N_2. Virtually no oxide remained after this treatment even though the etching was followed by a water rinse. However, exposure to air rapidly grew an oxide layer. Thus, it appears that the 5–20 Å thick oxide commonly observed on polished surfaces results primarily from exposure to air and not from water rinsing (or water in the etching solution).

The initial oxide growth rate due to air exposure is very rapid and appears to saturate after a few minutes. However, the oxides of GaAs,

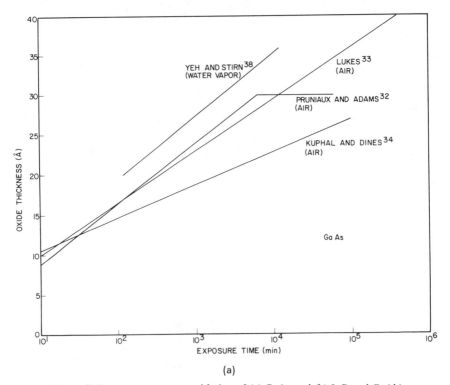

Figure 6. Room-temperature oxidation of (a) GaAs and (b) InP and GaAlAs.

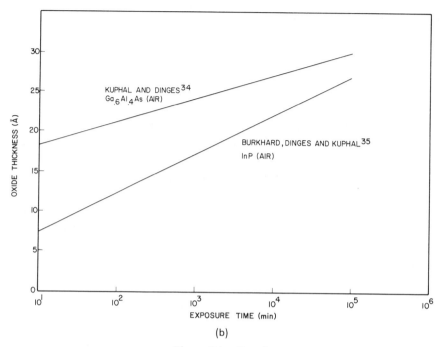

Figure 6 (*continued*)

$Ga_{0.6}Al_{0.4}As$, and InP have been shown to continue to thicken as a logarithmic function of time as shown in Fig. 6. For the oxide growths of Fig. 6, the samples were exposed to laboratory air with varying humidity and temperature. Note that the curve of Pruniaux and Adams[32] indicates that the oxide thickness on GaAs saturates after a few days, while that of Lukes[33] and Kuphal and Dinges[34] does not, even after one year. In fact in the later investigation, samples stored in air for as long as eight years still fit the logarithmic curve.

While the above ellipsometric measurements indicate a continuous oxide growth, certainly some collection of carbonaceous materials is also possible. Similar experiments on the other III–V's have not been reported, although, Rosenberg[36] has investigated the room-temperature oxidation of crushed InAs, InSb, GaSb, and AlSb. After an initial fast growth, the oxides appear to thicken logarithmically.

As discussed above, water rinsing in a nitrogen atmosphere does not grow an oxide layer, while exposure to water and O_2 can grow a substantial thickness of oxide. In fact GaAs and GaP will grow ≈ 1000 Å of oxide when placed in water with bubbling O_2.[37] Humid air is known to accelerate oxide growth for GaAs,[38] GaP, InSb,[39] and InP[40] and thus the thickness of oxides grown by exposure to air will vary with the humidity of the room. Yeh and Stirn[38] used water-saturated air to grow an oxide on GaAs in order to successfully fabricate MIS solar cells.

2.2. Chemical Etching and Growth of a Chemical Oxide

After polishing with Br–MeOH the III–V substrate must be removed from the polishing puck and often there is a need to store the substrate for a period of time. Both of these procedures allow the growth of an air–oxide layer. In some cases it is desirable to etch the surface just prior to use in order to remove the air-grown oxide and remove surface contaminants. However, for some applications, such as MIS Schottky diodes, it is necessary to grow a thin oxide layer in order to increase the barrier height. The growth of an oxide in a chemical solution or vapor is a simple way of accomplishing this. This section presents the basic concepts and results of oxide removal and chemical oxide growth.

When considering the possible oxide residues which may be left by an etch, the following factors are important:

1. Does the etch contain an oxidizing agent?
2. Some oxides, such as As_2O_3, P_2O_5, and Sb_2O_3, are highly soluble in water.
3. The solubility of Ga_2O_3 and In_2O_3 is greatest in strongly acidic or basic etches.
4. Equilibrium thermodynamics provides a guide to the final residue products.

The results for various etches on GaAs, InP, InSb, and InGaAs are given in the Tables 3–5 and are discussed below.

Thick oxides can be grown by simply dipping the substrate in an oxidizing solution, such as H_2O, H_2O_2, or HNO_3. The results on a number of III–V's are summarized in Table 5. One notes the wide variety of thickness obtained by these chemical oxidation methods. This is probably due to an apparent strong dependence of the growth rate on doping level, incident light, substrate conductivity type, and temperature. These factors point to the importance of free carriers in the chemical oxidation process as has been demonstrated by Schwartz and co-workers[37,54a,54b] on GaAs and GaP. On GaAs, a doping-level change from $N_D^- N_A = 1.5 \times 10^{18} - 2 \times 10^{17}$ reduces the growth rate significantly. There is not a complete understanding of this phenomenon at present, but the reader should be aware of such strong dependences and calibrate the process accordingly. The acidic etches such as HF, HCl, and H_2SO_4 which contain no oxidizer completely etch away the oxides on all of the III–V's. HNO_3 is an acidic etch but is itself an oxidizer and it will grow an oxide[48,52] while at the same time it dissolves any Ga_2O_3 and In_2O_3. Thus, on GaAs

Table 3. GaAs Etches

Etch	Contains an oxidizer?	Acidic	Results and remarks
HF[41,42]	No	Yes	Etches both the Ga_2O_3 and As_2O_3. Thus an oxide-free surface is formed until exposure to air. The air-grown oxide is composed of Ga_2O_3, As_2O_3, and probably some elemental As.
HCl[31,41,43,44]	No	Yes	Same as HF.
HNO$_3$[45]	Yes	Yes	Both Ga_2O_3 and As_2O_3 are formed, but the Ga_2O_3 has a tendency to be leached out by the acidic solution. Thus, an A_2O_3 rich layer is formed. Since the A_2O_3 is not stable in the presence of GaAs, there is a possibility of the As_2O_3 decomposing to form elemental As.
H_2SO_4 : H_2O_2 : H_2O[42,45–47]	Yes	Depends on the H_2SO_4/H_2O_2 ratio (R)	Both Ga_2O_3 and As_2O_5 are formed. The presence of As_2O_5 has also been reported. The As_2O_3 is leached out leaving a Ga_2O_3-rich layer. Since the pH is determined by the H_2SO_4/H_2O_2 ratio, the etch rate of the Ga_2O_3 and hence the thickness of the residual oxide varies with R. For $R = 10$, $t_{ox} = 10$ Å; for $R = 4$, $t_{ox} = 50$ Å.
Br-Methanol[31,43]	No	No	

Table 4. InP Etches

Etch	Contains an oxidizer?	Acidic	Results and remarks
HF[48]	No	Yes	Removes all of the oxide but an oxide regrows upon exposure to air. The air-grown oxide is made up primarily of $InPO_4$ and H_2O, but possibly include some $InO \cdot OH$, $In(OH)_3$, and/or H_3PO_4.
HCl[43]	No	Yes	Same as HF.
H_2SO_4[43]	No	Yes	Same as HF.
$H_2SO_4:H_2O_2:H_2O$[48,49]	Yes	Depends on the $H_2SO_4:H_2O_2$ ratio	This etch appears to remove all of the oxide even though there is an oxidizer present. This implies that H_2SO_4 readily dissolves $InPO_4$.
HNO_3[43,50–52]	Yes	Yes	On N-type thick oxide layers can be grown. The rate of oxidation is greatly accelerated by light and elevated temperature. The rate of oxide growth is highly dependent upon conductivity type, with the growth much faster on N-type. Growth on P-type is restricted to 30 to 40 Å.
Br–Methanol[31,43]	No	No	

Table 5. Chemical Oxidation

Semiconductor	Oxidizer	Conditions	Results	References
InP	HNO_3	10-s dip	100 Å in 14 s; $t_{ox} = At^{0.7\pm0.05}$	56
		73°C with light	Thin oxide ($InPO_4$)	50, 51
		65%/23°C/light		52
		5-min dip	20–40 Å ($InPO_4$)	43
	H_2O_2	Boiling, 30 min or longer	150–250 Å	56
	Br in H_2O	23°C, few seconds	20–60 Å	57
GaP	H_2O_2	Boiling (106°C)	Thickness ~ independent of temperature, 100 Å after 11 h	54a
		With large ohmic contact, boiling	1400 Å, P-type grows faster than N-type	54a, 54b
	H_2O	100°C, 8 h	100 Å	54a, 54b
	HNO_3	23°C, 7 days	90 Å	54b
GaAs	H_2O	Boiling	Large N_D–N_A dependence. The oxide is $Ga_2O_3 \cdot H_2O$.	37
	H_2O_2	Room temperature	115 Å after 6 days	54a
	H_2O	Room temperature	850 Å after 6 days	54a
InSb	H100[a]	30 s at room temperature	60–120 Å: 75% Sb oxide and 25% In oxide	58
	$H_2O:HNO_3:HCl$	Drops dispersed on the spinning wafer	Primarily In_2O_3	

[a] 70 g of KOH + 4 g of tartaric acid + 8 oz of ethylenediamine tetraacetic acid + 78 g of H_2O solution mixed before using with 30% H_2O_2 in volume ratio of 5:2.

and InGaAs, HNO_3 leaves a residue of As_2O_3 which can be removed with a water rinse. Similar results are expected on InAs. On InP, however, the HNO_3 grows an $InPO_4$ film which apparently is not readily dissolved by the acid and thus the HNO_3 can be used to grow a thick oxide layer on InP.[50–52] The oxide growth rate on InP in HNO_3 is strongly dependent upon conductivity type. Figure 7(a) shows that 1000 Å or more can be grown in a few minutes on N-type,[50,51] while under the same conditions only 40 Å will grow on P-type. Michel and Ehrhardt[52] used pure HNO_3 vapor to grow oxides on InP. They conclude that the absence of H_2O improves the quality of the oxide for MIS applications. For oxides

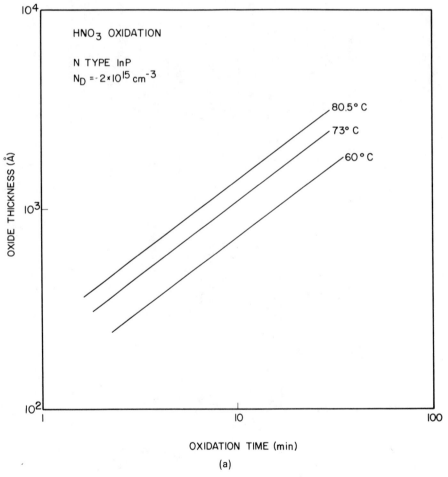

Figure 7. Chemical oxidation of (a) InP in HNO_3 (Wada et al.[50]) and (b) GaP in H_2O_2 (Schwartz and Sundbung[54b]).

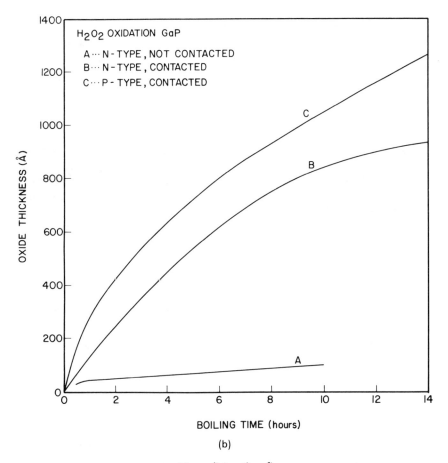

Figure 7 (*continued*)

thinner than ≈7 Å, Guivarc'H et al.[53] determined that the oxide was not continuous. This could imply an island growth or just extensive pinholes. H_2O_2 seems to oxidize InP in a manner similar to HNO_3, however, this oxidant has not been as well characterized for InP. On GaP, Schwartz and Sundburt[54b] studied the oxide growth in H_2O_2; Fig. 7(b) gives the growth curves. A thick oxide can also be grown on GaAs using H_2O_2[54] and presumably the other III–V's will also grow similar oxides in H_2O_2 and HNO_3.

$N_2SO_4:H_2O_2:H_2O$ etches the substrate in a normal mode, that is, the H_2O_2 oxidizes the surface and the H_2SO_4 etches the oxide. However, the H_2O_2 tends to neutralize the etching solution which reduces the etch rate of Ga_2O_3 and/or In_2O_3, and thus a film of Ga_2O_3 and/or In_2O_3 can be grown on GaAs with this etch. The thickness of the Ga_2O_3/In_2O_3 layer depends on the temperature and the $H_2SO_4:H_2O_2$ ratio since this determines the pH of the solution. The solubility of $InPO_4$ in H_2SO_4 appears to be less dependent upon the pH and thus an oxide layer is not grown on InP by this etch.

3. Thermal Oxides

3.1. General Overview

Table 2 indicates that the chemical composition of most interfaces resulting from thermal oxidation of III–V compounds are the products predicted by equilibrium thermodynamics. This undoubtedly is due to the increased thermal energy which allows kinetic barriers to be overcome.

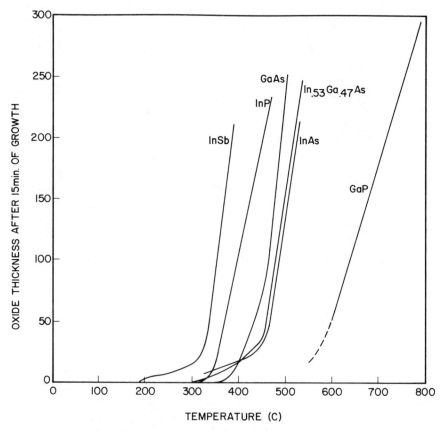

Figure 8. Thermal oxide thickness on III–V compounds after 15 min in dry O_2.

The interface of InP for $T \leq 600°C$ is a notable exception to equilibrium composition since no elemental P should be present.[7] It is thought that the P cannot diffuse through the oxide layer and is thus trapped at the interface.[60]

The temperatures at which rapid oxidation begins on the III-V compounds are summarized in Fig. 8, which plots the thickness of oxide grown after 15 min vs. the growth temperature.[59-63] Rapid oxidation begins at nearly the same temperature for InP, InAs, GaAs, and

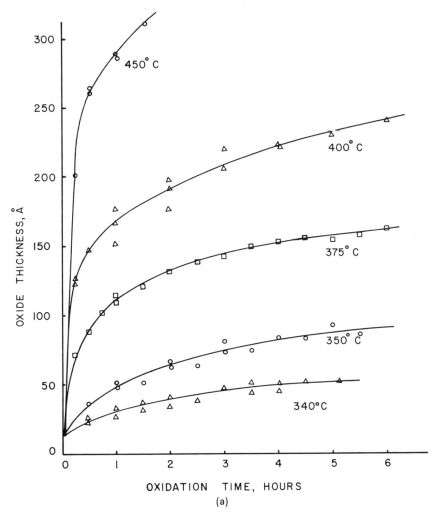

Figure 9. Thermal oxide thickness vs. time for (a) InP (Wager and Wilmsen[60]) and (b) GaAs (Muraka[30]).

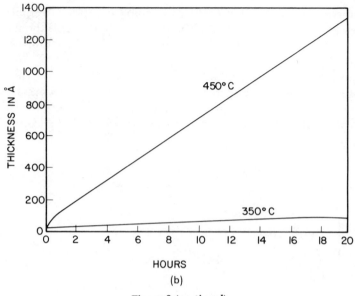

Figure 9 (*continued*)

In$_{0.53}$Ga$_{0.47}$As, but is considerably higher for GaP. More complete growth curves for InP and GaAs are given in Fig. 9.

For Si, thermal oxidation proceeds by the diffusion of oxygen through the oxide where it reacts with the Si substrate.[64] For thin oxides the process is reaction limited and for thick oxides diffusion limits the oxide growth rate. The thermal oxidation mechanisms for the III–V compounds are not fully understood or characterized, however, both diffusion and reaction processes must be involved. It is not known if the substrate ions diffuse out or if the oxygen diffuses in, however. O^{18}/O^{16} experiments rule out oxygen diffusion through interstitial sites.[65,66] This result plus the fact that unoxidized column V elements collect at the interface,[67] suggests that oxygen does not diffuse inward to the oxide/substrate interface. GaP may be an exception[63] since elemental P does not collect at the interface.

The oxidation model for the III–V's is complicated by the evaporation of the column V elements and their oxides, the difference in diffusion rate, and thermodynamic stability of the oxides and elements. However, the thermal oxide growth rates of the III–V's do follow a trend similar to Si, that is, there is a rapid initial growth which slows down considerably as the oxide thickens. This implies that there is a change in the rate-controlling process that depends on the oxide thickness. Without further knowledge these rate-controlling steps cannot be accurately defined. However, they are probably a complex form of diffusion and reaction processes.

3.2. InP

The thermal oxidation of InP begins with the formation of a layer of $InPO_4$,[68] which is the compound expected to form during equilibrium growth. For a growth temperature greater than $\approx 650°C$, the entire oxide film has been found[69] to be predominantly $InPO_4$ with some admixed In_2O_3. The interface of the high-temperature oxides does not contain elemental P.[70] For growth temperatures below $\approx 650°C$, an initial $InPO_4$ layer grows. With further growth an outer layer containing In_2O_3 forms with an $InPO_4$ inner layer and elemental P at the interface.[60,67,70] Changes in the topography also occur as the growth temperature increases above 600°C.[71] Figure 10 illustrates this change with SEM micrographs of oxides grown at 600, 625, 650, and 780°C. The oxide is observed to change from relatively smooth to one with large bubbles. The bubbles indicate that the oxide has softened and has then been pushed away from the InP surface by pressure from beneath the oxide. It is believed that P vapor provides this pressure. For the oxides grown at $T \leq 650°C$, the elemental P at the interface vaporizes and inflates the oxide bubbles. Above 650°C, the concentration of elemental P at the interface is greatly decreased and thus decomposition of the InP substrate becomes the source of P vapor. For the low temperature oxidation, the InP substrate below the oxide remains smooth but for higher temperatures the decomposition causes the substrate to pit. In order to form the bubbles, the oxide must soften. This softening provides the key to understanding the composition and the compositional changes of the oxide as explained in the following discussion.

For low growth temperatures, the diffusion rate of P through the oxide is smaller than that of In. The P that does diffuse to the surface either evaporates or reacts with In and O to form $InPO_4$. The remaining In reacts with O to form In_2O_3. Much of the P is trapped at the interface, where it remains unoxidized. Had the oxide crystallized at these low temperatures then the P may have escaped to the surface by diffusion along the grain boundaries. However, crystallization does not occur below 700°C.[69]

At a growth temperature of $T > 650°C$ the oxide softens allowing the diffusion rate of P to greatly increase. Thus, the elemental P is no longer trapped at the interface but diffuses to the surface where it reacts to form $InPO_4$. Some P still evaporates resulting in excess In at the surface which is incorporated into growing films as In_2O_3.

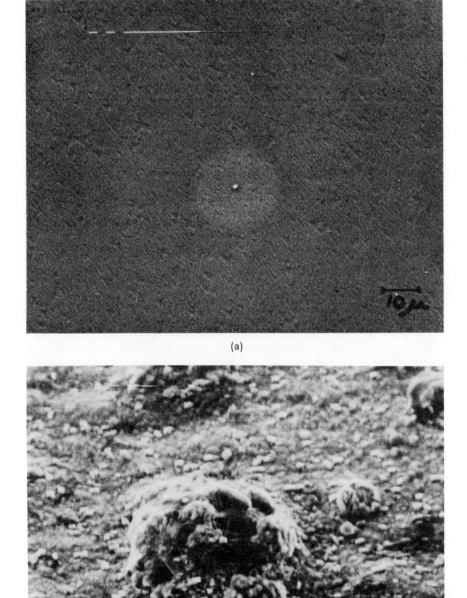

Figure 10. Surface topography of thermal oxides of InP grown in dry O_2 at (a) 600°C, (b) 625°C, (c) 650°C, and (d) 780°C.

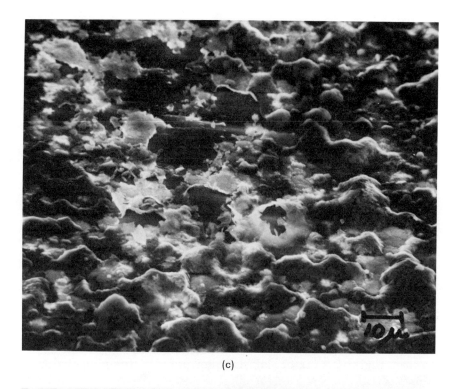

(c)

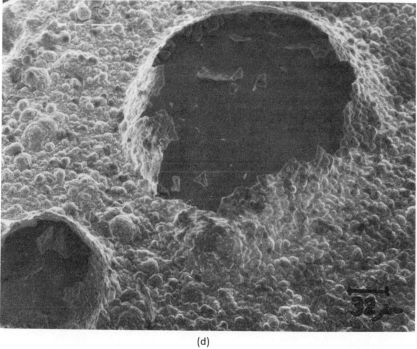

(d)

Figure 10 (*continued*)

The above P-diffusion-limited model along with the volatility of P and thermodynamics explains the main observations of the bulk oxide composition and surface topography of InP thermal oxides. These factors lead to a complex oxide/InP interface which changes with oxide thickness and temperature. For low temperature ($T \leq 600°C$) and thin layers ($t_{ox} \leq 40$ Å), the interface is relatively smooth and composed primarily of $InPO_4$ with only low concentration of elemental P and In_2O_3. As the oxide grows thicker, elemental P is trapped at the interface and In_2O_3 appears in greater concentration in the bulk of the oxide layer.

For high-temperature growth ($T \geq 650°C$), the oxide softens allowing much of the interfacial P to diffuse out but it also allows the P pressure to severely blister the oxide. Thus, the oxides grown above $\approx 625°C$ are not in physical contact with the InP substrate over a large percentage of the area. At even higher temperatures, the substrate begins to pit under the bubbles. High-pressure growth may prevent blistering but this has not yet been examined.

3.3. GaP

The thermal oxide of GaP does not have the compositional complexity of InP oxides since for all growth temperatures studied (185–>1100°C), the oxide is a uniform layer of $GaPO_4$ with no elemental P at the interface.[72-74] This implies that the kinetic barriers (primarily diffusion) to the oxidation of GaP are small and that the growth occurs near equilibrium. SEM micrographs show the oxide surface to remain smooth up to $\approx 950°C$. A connected void structure forms beneath the oxide even for 600 Å of oxide grown at 950°C (see Fig. 11). Above 1140°C, GaP oxidizes very rapidly. This is in part due to the flaking off of the oxide which exposes the bare substrate. Upon cooling, the thick oxide layers develop cracks due to stress. For oxides grown in steam, no interfacial voids are observed and the surface is much smoother than with dry-oxygen oxidation. The oxidation rate in steam is approximately 10 times that in dry oxygen and 2μ of oxide can be grown in 6 h at 850°C. These thick oxides crack into small tiles when pulled from the furnace.[63]

3.4. GaAs, GaSb, and InSb

The thermal oxides of these three compounds all follow a similar pattern, that is, the oxide layer is primarily Ga_2O_3 or In_2O_3 with elemental As or Sb collected at the interface. This is to be expected[19] from the phase diagrams of these materials but also from evidence that the diffusion rate of As and Sb is low.[67]

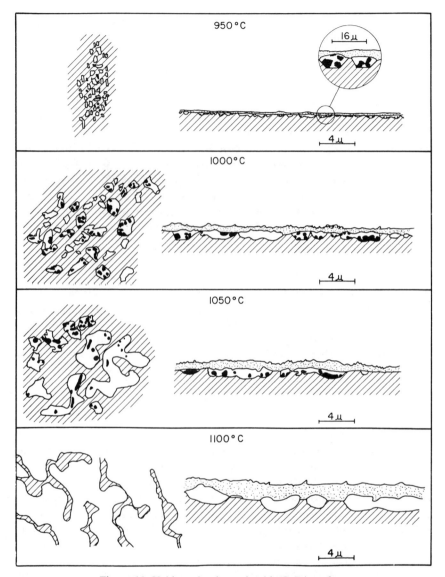

Figure 11. Voids at the thermal oxide/GaP interface.

Rosenberg and Lavine[75] and Rosenberg[76] were the first to detect the elemental Sb at the interface of thermally oxidized InSb. They angle lapped the samples and then traversed the lap with electron diffraction. The presence of elemental Sb was later confirmed by a number of other researchers[77,83] using other techniques for oxides grown at temperatures as low as 250°C. Korwin-Pawlawski and Heasell[78] stripped away oxides grown at 450 and 500°C and found that there were large In_2O_3 dendrites and hillocks of element Sb. This caused both the interface and the bulk oxide to be very rough.

The interface of the thermal oxide on GaAs is also quite rough. At lower temperatures, $T \leq 500°C$ Navratil and co-workers[59,79] found that the roughness began with the initial oxidation which indicated that evaporation of the substrate or the initial oxidation products could be the cause. The GaAs thermal oxide crystallizes into polycrystalline β Ga_2O_3 above a growth temperature of 500°C.[30] This also leads to surface roughness as seen in the photomicrograph of Fig. 12. The GaAs thermal oxide is composed primarily of Ga_2O_3 with some $GaAsO_4$ and As_2O_3 on the outer surface. At the interface, however, only Ga_2O_3 and elemental As are found.[80,81] This follows from the phase diagram which shows that As_2O_3 and $GaAsO_4$ are reduced by the GaAs substrate to form Ga_2O_3 and As. As_2O_3 and As are volatile and thus evaporation causes a loss of these compounds from the film if they can reach the surface. At high growth temperatures ($T > 500°C$), the polycrystalline grain boundaries provide an easier diffusion path for the elemental As and there is evidence that some of the interfacial As escapes, probably by this route.[80]

The thermal oxides of GaSb have not been extensively investigated, but the available evidence[19,82,83] indicates that elemental Sb collects at the interface and that the oxide bulk is composed of Ga_2O_3. Little Sb_2O_3 is found in the oxide and no evidence of $GaSbO_4$ has been reported.

3.5. InAs

While the In–As–O phase diagram is identical to that of Ga–As–O,[22] the thermal oxidation of InAs produces a somewhat different oxide and interface than that of GaAs. These oxide layers have been found to be In_2O_3 containing a large concentration of As_2O_3 even when grown at $T > 500°C$.[84,85] Thus, less evaporation of As_2O_3 takes place. In addition the elemental As is distributed throughout the oxide film and not concentrated at the interface. Yamaguchi et al.[85] have also reported a metamorphic layer in the substrate that increases in thickness with oxidation time and temperature. This layer was reported to be $\sim 0.3~\mu$ thick for growth at 450°C for 1 h and $\sim 100~\mu$ thick at 600°C for 20 h. The cause of the metamorphic layer is not presently known.

3.6. $In_{0.53}Ga_{0.47}As$

The thermal oxide growth rate of $In_{0.53}Ga_{0.47}As$ is between that of InAs and GaAs, although all three are very close.[62] At low temperature ($T = 350°C$), the oxide is composed of Ga_2O_3, In_2O_3, and As_2O_3. For

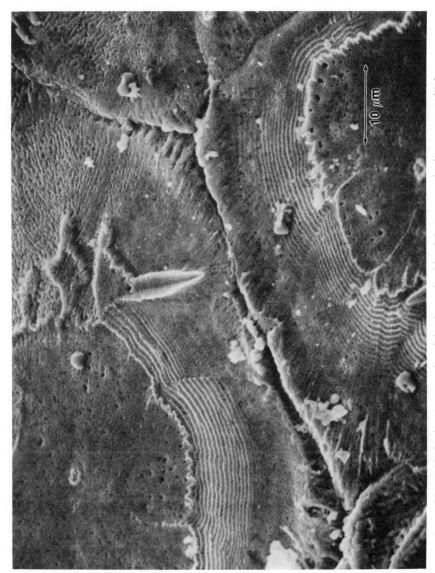

Figure 12. Surface topography of the thermal oxide of GaAs grown in dry oxygen at high pressure at 500°C. (R. Zeto, private communications.)

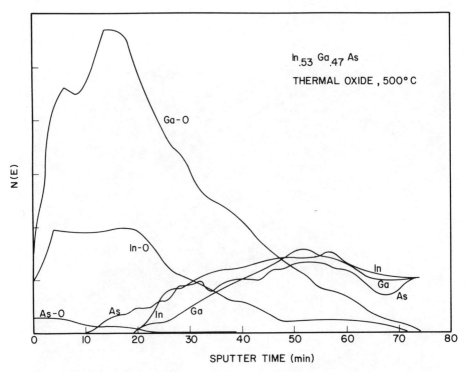

Figure 13. XPS profile of a thermal oxide grown on $In_{0.53}Ga_{0.47}As$ at 500°C in dry oxygen.

$T = 500°C$, the grown film contains a large concentration of Ga_2O_3 but almost no As_2O_3, as shown in Fig. 13. From Raman scattering it is known that elemental As has collected in films grown at 400°C and possibly lower.[22]

4. Anodic Oxides

The growth of an anodic oxide on the III–V compounds is quick and easy to accomplish. However, the process of anodization is complex and many details of the process are not well understood. Even so, many people have grown and analyzed anodic films for MOSFET application. In this section, a simplified model for anodization is presented. The model is not fully rigorous in an electrochemical sense but it provides a basic level of understanding for the user.

4.1. Anodic Oxidation Process

In principle anodization is a very simple process; one simply connects a metal or semiconductor to the positive terminal of a battery and places is along with an inert electrode in a suitable conducting liquid solution (called an electrolyte). The voltage drop around the loop can be modeled by the series circuit of Fig. 14.

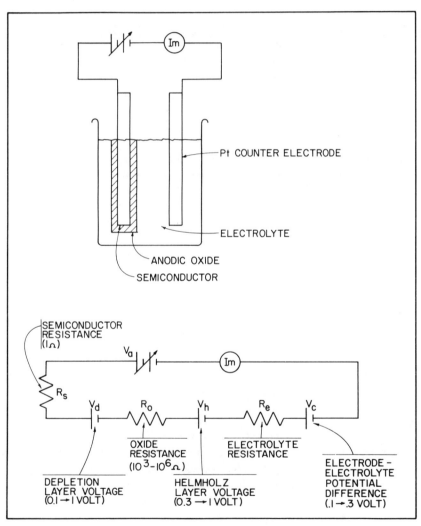

Figure 14. Anodic oxidation setup and equivalent circuit.

The applied voltage V_a is equal to the sum of the internal voltages, resulting in the following equation:

$$V_a = V_d + V_h + V_c + I(R_s + R_0 + R_e)$$

$$V_a = V_{\text{cont}} + IR(t)$$

$R(t)$ is a function of time because the oxide is growing. Note that there are voltages associated with the electrodes which can be considered to be contributing to a net contact potential difference, much the same as found in a battery. These voltages will change with electrolyte, semiconductor doping, surface preparation, light intensity, and so on. This light sensitivity is similar to what is found in electrochemical solar cells. In addition to these voltages, which have been lumped together as V_{cont}, there are a number of resistances associated with the semiconductor, oxide, and electrolyte. Normally, the oxide resistance is much larger than the other two except perhaps during the initial stages of oxide growth. This oxide resistance is often highly nonlinear and decreases with applied voltage. The growth of a thick oxide at room temperature requires a very large electric field across the oxide and for these conditions the oxide resistance is relatively low compared to that of a good insulator.

Since anodic oxidation is carried out near room temperature, the diffusion process cannot depend solely upon thermal activation of the atoms as occurs with thermal oxidation. In thermal oxidation, the diffusion process is modeled as a series of energy barriers as shown in Fig. 15.

The diffusion flux is given by

$$\text{flux} = -D\frac{dC}{dx} + \frac{q}{kT}D\varepsilon C$$

where

$$D \propto \exp[-(\Delta E - b\varepsilon)/kT]$$

ε = electric field
C = concentration
D = diffusion constant
q = electronic charge
k = Boltzmann's constant
T = absolute temperature

and b = the product of the jump distance and the electronic charge and is usually $\approx 20q$ Å.

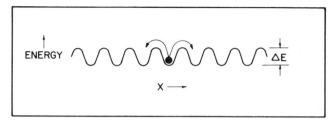

Figure 15. Energy barrier model for thermal diffusion.

For thermal oxidation, $\varepsilon \sim 0$, thus

$$\text{flux} \propto e^{-\Delta/kT} \frac{dC}{dx}$$

For anodic oxidation, however, ε is very large, usually $\sim 5 \times 10^6$ V/cm. Thus

$$\text{flux} \propto \frac{q}{kT} \exp[-(\Delta E + b\varepsilon)/kT]\varepsilon C$$

The diffusion barrier in the forward direction is reduced by $b\varepsilon$ as illustrated in Fig. 16. For $b \approx 20q$ Å, $b\varepsilon \approx (20 \times 10^{-8})(5 \times 10^6) = 1$ eV. Note that $b\varepsilon$ is many kT, even for $T = 500°C$ (773 K). Thus anodic oxidation is much faster than thermal oxidation even though the process is carried out near room temperature. As an example, several thousand angstroms of oxide can be grown on GaAs in a few seconds if the current from the battery is not limited. This is called constant voltage anodization. Normally, however, the current is limited to the range 0.1–5.0 mA/cm² and several minutes are required to grow the anodic film. This is still fast compared to thermal oxidation.

The chemical/electrical processes which take place during anodization are summarized in Fig. 17. The electrolyte must contain charge

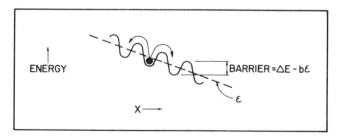

Figure 16. Energy barrier model for field-aided diffusion.

carriers since it must carry current. Undoubtedly, this current is carried, at least in part, by the anodizing ions and thus there is both mass and charge transport in the electrolyte. For continuity, there must also be mass and charge transport or generation/recombination at the two electrodes. Assuming that the OH^- ion is the primary oxidizing/transport specie, Fig. 17 illustrates these anodization processes for GaAs.

Note that holes are required in the GaAs substrate in order to form the Ga^{+3} and As^{+3} ions. Since GaAs is a wide-band-gap semiconductor, the density of holes is very low in n-type substrates. When anodizing n-type GaAs or InP, a strong light with energy in the range $1.4 < E_{light} < 4.5$ eV is required. The light also effects the depletion-layer width and voltage. The chemical equations for the above processes are:

At the Anode

$$12H^+ + 2GaAs \rightarrow 2Ga^{3+} + 2As^{3+}$$

$$2Ga^{3+} + 3(OH^-) \rightarrow Ga_2O_3 + 3H^+$$

$$2As^{3+} + 3(OH^-) \rightarrow As_2O_3 + 3H^+$$

At the Cathode

$$6H^+ + 6e^- \rightarrow 3H_2$$

$$6H_2O + 6e^- \rightarrow 6(OH^-) + 3H_2$$

4.2. Anodization Parameters

4.2.1. The Electrolyte

As shown in the previous section the electrolyte is simply a source of OH^- ions. The oxidation can also take place through H_2O or H_3O^+. Thus water (H_2O, H_3O^+, or OH^-) is the oxidizing agent and the oxidation–reduction reaction can be simply written as $2GaAs + 6H_2O \rightarrow Ga_2O_3 + As_2O_3 + 6H_2$. The kinetic reaction at the surface is, however, far more complex and there is also a need to increase the conductivity of the aqueous solution. For the anodization of GaAs, a few percent tartaric or citric acid is added to the water. For InP, phosphoric acid is often used. This changes the conductivity and the pH of the electrolyte. In water, the acid disassociates, that is, gives up a proton. For example:

phosphoric acid $\quad H_3PO_4 + H_2O \rightarrow H_3O^+ + H_2PO_4^-$

citric acid $\quad H_3C_6H_5O_7 + H_2O \rightarrow H_3O^+ + H_2C_6H_5O_7^-$

A base gives up an electron:

$$KOH + H_2O \rightarrow K^+ + OH^- + H_2O$$

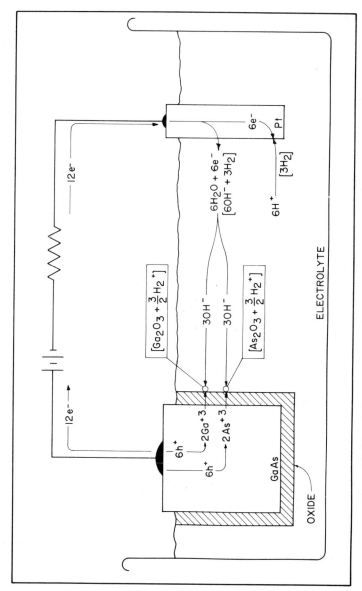

Figure 17. Chemical processes, mass flow, and current during anodization of GaAs.

Obviously, the charge carrier concentrations (+ and − ions) increase with the addition of the acid or base and hence the conductivity increases with the pH change. These added acids or bases also may dissolve part or all of the growing oxide layers, for example, the entire oxide layer may dissolve away (this is electrochemical etching) or the etching may be selective and remove only one of the mixed oxide species, or both oxides may be etched. This etching process appears to be complex and is not well understood, however, the following four parameters appear to be important:

> Viscosity of the electrolyte.
> Current density during growth.
> The anion specie.
> pH of the electrolyte.

Their role in the etching process can be understood in a simplified way as discussed below.

4.2.2. Viscosity

An increase in the electrolyte viscosity decreases both the etch and the growth rate of the growing oxide layer. This is caused by the reduced mobility of the solution anions which prevents the replenishment of anions at the surface.[86] Note that anions other than OH^- which participate in the oxidation process are usually large compared to H_2O^+ and OH^-, for example, citric acid ($H_2C_6H_5O_7^-$). Since the citric ion (an etchant) is impeded more than the OH^- oxidizer, oxide growth is enhanced over etching. To increase the electrolyte viscosity, propylene glycol may be added to the aqueous solution. The larger the glycol/water ratio, the slower the oxide growth and the more differential the dissolution.

4.2.3. Current Density

The effect of current density can be similarly explained. Consider the simplified case where the primary electrolyte current is carried by the OH^- and/or H_3O^+ ions and the anion current does not change with variation in the total current. For this example the oxide growth rate increases with total current but the dissolution rate remains approximately constant. From this it is postulated that for high current density, there is rapid oxide growth with little or no dissolution. As the current density decreases, the percentage of dissolution increases. Thus, if the dissolution is selective, for example, In_2O_3 is dissolved faster than P_2O_5, in the anodization of InP, then the composition may change with current density, particularly at low current densities. It is also evident that when the current density is reduced to such a low level that the growth rate equals the dissolution rate, no oxide layer remains on the substrate surface.

4.2.4. pH

The pH of the electrolyte is a measure of the hydrogen ion (H^+) concentration, that is, for a neutral solution the pH = 7 and the H^+ and OH^- concentrations are both equal to 10^{-7}. For a pH = 2 the solution is acidic and the H^+ concentration = 10^{-2} and the OH^- concentration = 10^{-12}. The H^+ and OH^- ions participate in intermediate reactions which

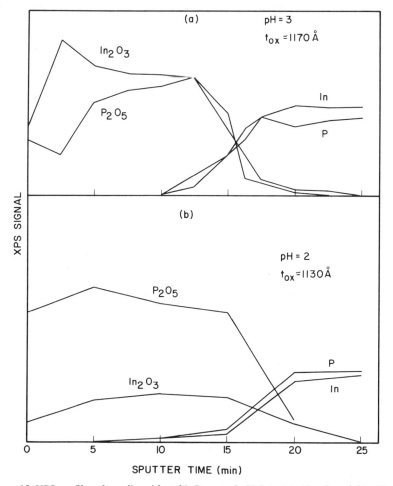

Figure 18. XPS profiles of anodic oxides of InP grown in H_3PO_4: (a) pH = 3, and (b) pH = 2.

control or catalyze the dissolution process. For example, consider the reaction between In and H_2O:

$$2In^{3+} + 3H_2O \rightleftarrows In_2O_3 + 6H^+$$

The reaction to the right results in an oxide layer, but the one to the left is a dissolution/reduction reaction; the reversal of the reaction direction occurs at a pH = 2.5. Thus, for pH < 2.5, the In_2O_3 is dissolved rapidly, while at a pH > 2.5 the In_2O_3 film grows. There are many such equilibrium dissolution equations for the various elements. These are summarized in the Pourbaix diagrams.[87,88]

The variation in oxide-layer composition as a function of pH is clearly demonstrated by the anodic layers on InP grown in a 3% phosphoric acid electrolyte.[61] XPS composition profiles for a pH of 3 and 2 are shown in Fig. 18. Note the drastic reduction in the In_2O_3 concentration when the pH is reduced from 3 to 2. This is predicted by the Pourbaix diagram for In as discussed above. While the Pourbaix diagram works very well for In, the results for P are somewhat misleading in that no solid phases are predicted. However, the diagrams assume equilibrium. Thus, it appears that the oxide films shown in Fig. 18 are grown fast enough to prevent total dissolution. In fact, it has been found that the oxides grown in a pH of 2 must be grown very fast in order to prevent complete dissolution of the anodic films.

4.3. Initial Growth

Examination of the cell potential (V_{net}) vs. growth time for a constant current source (Fig. 19) is very instructive. The V_{cont} and $Ir(t)$ regions

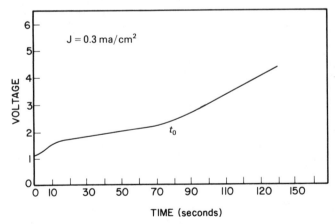

Figure 19. Cell voltage vs. time during the anodization of GaAs.

were discussed earlier. V_{cont} can be thought of as the net contact potential and the $Ir(t)$ as the voltage drop across the growing oxide. This later voltage increases linearly with time (for a constant current) and thus the voltage is a measure of the oxide thickness. With InP the V vs. t curve is often much more complex than the simple curve shown in Fig. 19. This could be caused by the lack of holes in the InP, resulting in a large voltage drop across the space-charge layer.[89]

The $V - t$ curve in the passivation region is usually characterized by a slow increase in voltage with time. During this time, oxide islands are nucleated and grown. At approximately t_0, the islands have coalesced into a continuous layer over the surface. The slow voltage increase during this period indicates that the islands are getting somewhat thicker as they grow laterally along the surface. This nucleation and island-growth process is illustrated for GaAs[2,3] and InP[4] in the high magnification transmission electron micrographs of Figs. 20–22. One can observe that the islands do indeed grow thicker as they grow laterally.

For the GaAs anodic oxide, the islands are generally circular in shape except when altered by coalescence. The islands are seen to be of various sizes from very small, ≈ 100–200 Å in diameter, to very large, $\approx 4~\mu$. The shape is generally round until significant coalescence has occurred. Some islands have clearly coalesced, while others appear to be near the nucleation stage. This large variation in the stage of island development suggests that the nucleation, island growth, and coalescence occur continuously during the passivation time and not in separate stages.

The photomicrographs of Fig. 20 show some islands which have clearly coalesced into irregular shapes. In other areas the islands are circular even though some are quite large. This would imply that the lateral growth of the islands is uniform in the radial direction and that after coalescence the islands are forced back into a circular shape. The islands of a deposited metal or water droplets behave in a similar fashion. For those systems there is surface migration of the atoms or molecules to minimize the surface-to-volume ratio in what is often called liquidlike coalescence. For the anodic oxide, it is noted that some coalescing islands have a very distinct boundary [see Fig. 21(a)], while others show no boundary [Fig. 21(b)]. The islands also seem not to move [Fig. 21(c)] but rather touch and then fill in. The dashed lines indicate the apparent shape and location of the islands before touching. The boundary appears not to move as the neck region is filled in. Thus, no liquidlike rapid coalescence appears to occur. Instead, it is believed that there is preferential growth in the neck regions caused by field intensification. Also, since the photographs show islands either well into coalescence or just touching, it is believed that the oxide growth in the neck regions is very fast.

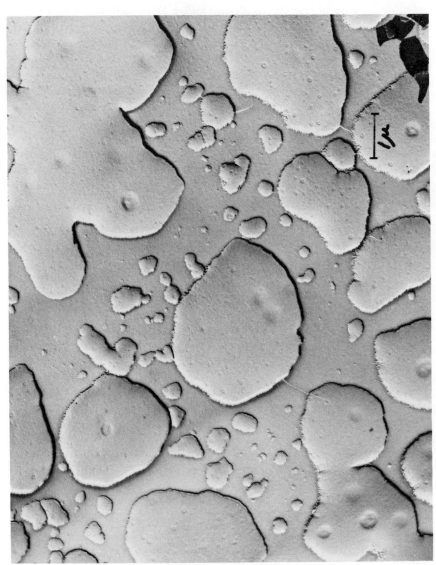

Figure 20. TEM photomicrograph of anodic oxide islands on GaAs.

The length of the shadow cast by the islands provides a convenient method of obtaining a relative measure of the island height. The shadow lengths were calibrated with replicas from SiO_2 islands formed in thermally oxidized SiO_2 layers by photolithography. A plot of the shadow length vs. island area for one sample is given in Fig. 23 for island areas ranging from $\approx 10^{-3}$ to $3\,\mu^2$. The island height is seen to increase

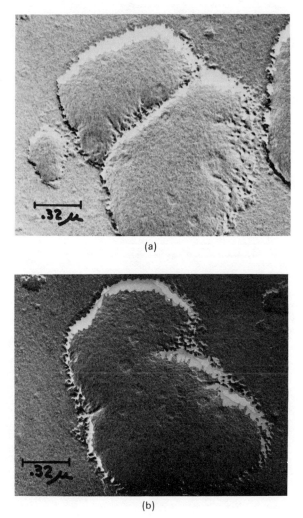

(a)

(b)

Figure 21. TEM photomicrographs of GaAs anodic oxide islands: (a) coalesced islands with a distinct boundary, (b) coalesced islands with most of the boundary filled in, and (c) (page 442) during coalescence.

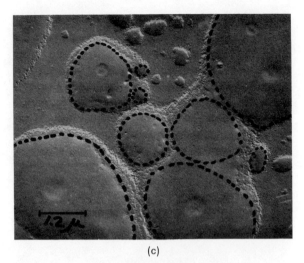

(c)

Figure 21 (*continued*)

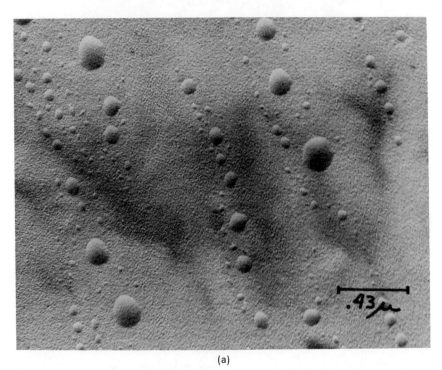

(a)

Figure 22. Anodic oxide islands on InP: (a) formed by dipping in the electrolyte and (b) growth in tartaric acid (pH = 7), strong light.

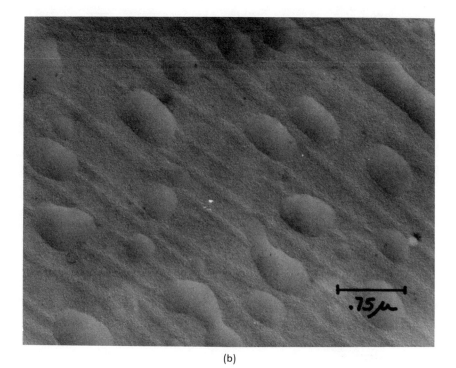

(b)

Figure 22 (*continued*)

monotonically with areas up to an area of ≈0.05–0.1 μ^2 after which the height remains approximately constant. The islands initially grow outward and horizontally until a critical size is reached, after which the islands grow primarily horizontally, although some outward growth is possible. The very smooth curve of Fig. 23 implies that there is a near one-to-one relationship between island area and height.

The photomicrographs of Fig. 22 illustrate the primary island structure of anodic oxides on InP. While the samples were prepared in different ways, each one has the same smooth, round islands which is believed to be the basic island shape. The island characteristics did not change with electrolyte pH equal to 2.7 and 7.0, the amount of illumination, the electrolyte, and rinsing with water or methanol. The smooth-edged islands of the InP oxide are much different than the ragged and pitted edges of the GaAs anodic oxide island shown in Figs. 20 and 21. The InP islands often (but not always) form rows ≈0.2–0.6 μ apart. The lining up of the islands imply that there are lines of preferred nucleation sites which could be the result of surface steps.

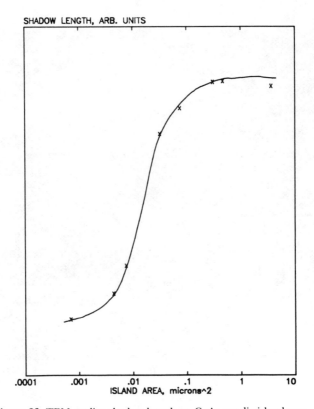

Figure 23. TEM replica shadow length *vs.* GaAs anodic island area.

The approximate height of the islands was estimated by measurement of the shadow length as discussed above. The shadows on the InP replicas were not as distinct as those on GaAs, which reduces the accuracy of the measurement and prevents estimation of the shadow length of the smaller islands. For the larger islands this method yields a height of ≈ 200 Å, which is similar to that obtained on GaAs.

Some coalescence is seen in the islands of Fig. 22(b) in which islands in the same line form elongated islands. The coalesced islands are all smooth with no distinct boundary between them, in contrast to the case of GaAs. However, as with GaAs, the islands appear stationary with the neck regions filling in by a fast-growth process due to electric field enhancement in these regions. The elongated and irregular island shapes argue against a liquidlike coalescence due to surface migration of molecules, as discussed previously for GaAs.

4.4. Chemical Composition of Anodic Oxides and Interfaces

In this section, the composition of anodic oxides grown on various substrates is presented in summary form.

4.4.1. GaAs

Feldman and his co-workers[90] appear to be the first researchers to have investigated the composition of anodic oxides on GaAs. Using Rutherford backscattering (RBS), they determined that the oxide bulk had a Ga:As ratio of ≈ 1 while the surface was depleted of As. The composition of the anodic oxide/GaAs interface has been the subject of some debate, primarily related to the presence or absence of elemental As and to a lesser extent if a thin layer of Ga_2O_3 exists.

The composition of thin anodic oxide layers on GaAs is illustrated by the XPS profile of Fig. 24.[91,92] The profile indicates an inner layer of Ga_2O_3 containing little or no As_2O_3. While it is possible that the sputter beam used to profile these oxides could reduce the As_2O_3, it does not seem likely that the same interfacial Ga_2O_3 layer would be observed for all thicknesses. Similar results for thick oxides have been reported by Mizokawa et al.[93] and Breeze et al.[94] Raman scattering[18] and spectroscopic ellipsometer[95] measurements have not detected this Ga_2O_3 inner layer. However, the dielectric constant and band gap of Ga_2O_3 are very close to those of As_2O_3 and thus these techniques may not be sensitive enough to detect such a thin inner layer. In addition recent TEM cross sections of anodic films show a narrow Ga rich band at the interface.[119] These data tend to favor the existence of the Ga_2O_3 inner layer.

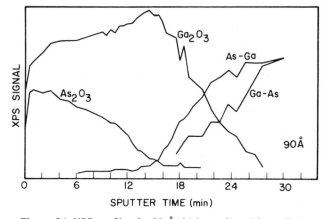

Figure 24. XPS profile of a 90-Å-thick anodic oxide on GaAs.

The XPS profiles also indicate a small amount of elemental As at the interface.[91,93] However, optical techniques and RBS channeling do not detect any As.[95,96] These latter techniques are much more sensitive to elemental As than is XPS and thus it would seem that the evidence is weighted against the presence of elemental As at the interface.

The bulk oxides grown in most electrolytes contain Ga_2O_3 and As_2O_3 in approximately equal proportion and are uniformly distributed, except at the interface and in a thin surface layer. Mizokawa *et al.*[93] report evidence of $GaAsO_4$ in anodic layers grown in an electrolyte of $K_2Cr_2O_7$. The profiles of these oxide layers are also nonuniform which is different from most other reports.

4.4.2. InP

The anodic oxides of InP are composed of a mixture of In_2O_3 and P_2O_5 in contrast with the In_2O_3 and $InPO_4$ mixture found for the thermal- and plasma-grown oxides.[91,97] The presence of P_2O_5 indicates that the oxide is not grown under equilibrium conditions since $InPO_4$ is the expected oxidation product. Investigation of oxides grown in various electrolytes, current densities, and pH's (1.5–13) indicates that the depth profile of the anodic oxides is strongly dependent upon the growth conditions. An example of this is shown in the XPS profiles of Fig. 18 which illustrate the effect of pH on the dissolution of In_2O_3 from the growing oxide film. However, while the bulk oxide compositional profile varies with growth conditions, the interfacial composition contains an approximately constant P_2O_5/In_2O_3 ratio. The ratio is usually ~1 but increases for bulk oxides which have a high concentration of P_2O_5. This constancy of the interfacial compositional ratio begins with the island stage of growth as illustrated by the XPS profiles of two very thin oxides in Fig. 25.[91] For these thin InP anodic oxides no inner layer of a single compound is observed as was shown to be the case with GaAs (Fig. 24). The composition of the InP interface is indeed similar to the bulk oxide and no evidence for elemental P has been reported.

Electrical measurements of InP anodic oxide have been used to determine some structural features.[98] First, the resistivity of the oxide has been observed to increase dramatically with the P_2O_5/In_2O_3 ratio (Fig. 26). In addition, temperature data on the high-resistivity oxides show very little resistance change with temperature.[99] These results strongly suggest that the In_2O_3 and P_2O_5 are phase separated within the oxide layer into small volumes of high and low resistivity. A simplified model for oxide structure is shown in Fig. 27. It is not known if the interfacial region also contains this phase-separated structure, but there is no reason to believe that it does not.

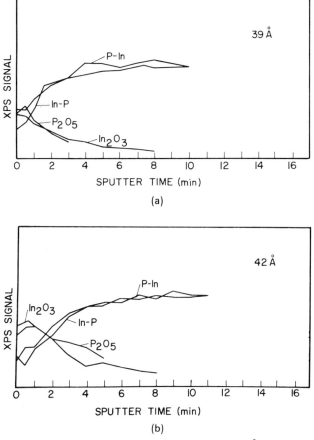

Figure 25. XPS profiles of anodic oxides on InP: (a) 39 Å and (b) 42 Å.

Similar phase separation of the oxide components does not appear to occur with the anodic oxides of other III–V semiconductors except perhaps GaP. In fact, the composition of all other anodic oxides except GaP do not change much with growth conditions. This suggests that it is the oxidation of the P and not the dissolution of In_2O_3 or Ga_2O_3 that plays the lead role in determining the resulting oxide composition and structure. However, this is only speculation at the present time.

4.4.3. GaP

The anodic oxides of GaP appear similar to those of InP in that the composition of the oxide bulk is strongly dependent on the pH of the electrolyte and is nonuniform except when the pH is small. Poate *et*

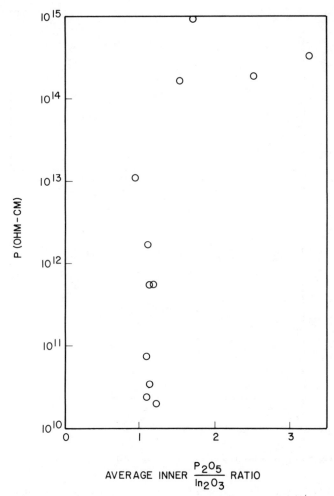

Figure 26. Resistivity of InP anodic oxides as a function of P_2O_5/In_2O_3 ratio.

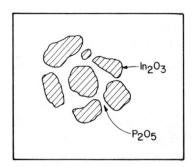

Figure 27. Phase-separated model for the InP anodic oxide.

al.[100-102] showed this to be the case using RBS. XPS and RBS profiles show that the interface is composed of both Ga and P oxides with no apparent elemental P. Bilz et al.[103] report a 5.8–6.5 eV shift of the P(2p) binding energy which indicates that the P is probably in the form of P_2O_5 and not a phosphate, whereas just the opposite was found by other workers for the thermal oxides.[72-74] These binding-energy shifts of the P(2p) line are similar to those reported on the thermal and anodic oxides of InP, respectively. No elemental P is detected at the anodic oxide/GaP interface.

4.4.4. InAs

Chemical analysis of InAs anodic oxides has been reported by Baglee et al.[104,105] using XPS sputter profiling. Unlike the InP anodic oxides, it appears that relatively small changes occur in the InAs anodic oxide profiles with various electrolytes and pH values. The profiles indicate a uniform distribution of In_2O_3 and As_2O_3 in the oxides with the possibility of a small concentration of elemental As in the oxide and at the interface. However, the latter is probably due to reduction of the As_2O_3 by the ion milling. There appears to be a small change in interface width with electrolyte but this could also be due to the ion milling.

4.4.5. GaSb

Bilz and co-workers[103] have reported the composition of anodic oxides of GaSb. Their XPS profiles of oxides grown in three different electrolytes (0.1 N $KMnO_4$, 0.1 N KOH, and 30% H_2O_2) indicated large differences in the composition between the $KMnO_4$ grown oxide and the other two. In the oxide grown in $KMnO_4$ the Sb oxide level greatly exceeds that of the Ga oxide and appears to form a Sb oxide rich layer at the interface. They also report a Sb-oxide-rich interfacial layer for the anodic oxide grown on InSb in the same electrolyte. The oxides grown in the other two electrolytes appear to have an Sb oxide/Ga oxide ratio of ≈1, except at the surface where some Sb oxide is lost to the electrolyte.

4.4.6. InSb

Dewald[106] was the first to report on the chemical composition of an anodic oxide of InSb. He used a microtechnique to chemically analyze anodic oxides of InSb grown in an 0.1 N KOH electrolyte. Sb was found to be deficient at the surface but the Sb:In ratio was ~1 throughout the oxide bulk and at the interface. This was independent of growth rate. Dewald[107] applied his previously developed model to explain the kinetics of the anodic growth on InSb. This model assumed that the oxide grew by the outward movement of cations and was limited by either the ion space-charge current within the oxide or the entrance of the ions into the film at the film/InSb interface. Since pure Sb oxides readily dissolve in KOH, the presence of a large concentration of Sb oxide in the anodic film suggested that a simple interstitial ion transport model was not sufficient to explain the observed growth. Later sputter–Auger profiles of Wilmsen[77] agreed with the chemical profiles of Dewald. The Auger results indicated a low Sb concentration at the surface, uniform Sb concentration in the oxide bulk, and no elemental Sb at the interface.

4.5. Thermal Annealing of the Anodic Oxides

Thermal annealing of Si/SiO_2 MOS devices significantly reduces the interface state density. The annealing ties up bond states by either attaching an H atom or by more completely oxidizing the interface. Similar improvements are rarely achieved for the oxide/III–V interface. However, some very useful information about the composition, thermodynamics, and stability of the oxides has been obtained by annealing the anodic oxides.

The results of thermal annealing the III–V anodic oxides are summarized in Table 6. From this table it is seen that the anodic oxide interfaces of GaAs, InAs, GaSb, and InSb approach the equilibrium composition at relatively low annealing temperatures. For InP and GaP, the oxides either do not change or they combine to form a phosphate,[18,24] but no evaporation or collection of elemental P has been detected. All the other oxides decompose to form elemental Sb or As. The evaporation of the Sb_2O_3 and As_2O_3 is also possible. For GaAs, the oxide layer has been found to begin to crystallize at approximately the same temperature at which the As_2O_3 begins to evaporate, namely 450°C.[112]

The above discussion emphasizes the chemical changes which occur when the anodic oxides are heated. Most of these changes will occur independent of the annealing gas. Although using an As_2O_3 or Sb_2O_3 vapor will help prevent evaporation of these oxides, it will not prevent oxide decomposition at the interface. Below the critical decomposition temperatures, annealing may be of benefit to the electrical properties, mainly through the evaporation of H_2O.

Upon heating the GaAs anodic oxide, As in some form was found to diffuse out of the layer at a temperature of 450°C.[90] This also caused the oxide layer to become thinner. Spitzer et al.[108,109] showed that baking in N_2 caused evaporation of water from the anodic oxide but also showed that there was no apparent reduction in oxide thickness for baking temperatures ≤400°C. Etching away an oxide layer baked at 400°C

Table 6. Annealed Anodic Oxide Interface

Substrate	Temperature at which the interface changes	Reaction products	Remarks	References
GaAs	300°C	Ga_2O_3, As	At 450°C, H_2O and As_2O_3 evaporate	18, 109
GaSb	300°C	Ga_2O_3, Sb		18
GaP	Not reached	None		18
InP	600°C	$InPO_4$	P_2O_5 does not evaporate. There is considerable surface topography change in the oxide above $T \approx 625°C$. H_2O is evaporated at $T \approx 200°C$.	24, 120
InSb	240–260°C	In_2O_3, Sb	Evaporation of Sb_2O_3 is possible at 400°C.	23
InAs	500°C	In_2O_3, As		22

exposed a granular interface material which did not readily etch. Baking at 600°C greatly increased the etch resistance of the entire film. It was also reported that the interface became rough after anodizing to a high voltage. Ishii and Jeppson[110,111] used RBS and mass analysis of the species evaporating from the surface to investigate the effects of annealing on the composition of GaAs anodic oxides. Up to 300°C, no changes were observed. Above 300°C, both H_2O and As oxide began to evaporate. At 450°C, the evaporation was rapid and essentially complete within 15 min.

Weiss and Hartnagel[112] also reported As evaporation from GaAs anodic films at 400°C and showed that the evaporation was accompanied by the appearance of β-Ga_2O_3 crystallites. The rate of crystallization was found to be a function of the current density used to grow the oxides, that is, the higher the current density the more readily the oxide layer crystallized when heated.

Nakagawa et al.[23] investigated the annealing of the InSb anodic oxide using Raman scattering. Anodic layers examined before annealing indicated little or no elemental Sb. However, after heating in air, Raman scattering indicated the presence of elemental Sb which begins to appear at a temperature between 240 and 260°C. They suggest that the formation of elemental Sb during annealing is not caused by O_2 in the air but rather by the decomposition of Sb oxide through a reduction reaction.

Schwartz et al.[18] examined the anodic oxides of GaSb with Raman scattering before and after annealing. They found no evidence for elemental Sb in the "as-grown films", but after annealing at $T \leq 300$°C the presence of Sb was clearly seen in the spectra. At an annealing temperature of 300°C, the Sb appeared to be amorphous. At 350°C or above, the Sb deposits were crystalline. Coupled with pseudobinary reaction experiments they conclude that the oxides are not grown near equilibrium conditions and that the Sb_2O_3 reacts with the substrate during annealing.

InP anodic oxides appear to be composed of P_2O_5 and In_2O_3 which are not the equilibrium products. Schwartz et al.[7] annealed an anodic film at 600°C for 1 h and later[113] at 850°C in order to observe any changes in the film composition. In particular they sought to observe the elemental P which would indicate that the P_2O_5 reacted with the InP substrate. Using Raman scattering no elemental P was detected in these annealed films. Fathipour et al.[24] used XPS, X-ray diffraction, and scanning electron microscopy to investigate the annealing. XPS sputter profiles of oxides annealed at 500 and 600°C clearly show a large concentration of P oxide distributed throughout the annealed film. Thus, P oxide does not evaporate from the anodic films annealed at $T \leq 600$°C. Above 600°C the oxide film becomes very rough with overgrowths and eventually the film peels off. Accurate profiles of these films could not be obtained, but the XPS data clearly indicated the presence of P oxide. X-ray dispersive data also indicated P in the 650°C-annealed films. Annealing the oxides at 600°C also caused an InPO$_4$ peak to appear in the X-ray diffraction spectra. Thus, it seems that at least some of the P$_2$O$_5$ and In$_2$O$_3$ convert to InPO$_4$. Water was found to evaporate from the anodic oxide at $\approx$200°C.[120]

5. Plasma-Grown Oxide

The surface of a metal or semiconductor is oxidized when exposed to a gaseous plasma containing oxygen. The gas could be O$_2$, N$_2$O, or CO$_2$ and is usually excited by a rf coil. Placing InP in or near such a plasma at room temperature quickly grows 20–50 Å of oxide.[26] To grow

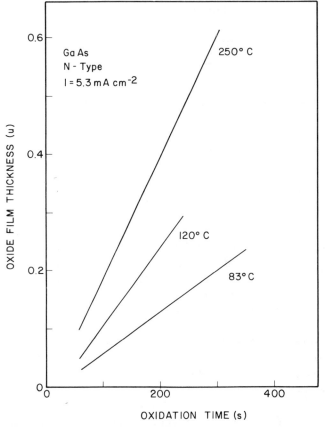

Figure 28. Plasma oxide thickness on GaAs as a function of time (Sugano[115]).

thicker layers usually requires a dc bias on the substrate in much the same way as with wet anodization. In fact, the process of plasma oxidation or plasma anodization is very similar to that of wet anodization. In this case, the plasma serves as a source of charged oxygen species which are drawn to the oxide surface. The impinging ions can sputter the surface and thus reduce the net growth rate.[114]

Most of the applied dc voltage appears across the oxide and field-aided diffusion of either the oxygen or the substrate atoms occurs. The growth direction is not known but it is believed that for GaAs the substrate atoms diffuse to the surface where they oxidize.[115]

For GaAs, Sugano[115] reported that the oxide thickness increases approximately linearly with time if the current is held constant, as shown in Fig. 28. This is similar to wet anodization. The growth rate is also relatively independent of gas pressure as would be expected for a diffusion-controlled process. XPS profiles of a plasma oxide, before and after annealing, were reported by Watanabe et al.[66] as shown in Fig. 29. From these profiles it is seen that the as-grown film is uniformly composed of Ga_2O_3 and As_2O_3 in approximately equal concentrations. The interface appears abrupt and may contain a low concentration of elemental As. Annealing at 600°C causes an evaporation of As_2O_3 from the surface and decomposition of As_2O_3 at the interface into elemental As. Chang et al.[25] used TEM to show that, after annealing, the interfacial As is in the form of crystalline clusters and is not uniformly distributed. The element As results from the decomposition of As_2O_3 by the GaAs substrate as predicted by thermodynamics.[17-19]

The plasma oxidation of InP is less well characterized even though there have been several independent studies. Clark and Fok[26] investigated the surface modification of InP using 10 W of rf power in an oxygen plasma. Using XPS they found that the grown oxide was composed of $InPO_4$ and that the oxide reached a self-limiting thickness of about 20 Å after an oxidation time of 60 s. Kanazawa and Matsunami[116] explored the possibility of using plasma oxidation of InP for metal/oxide/semiconductor applications. They capacitively coupled 80 W of rf power into a magnetically confined oxygen plasma. Their plasma oxides grew to a saturated thickness of about 700 Å in less than 1 h. The oxidation rate was found to depend upon whether the substrate was grounded or allowed to float. Auger depth profiles of these thick oxides indicated deposits of phosphorus at the oxide-semiconductor interface, similar to that observed for thermal oxides of InP. Imai et al.[117] applied plasma oxidation in an attempt to improve the reverse leakage current of InP Schottky diodes. 100 W of rf power was used to excite the oxygen plasma. They found that the reverse leakage current was substantially reduced for oxidation times of less than 2 min, however, the current

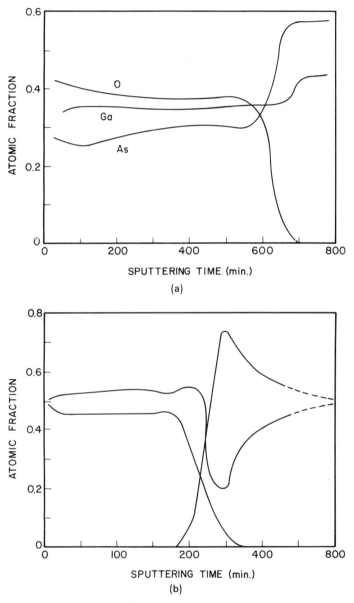

Figure 29. XPS profile of a plasma-grown oxide on GaAs: (a) as grown and (b) after annealing at 600°C (Watanabe et al.[66]).

dramatically increased for longer oxidation times. No saturation of the oxide thickness was observed for oxidation times of up to 8 min. The oxide thicknesses were estimated to be less than 100 Å.

The above oxides were grown with the substrate in the same chamber as the plasma and with no substrate heating. Wager et al.[118] measured the oxide thickness grown after 5 min at various temperatures with the substrate in a chamber separate from that of the plasma. Figure 30 compares the oxide thickness resulting from no plasma (thermal oxidation) and at two different distances from the oxygen plasma source. For all temperatures, the oxide thickness decreases with a reduction in plasma intensity. At $T \approx 350°C$ the oxide growth appears to be dominated by the thermal oxidation process, that is, above 350°C the plasma oxidation is no longer the dominant mechanism of oxide growth. XPS profiles indicate that the plasma oxides of InP are similar to the thermal oxides, with the initial layer composed of $InPO_4$, and as the oxide thickens the P is trapped at the interface and the outer surface becomes P depleted. This leads to an outer layer of In_2O_3 mixed with the $InPO_4$.

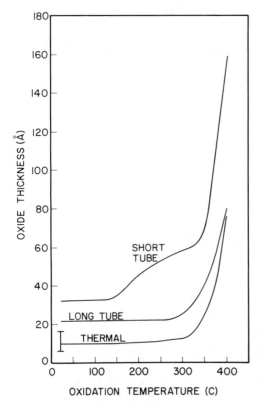

Figure 30. Plasma-grown oxide on InP after 5 min. of growth as a function of substrate temperature in two different plasma intensities. The thermal oxide thickness is included for comparison.

Acknowledgments

I wish to thank my students whose research made this work possible. Art Nelsen deserves special thanks for his help with the tables and Kent Geib, Steve Goodnick, and John Wager for reading the manuscript. This work was supported by ONR and ARO.

References

1. P. Pianetta, I. Lindau, C. M. Garner, and W. E. Spicer, Determination of the oxygen binding site on GaAs (110) using soft X-ray photoemission spectroscopy, *Phys. Rev. Lett.* 35, 1356–1359 (1975).
2. S. Szpak, Electro-oxidation of gallium arsenide: I. Initial phase of film formation in tartaric acid–water–propylene glycol electrolyte, *J. Electrochem. Soc.* 124, 107–112 (1977).
3. W. H. Makky, F. Cabrera, K. M. Geib, and C. W. Wilmsen, Initial stages of anodic oxidation GaAs, *J. Vac. Sci. Technol.* 21, 417–421 (1982).
4. W. H. Makky and C. W. Wilmsen, Island stage of InP anodization, *J. Electrochem. Soc.* 130, 569–662 (1983).
5. ASTM index of Powder Diffraction File.
6. G. V. Samsonov, *The Oxide Handbook*, Plenum Press, New York (1973).
7. G. P. Schwartz, W. A. Sunder, and J. E. Griffiths, The In–P–O phase diagram: Construction and applications, *J. Electrochem. Soc.* 129, 1361–1367 (1982).
8. D. D. Wagman, W. H. Evans, V. B. Parker, I. Halaw, S. M. Baily, and R. H. Schumm, *Selected Values of Chemical Thermodynamic Properties*, NBS Technical Note 270-3 (January 1968).
9. Paul G. Stecher, *The Merck Index*, 8th ed., Merck & Co., Rahway, N. J. (1968).
10. NBS Circular 539, Vol. 8 (1958).
11. R. C. Weast, *CRC Handbook of Chemistry and Physics*, CRC Press, Cleveland, Ohio (1975).
12. P. J. Harrop and D. S. Campbell, Selection of thin film capacitor dielectrics, *Thin Solid Films* 2, 273–292 (1968).
13. J. F. Wager, C. W. Wilmsen, and L. L. Kazmerski, Estimation of the bandgap of $InPO_4$, *Appl. Phys. Lett.* 42, 589–590 (1983).
14. D. H. Laughlin, The Correlation between the Composition and Conduction of InP Anodic Oxides, M. S. Thesis, Colorado State University (1980).
15. W. P. Doyle, Absorption spectra of solids and chemical bonding—I, Arsenic, Antimony and Bismuth trioxides, *J. Phys. Chem. Solids* 4, 144–147 (1958).
16. R. L. Weiher, Electrical properties of single crystal indium oxide, *J. Appl. Phys.* 33, 2834–2839 (1962).
17. C. D. Thurmond, G. P. Schwartz, G. W. Kammlott, and B. Schwartz, GaAs oxidation and the Ga–As–O equilibrium phase diagram, *J. Electrochem. Soc.* 127, 1366–1371 (1980).
18. G. P. Schwartz, G. J. Gaultieri, J. E. Griffiths, C. D. Thurmond, and B. Schwartz, Oxide–substrate and oxide–oxide chemical reactions in thermally annealed films on GaSb, GaAs and GaP, *J. Electrochem. Soc.* 127, 2488–2499 (1980).
19. G. P. Schwartz, Analysis of native oxide films and oxide–substrate reactions on III–V semiconductors using thermochemical phase diagrams, *Thin Solid Films*, 103, 3–16 (1983).

20. T. P. Smirnova, A. N. Golubenko, N. F. Zackarchals, V. I. Belyi, G. A. Kokovin, and N. A. Valiskeva, Phase composition of thin oxide films on InSb, *Thin Solid Films* 76, 11–21 (1981).
21. C. W. Wilmsen, Chemical composition and formation of thermal and anodic oxide/III–V compound semiconductor interfaces, *J. Vac. Sci. Technol.* 19, 279–289 (1981).
22. G. P. Schwartz, J. E. Griffiths, and G. J. Gaultieri, Thermal oxidation and native oxide–substrate reactions on InAs and $In_xGa_{1-x}As$, *Thin Solid Films* 94, 213–222 (1982).
23. T. Nakagawa, K. Ohta, and N. Koshizuka, Raman scattering study of unoxidized antimony in anodic oxide-films of InSb, *Japan. J. Appl. Phys.* 19, L339–L341 (1980).
24. M. Fathipour, W. H. Makky, J. McLaren, K. M. Geib, and C. W. Wilmsen, High temperature annealing of InP anodix oxides, *J. Vac. Sci. Technol. A* 1, 662–666 (1983).
25. R. P. H. Chang, T. T. Sheng, C. C. Chang, and J. J. Coleman, The effect of interface arsenic domains on the electrical properties of GaAs MOS structures, *Appl. Phys. Lett.* 33, 341–342 (1978).
26. D. T. Clark and T. Fok, Surface modification of InP by plasma techniques using hydrogen and oxygen, *Thin Solid Films* 78, 271–278 (1981).
27. R. Glang, in: *Handbook of Thin Film Technology* (L. I. Maissel and R. Glang, Eds.), pp. 1–16, 1–17, McGraw-Hill, New York (1970).
28. K. K. Kelly, *U.S. Bureau of the Mines*, Bulletin 383 (1935).
29. R. P. Burns, Systematics of the evaporation coefficients of Al_2O_3, Ga_2O_3 and In_2O_3, *J. Chem. Phys.* 44, 3307–3319 (1966).
30. S. P. Muraka, Thermal oxidation of GaAs, *Appl. Phys. Lett.* 26, 180–182 (1975).
31. D. E. Aspnes and A. A. Studna, Chemical etching and cleaning procedures for Si, Ge and some III–V compound semiconductors, *Appl. Phys. Lett.* 39, 316–318 (1981).
32. B. R. Pruniaux and A. C. Adams, Dependence of barrier height of metal semiconductor contact (Au–GaAs) on thickness of semiconductor surface layer, *J. Appl. Phys.* 43, 1980–1983 (1972).
33. F. Lukes, Oxidation of Si and GaAs in air at room temperature, *Surf. Sci.* 30, 91–100 (1972).
34. E. Kuphal and H. W. Dinges, Composition and refractive index of $Ga_{1-x}Al_xAs$ determined by ellipsometry, *J. Appl. Phys.* 50, 4196–4200 (1979).
35. H. Burkhard, H. W. Dinges, and E. Kuphal, Optical Properties of $In_{1-x}Ga_xAs_y$, InP, GaAs and GaP determined by ellipsometry, *J. Appl. Phys.* 53, 655-662 (1982).
36. A. J. Rosenberg, The oxidation of intermetallic compounds-III, The room-temperature oxidation of $A^{III}B^V$ compounds, *J. Phys. Chem. Solids* 14, 175–180 (1960).
37. B. Schwartz, S. E. Haszko, and D. R. Wonsidler, The influence of dopant concentration on the oxidation os N–type GaAs in H_2O, *J. Electrochem. Soc.* 118, 1229–1231 (1971).
38. Y. C. M. Yeh and R. J. Stirn, Single crystal and polycrystalline GaAs solar cells using AMOS technology, 11th IEEE Photovoltaic Specialist Conference, New York (1975).
39. H. Iwasaki, Y. Mizokawa, R. Nishitani, and S. Nakamura, Effects of water vapor and oxygen excitation on oxidation of GaAs, GaP and InSb surfaces studied by X-ray photoemission, *Japan. J. Appl. Phys.* 18, 1525–1529 (1979).
40. J. F. Wager, D. L. Ellsworth, S. M. Goodnick, and C. W. Wilmsen, Composition and thermal stability of thin native oxides on InP, *J. Vac. Sci. Technol.* 19, 513–518 (1981).
41. J. L. Zilko and R. S. Williams, Auger electron spectroscopy study of GaAs substrate cleaning procedures, *J. Electrochem. Soc.* 129, 406–409 (1982).
42. I. Shiota, K. Motoya, T. Ohmi, N. Miyamoto, and J. Nishizawa, Auger characterization of chemically etched GaAs surfaces, *J. Electrochem. Soc.* 124, 155–157 (1977).
43. P. A. Bertrand, XPS study of chemically etched GaAs and InP, *J. Vac. Sci. Technol.* 18, 28–33 (1981).

44. C. C. Chang, P. H. Citrin, and B. Schwartz, Chemical preparation of GaAs surfaces and their characterization by Auger electron and X-ray photoemission spectroscopies, *J. Vac. Sci. Technol. 14*, 943–952 (1977).
45. T. Oda and T. Sugano, Studies on chemically etched silicon, Gallium arsenide, and gallium phosphide surfaces by Auger electron spectroscopy, *Japan. J. Appl. Phys. 15*, 1317–1327 (1976).
46. R. P. Vasquez, B. F. Lewis, and F. J. Grunthaner, X-ray photoelectron spectroscopy study of the oxide removal mechanism of GaAs (100) molecular beam epitaxial substrates in situ heating, *Appl. Phys. Lett. 42*, 293–295 (1983).
47. R. P. Vasquez, B. F. Lewis, and F. J. Grunthaner, Cleaning chemistry of GaAs (100) and InSb (100) substrates for molecular beam epitaxy, *J. Vac. Sci. Technol. B 1*, 791–794 (1983).
48. D. T. Clark, T. Fok, G. G. Roberts, and R. W. Sykes, An investigation by electron spectroscopy for chemical analysis of chemical treatments of the (100) surface of n-type InP epitaxial layers for Langmuir film ay redeposition, *Thin Solid Films 70*, 261–283 (1980).
49. S. Singh, R. S. Williams, L. G. VanUitent, A. Schlierr, I. Camlibel, and W. A. Bonner, Analysis of InP surface prepared by various cleaning methods, *J. Electrochem. Soc. 129*, 447–448 (1982).
50. O. Wada, A. Majerfeld, and P. N. Robson, InP Schottky contacts with increased barrier height, *Solid-State Electron. 25*, 381–387 (1982).
51. O. Wada and A. Majerfeld, Low leakage nearly ideal Schottky barriers to n-InP, *Electron. Lett. 14*, 125–126 (1978).
52. C. Michel and J. J. Ehrhardt, Oxidation of (n)-InP by nitric acid, *Electron. Lett. 18*, 305–307 (1982).
53. A. Guivar'H, H. L'Haridon, G. Pelous, G. Hollinger, and P. Pentosa, Chemical cleaning of InP surfaces: Oxide composition and electrical properties, *J. Appl. Phys.* (to be published).
54a. B. Schwartz, Preliminary results on the oxidation of GaAs and GaP during chemical etching, *J. Electrochem. Soc. 118*, 657–658 (1971).
54b. B. Schwartz and W. J. Sundburg, Oxidation of GaP in an aqueous H_2O_2 solution, *J. Electrochem. Soc. 120*, 576–580 (1972).
55. M. Invishi and B. W. Wessels, Deep level transient spectroscopy of interface and bulk trap states in InP MOS structures, *Thin Solid Films 103*, 141–153 (1983).
56. H. Lim, G. Sagnes, and G. Bastide, A study of the chemical oxide/InP interface states, *J. Appl. Phys. 53*, 7450–7453 (1982).
57. K. Kamura, T. Suzuki, and A. Kunioka, InP metal–insulator–semiconductor Schottky contacts using surface oxide layers prepared with bromine water, *J. Appl. Phys. 51*, 4905–4907 (1980).
58. F. D. Auret, An AES evaluation of cleaning and etching methods for InSb, *J. Electrochem. Soc. 129*, 2752–2755 (1982).
59. K. Navratil, Thermal oxidation of gallium arsenide, *Czech. J. Phys. 18*, 266–274 (1968).
60. J. F. Wager and C. W. Wilmsen, Thermal oxidation of InP, *J. Appl. Phys. 51*, 812–814 (1980).
61. M. L. Korwin-Pawlowski and E. L. Heasell, Thermal oxide layers on indium antimonide, *Phys. Stat. Sol. (a) 27*, 339–346 (1975).
62. T. Hwang, K. M. Geib, C. W. Wilmsen, A. R. Clawson, and D. I. Elder, Thermal oxidation of $In_{0.53}Ga_{0.47}As$, *J. Appl. Phys.* (to be published).
63. K. Kato, K. M. Geib, R. G. Gann, P. Brusenback, and C. W. Wilmsen, Thermal oxidation of GaP, *J. Vac. Sci. Technol. A2*, 588–592 (1984).

64. A. S. Grove, Physics and Technology of Semiconductor Devices, Chapter 2, Wiley, New York (1967).
65. A. J. Nelson, Composition, Structure and Growth Kinetics of Thermal Oxides of InP, M.S. Thesis, Colorado State University (1982).
66. K. Watanabe, M. Hashiba, Y. Hirahota, M. Nishino, and T. Yamashina, Oxide layers on GaAs prepared by thermal, anodic and plasma oxidation: In-depth profiles and annealing effects, Thin Solid Films 56, 63–73 (1979).
67. C. W. Wilmsen, Oxide layers on III–V compound semiconductors, Thin Solid Films 39, 105–117 (1976).
68. J. F. Wager and C. W. Wilmsen, Plasma-enhanced chemical vapor deposited SiO_2/InP interface, J. Appl. Phys. 53, 5789–5797 (1982).
69. A. Nelson, K. Geib, and C. W. Wilmsen, Composition and structure of thermal oxides of indium phosphide, J. Appl. Phys. 54, 4134–4140 (1983).
70. G. P. Schwartz, W. A. Sander, and J. E. Griffiths, Raman scattering study of the thermal oxidation of InP, Appl. Phys. Lett. 37, 925–927 (1980).
71. J. J. McLaren, A. Nelson, K. Geib, R. Gann, and C. W. Wilmsen, Surface topography of oxides on InP thermally grown at high temperatures, J. Vac. Sci. Technol. A1, 1486–1490 (1983).
72. M. Rubenstein, The oxidation of GaP and GaAs, J. Electrochem. Soc. 113, 540–542 (1966).
73. H. Iwasaki, Y. Mizokawa, R. Nishitani, and S. Nakamura, X-ray photoemission study of the oxidation process of cleaned (110) surface of GaAs, Gap and InSb, Japan. J. Appl. Phys. 17, 1925–1933 (1978).
74. R. Nishitani, H. Iwasaki, Y. Mizokawa, and S. Nakamura, An XPS analysis of thermally grown oxide film on GaP, Japan. J. Appl. Phys. 17, 321–327 (1978).
75. A. J. Rosenberg and M. C. Lavine, The oxidation of intermetallic compounds: I. High temperature oxidation of InSb, J. Phys. Chem. 64, 1135–1142 (1960).
76. A. J. Rosenberg, Oxidation of intermetallic compounds: II. Interrupted oxidation of InSb, J. Phys. Chem. 64, 1143–1150 (1960).
77. C. W. Wilmsen, Correlation between the composition profile and electrical conductivity of the thermal and anodic oxides of InSb, J. Vac. Sci. Technol. 13, 64–67 (1976).
78. M. L. Korwin-Pawlawski and E. L. Heasell, Thermal oxide layers on indium antimonide, Phys. Status Solid A 27, 339–346 (1975).
79. K. Navratil, I. Ohlidal, and F. Lukes, The physical structure of the interface between single-crystal GaAs and its oxide film, Thin Solid Films 56, 163–171 (1979).
80. G. P. Schwartz, G. J. Gualtieri, G. W. Kammlott, and B. Schwartz, An X-ray photoelectron spectroscopy study of native oxides on GaAs, J. Electrochem. Soc. 126, 1737–1749 (1979).
81. Y. Mizokawa, H. Iwasaki, R. Nishitani, and S. Nakamura, In depth profiles of oxide films on GaAs studied by XPS, Japan. J. Appl. Phys. 17, Suppl., 327–333 (1978).
82. K. Loschke, G. Kuhn, H. J. Bitz, and G. Leonhardt, Oxide films and $A^{III}B^V$ Halbleiter, Thin Solid Films 48, 229–236 (1978).
83. R. L. Farrow, R. K. Chang, S. Mroezkawski, and F. H. Pollak, Detection of excess crystalline As and Sb in III–V oxide interfaces by Raman scattering, Appl. Phys. Lett. 31, 768–770 (1977).
84. D. H. Laughlin and C. W. Wilmsen, Thermal oxidation of InAs, Thin Solid Films, 70, 325–332 (1980).
85. M. Yamaguchi, A. Yamamoto, H. Sagivra, and C. Vemura, Thermal oxidation of InAs and characterization of the oxide film, Thin Solid Films 92, 361–369 (1982).
86. H. L. Hartnagel, MOS-gate technology on GaAs and other III–V compounds, J. Vac. Sci. Technol. 13, 860–867 (1976).

87. M. Pourbaix, *Atlas of Electrochemical Equilibria in Aqueous Solutions*, Pergamon Press, New York (1966).
88. B. Schwartz, GaAs surface chemistry: A review, *CRC Crit. Rev. Solid-State Sci.* **5**, 609–624 (1975).
89. A. Yamamoto, M. Yamaguchi, and C. Vemura, Preparation and electrical properties of an anodic oxide of InP, *J. Electrochem. Soc.* **129**, 2795–2801 (1982).
90. L. C. Feldman, J. M. Poate, F. Ermanis, and B. Schwartz, Combined use of helium back-scattering and helium-induced X-rays in the study of anodically grown oxide films on gallium arsenide, *Thin Solid Films* **19**, 81–89 (1973).
91. K. M. Geib and C. W. Wilmsen, Anodic oxide/GaAs and InP interface formation, *J. Vac. Sci. Technol.* **17**, 952–957 (1980).
92. C. W. Wilmsen and R. W. Kee, Analysis of the oxide/semiconductor interface using Auger and ESCA as applied to InP and GaAs, *J. Vac. Sci. Technol.* **15**, 1513–1517 (1978).
93. Y. Mizokawa, H. Iwasaki, R. Hishitani, and S. Makamura, Quantitative chemical depth profiles of anodic oxide on GaAs obtained by X-ray photoemission spectroscopy, *J. Electrochem. Soc.* **126**, 1370–1374 (1979).
94. P. A. Breeze, H. L. Hartnagel, and P. M. A. Sherwood, An investigation of anodically grown films on GaAs using X-ray photoemission spectroscopy, *J. Electrochem. Soc.* **127**, 454–461 (1980).
95. D. E. Aspnes, G. P. Schwartz, G. J. Gualtieri, A. A. Studna, and B. Schwartz, Optical properties of GaAs and its electrochemically grown anodic oxide from 1.5 to 6.0 eV, *J. Electrochem. Soc.* **128**, 590–597 (1981).
96. C. J. Maggiore and R. S. Wagner, Ion beam characterization of the GaAs–GaAs oxide interface for plasma and anodic oxides, *J. Vac. Sci. Technol.* **19**, 463–466 (1981).
97. C. W. Wilmsen and R. W. Kee, Auger analysis of the anodic oxide/InP interface, *J. Vac. Sci. Technol.* **14**, 953–956 (1977).
98. D. A. Baglee, D. H. Laughlin, B. T. Moore, B. L. Eastep, D. K. Ferry, and C. W. Wilmsen, *Inst. Phys. Conf. Ser. No. 56*, Chap. 5 (1980).
99. W. H. Makky, Structural and Electrical Properties of InP Anodic Oxides, Ph.D. Thesis, Colroado State University (1983).
100. J. M. Poate, T. M. Buck, and B. Schwartz, Rutherford scattering study of the chemical composition of native oxides on gallium phosphide, *J. Phys. Chem. Solids* **34**, 779–786 (1973).
101. J. M. Poate, P. J. Silverman, and J. Yahalom, The growth and composition of anodic films on GaP, *J. Phys. Chem. Solids* **34**, 1847–1857 (1973).
102. J. M. Poate, P. J. Silverman, and J. Yahalom, Anodic oxide films on gallium phosphide, *J. Electrochem. Soc.* **120**, 844–845 (1973).
103. H. J. Bilz, G. Leonhardt, G. Kunn, K. Loschke, and A. Meisel, ESCA—untersuchungen an anodisch oxydierten 3–5 verbindungen, *Krist. Technik.* **13**, 363–368 (1978).
104. D. A. Baglee, D. K. Ferry, C. W. Wilmsen, and H. H. Wieder, Inversion layer transport and properties of oxides on InAs, *J. Vac. Sci. Technol.* **17**, 1032–1036 (1980).
105. D. A. Baglee, D. H. Laughlin, C. W. Wilmsen, and D. K. Ferry in: *The Physics of MOS Insulators* (G. Lucovsky, S. T. Pantelides, and F. L. Galeener, eds.) Pergamon Press, New York (1980).
106. J. F. Dewald, The kinetics and mechanism of the formation of anodic films on single crystal InSb, *J. Electrochem. Soc.* **104**, 244–251 (1957).
107. J. F. Dewald, A theory of the kinetics of formation of anodic films at high fields, *J. Electrochem. Soc.* **102**, 1–6 (1955).
108. S. M. Spitzer, B. Schwartz, and G. D. Weigle, Preparation and stabilization of anodic oxides on GaAs, *J. Electrochem. Soc.* **121**, 92C (1974).

109. S. M. Spitzer, B. Schwartz, and G. D. Weigle, Preparation and stabilization of anodic oxides on gallium arsenide, *J. Electrochem. Soc. 122*, 397–402 (1975).
110. T. Ishii and B. Jeppsson, Influence of temperature on anodically grown native oxides on gallium arsenide, *J. Electrochem. Soc. 124*, 1784–1794 (1977).
111. T. Ishii and B. Jeppsson, Influence of temperature on the structure and properties of an anodized native GaAs oxide, *Japan. J. Appl. Phys. 16*, 471–474 (1977).
112. B. L. Weiss and H. L. Hartnagel, Crystallization dynamics of native anodic oxides on GaAs for device applications, *Thin Solid Films 56*, 143–152 (1979).
113. G. P. Schwartz, *Springer Series in Electrophysics, 7*, 22a, Springer-Verlag, New York (1981).
114. A. T. Fromhold, Plasma oxidation, *Thin Solid Films 95*, 297–308 (1982).
115. T. Sugano, Oxide film growth on GaAs and silicon substrates of anodization in oxygen plasma and its application to devices and integrated circuit fabrication, *Thin Solid Films 72*, 9–17 (1980).
116. K. Kanazawa and Matsunani, Plasma-grown oxide on InP, *Japan. J. Appl. Phys. 20*, L211–L213 (1981).
117. Y. Imai, T. Ishibashi, and M. Ida, Characterization of Inp MIS Schottky diodes prepared by plasma oxidation, *J. Electrochem. Soc. 129*, 221–224 (1982).
118. J. F. Wager, W. H. Makky, C. W. Wilmsen, and L. G. Meiners, Oxidation of InP in a plasma-enhanced chemical vapor deposition reactor, *Thin Solid Films 95*, 343–350 (1982).
119. O. Krivanik and S. L. Fortner, HREM imaging and microanalysis of a III–V semiconductor/oxide interface, *Ultramicroscopy, 14*, 121–126 (1984).
120. A. A. Studna and G. J. Gualitieri, Optical properties and water absorption of anodically grown native oxides on InP, *Appl. Phys. Lett. 39*, 965–966 (1981).

Index

Acceptorlike surface states, 4
Adsorption
 chlorine, 28
 dependence on surface perfection, 25
 H_2, H_2S, H_2O, 29, 46, 54
 iodine, 29
 metals
  Al, Ga, In, 39
  Au, Ag, 44
  Cs, 35
 oxygen, 22, 52
 sticking coefficient, 25, 29
Annealing, 13, 19, 32, 104, 168, 169, 194, 200, 266, 450
Anodization, 430
Anodization parameters
 current density, 436
 electrolyte, 434
 pH, 437
 viscosity, 436
Antisite defect, 60
Autocorrelation, 301

Bandgap
 insulators, 170
 oxides, 174, 182, 204
 III-V compounds, 8, 124, 126–128, 328
Bardeen limit, 4, 77, 81
Bardeen model, 54, 57
Barrier height, 291; *see also* Schottky barriers

Charge coupled devices, 335, 347, 377, 381
Charge density contour, 16
Clean surface, 7, 237
Cluster formation, 50
Contacts
 with Al, 30, 32

Contacts (*contact*)
 ohmic, 51, 129
 resistance, 94, 147–151
 See also Ohmic contacts; Schöttky barriers
Coulomb scattering, 296
Cutoff frequency, 346, 368

Dangling bonds, 5, 15, 18, 40, 54
Deep depletion, 220, 223, 230
Defect energy, 184
Defect model, 58, 82, 112
Deposited insulators, 165, 239, 249, 269
Dielectric loss, 173
Dopants, 183
Drift, 199, 203, 241

Electron affinity, 2, 74
Electron mobility, 242, 254, 258, 261, 283, 303, 309, 314, 338, 356, 364, 366, 384
Electronegativity, 5, 28, 54
Etches, 415; *see also* Surface preparation
Evaporation, 168, 184, 408–411

Fermi level pinning, 4, 13, 18, 39, 46, 52, 58, 60, 80, 125, 180, 230
Frequency dispersion, 171, 230–236, 250–259, 358, 363

Green's function, 296
Gunn diode, 144

Heat of formation, 49, 55, 84, 116, 122, 176
Heterojunction transport, 285, 318
High field transport, 310

Ideality factor, 60, 90, 93
Image charge, 79

Image force, 79, 87
Imaging arrays, 354
Injecting contact, 3
In situ processing, 167, 180
Inversion layers, 251, 283, 307
Insulator
  band gap, 170, 174, 204
  deposited, 165, 239, 249, 269
  dielectric constant, 171
  traps, 188, 190, 202
Integrated circuits, 348, 369
Interdiffusion, 32, 45, 46, 56, 84, 182, 198
Interfacial oxide, 179
Interface reactions, 175, 178
Interfacial polarization, 172
Island growth, 34, 45, 47, 440, 442

Jellium, 55

Langmuir films, 249
Lattice constants, 8
Leed patterns, 36, 53

Memory cells, 355
MESFETs, 114, 343, 379
MIS-gated depletion devices, 335, 380
MIS
  capacitance, 189–205, 215, 224, 228–236, 250–259
  conductance, 226
  structures, 213, 327
MISFETs
  GaAs, 356
  InP, 290, 303, 333, 362
  InAs, 293, 309
  InGaAs, 387
  MISFET power gain, 370

Native oxide, 181, 192, 261
Neutral surface states, 4, 60

Ohmic contacts, 6, 56
  alloyed, 133, 144
  diffusion and implantation, 131
  epitaxy, 132
  field emission, 91
  heterojunction, 134
  resistance, 96
  resistance measurement
    Cox-Strack method, 137
    four point method, 139
    Shockley method, 140

Ohmic contacts (cont.)
  resistance measurement (cont.)
    transmission method, 142
  tunneling, 91
Oxidation
  anodic, 239, 243, 248, 261, 430
  chemical, 114, 414, 418
  with excitation, 22, 26, 31
  initial, 24, 26, 404, 412, 438
  mechanisms, 422
  plasma, 180, 453
  thermal, 201, 238, 248, 420
Oxide
  annealing, 450
  composition, 430, 437, 445, 447, 455
  islands, 440, 442, 448
  phase diagram, 407
  properties, 406
  topography, 440

Passivation, 351
Permittivity, 79
Phonon scattering, 304
Photoluminescence, 19, 27
Plasma deposition, 181, 197, 201, 256, 271, 453
Pseudopotential method, 16

Quantum well devices, 318

Remote optical phonons, 308
Richardson's constant, 86
Richardson's plot, 102

S parameter, 81, 111
Schöttky barrier
  capacitance, 57, 97, 105
  chemical reactivity, 84, 116
  current voltage, 85, 101
  energies
    GaAs, 109
    InP, 112
    other compounds, 55, 118
    quaternaries, 123
    ternaries, 123, 128
  formation, 2, 51, 57, 106, 114
  heat of formation, 55, 84, 112
  ideality factor, 90, 93
  interfacial layer, 107
  measurement, 99
  MIS diode, 98, 114, 335, 414
  technology, 98
  thermionic emission, 86, 101

Index

Schottky gate FET, 343
Schottky limit, 3, 76, 81
Schottky model, 54, 74
Screening, 297
Shubnikov-de Haas, 312, 338
Solar cells, 351
Sticking coefficient, 25
Surface
  crystallography, 7
  defects, 18, 49
  dipole, 3
  disorder, 13, 39, 50, 56, 60, 235, 296
  etched, 110, 117, 237, 414
  evaporation, 184
  ideal, 10
  preparation, 166, 179, 408
    GaAs, 88, 194, 237, 415
    GaP, 261
    InAs, 260
    InP, 112, 198, 250, 365, 372
    InSb, 189, 242
  quantization, 287
  recombination, 340
  reconstruction, 9
  relaxed, 9, 28, 54
  roughness, 299
  roughness parameters, 301, 316
  scattering, 295
  sputtered, 19, 22, 54

Surface (*cont.*)
  states, 13, 224, 262
  steps, 18–20
  subbands, 287, 311
Surface state calculation, 218

Terman's method, 229, 246
Thermal decomposition, 19
Thermal desorption, 37
Thermal oxidation
  GaAs, GaSb, InSb, 426
  GaP, 426
  InAs, 428
  InGaAs, 428
  InP, 423
Thermal oxidation rate, 420
Tight binding method, 15
Transport, 283
Trapping, 186, 188, 190, 200, 202, 332, 359, 371, 374
Trapping models, 336

UPS spectra, 23, 28, 41, 42, 45, 47, 48, 404

Vapor pressure
  oxides, 408, 411
  III-V elements, 410

Work function, 18, 38, 74, 111